Lecture Notes in Artificial Intelligence 10713

Subseries of Lecture Notes in Computer Science

More information about this series at http://www.springer.com/series/1244

Georg Rehm · Thierry Declerck (Eds.)

Language Technologies for the Challenges of the Digital Age

27th International Conference, GSCL 2017
Berlin, Germany, September 13–14, 2017
Proceedings

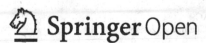

Editors
Georg Rehm
DFKI GmbH
Berlin
Germany

Thierry Declerck
DFKI GmbH
Saarbrücken
Germany

ISSN 0302-9743   ISSN 1611-3349   (electronic)
Lecture Notes in Artificial Intelligence
ISBN 978-3-319-73705-8   ISBN 978-3-319-73706-5   (eBook)
https://doi.org/10.1007/978-3-319-73706-5

Library of Congress Control Number: 2017963770

LNCS Sublibrary: SL7 – Artificial Intelligence

Preface

GSCL 2017 was the 27th edition of the biennial conference of the German Society for Computational Linguistics and Language Technology (GSCL e.V.). It was also the first to take place in Berlin. The conference offered, once again, high-quality articles covering a wide spectrum of computational linguistics research.

The unprecedented social, political, and economic relevance of online media and digital communication technologies inspired this year's main conference topic: "Language Technologies for the Challenges of the Digital Age." This relevance has, in turn, given rise to the still ongoing debate around several phenomena relating to online media, most importantly, around the topic of false news ("fake news"), online rumors, abusive language, and hate speech as well as filter bubbles and echo chambers. We have been actively engaged in several research activities around these topics, most notably in the project Digitale Kuratierungstechnologien (DKT; Digital Curation Technologies, funded by BMBF and coordinated by DFKI) and the European project PHEME, in which the University of Saarland participated. The two invited talks of GSCL 2017 examined aspects of our main conference topic from different perspectives: Holger Schwenk (Facebook AI Research, Paris, France) talked about "Learning Deep Representations in NLP," especially with regard to the aspect of overcoming language barriers with the help of machine translation. Kalina Bontcheva (University of Sheffield, UK) focused on "Automatic Detection of Rumours in Social Media."

It would have been impossible to organize this conference without the help and commitment of many people. We would like to thank all speakers, participants, workshop organizers (details at http://gscl2017.dfki.de), as well as the members of the Program Committee. We would also like to thank the people who made it possible to organize GSCL 2017 at the Humboldt University of Berlin: Stefan Müller, Anke Lüdeling, Manfred Krifka, Steffen Hofmann, and Dagmar Oehler. Of crucial importance in the planning and organization of the conference was the Executive Committee of GSCL e.V., especially Heike Zinsmeister, Torsten Zesch, Andreas Witt, and Roman Schneider. Furthermore, we would like to thank Nieves Sande, who organized all the practical aspects of the event and coordinated the conference office, Julian Moreno Schneider, who not only created and maintained the conference website but who also played a central role in assembling this proceedings volume, as well as Stefanie Hegele, Charlene Röhl, and Peter Bourgonje, who helped on site at the conference desk and with the logistics. Thanks are also due to the three gold and seven silver sponsors without whom the conference would not have been possible.

This Open Access conference proceedings volume was made possible with the help of the BMBF project Digitale Kuratierungstechnologien (DKT). We are grateful to Alexandra Buchsteiner (Pt-J) for supporting the initial suggestion to publish this book.

November 2017

Georg Rehm
Thierry Declerck

Organization

GSCL 2017 was organized by Deutsches Forschungszentrum für Künstliche Intelligenz GmbH (DFKI) in cooperation with Humboldt-Universität zu Berlin.

Conference Chairs

Georg Rehm   DFKI, Germany
Thierry Declerck   DFKI, Germany

Organizing Committee

Georg Rehm   DFKI, Germany
Thierry Declerck   DFKI, Germany
Nieves Sande   DFKI, Germany
Julian Moreno-Schneider   DFKI, Germany
Stefan Müller   Humboldt-Universität zu Berlin, Germany

Program Committee

Michael Beisswenger   Universität Duisburg-Essen, Germany
Delphine Bernhard   Université de Strasbourg, France
Chris Biemann   Universität Hamburg, Germany
Aljoscha Burchardt   DFKI, Germany
Christian Chiarcos   Goethe-Universität Frankfurt, Germany
Philipp Cimiano   Universität Bielefeld, Germany
Bertold Crysmann   CNRS, France
Michael Cysouw   University of Marburg, Germany
Thierry Declerck (Chair)   DFKI, Germany
Stefanie Dipper   Ruhr-Universität Bochum, Germany
Christian Federmann   Microsoft, USA
Bernhard Fisseni   Uni Duisburg-Essen, Germany
Alexander Fraser   Ludwig-Maximilians-Universität München, Germany
Jost Gippert   Universität Frankfurt, Germany
Dagmar Gromann   Artificial Intelligence Research Institute, Spain
Iryna Gurevych   Technische Universität Darmstadt, Germany
Gerhard Heyer   Universität Leipzig, Germany
Roman Klinger   Universität Stuttgart, Germany
Brigitte Krenn   Austrian Research Institute for Artificial Intelligence,
                           Austria
Sandra Kübler   Indiana University, USA
Jonas Kuhn   Universität Stuttgart, Germany

Lothar Lemnitzer	BBAW, Germany
Piroska Lendvai	University of Göttingen, Germany
Nils Lenke	Nuance Communications, Aachen, Germany
Henning Lobin	Universität Gießen, Germany
Cerstin Mahlow	Universität Stuttgart, Germany
Alexander Mehler	Goethe-Universität Frankfurt, Germany
Friedrich Neubarth	Austrian Research Institute for Artificial Intelligence, Austria
Sebastian Padó	Universität Stuttgart, Germany
Karola Pitsch	Universität Duisburg-Essen, Germany
Uwe Quasthoff	Universität Leipzig, Germany
Georg Rehm (Chair)	DFKI, Germany
Frank Richter	Goethe-Universität Frankfurt, Germany
Manfred Sailer	Goethe-Universität Frankfurt, Germany
David Schlangen	Universität Bielefeld, Germany
Helmut Schmid	CIS München, Germany
Thomas Schmidt	IDS Mannheim, Germany
Ulrich Schmitz	Universität Duisburg-Essen, Germany
Roman Schneider	IDS Mannheim, Germany
Bernhard Schröder	Universität Duisburg-Essen, Germany
Sabine Schulte im Walde	Universität Stuttgart, Germany
Manfred Stede	Universität Potsdam, Germany
Benno Stein	Universität Weimar, Germany
Angelika Storrer	Universität Mannheim, Germany
Elke Teich	Universität des Saarlandes, Germany
Mihaela Vela	Universität des Saarlandes, Germany
Michael Wiegand	Universität des Saarlandes, Germany
Andreas Witt	Universität zu Köln, Germany
Marcos Zampieri	University of Cologne, Germany
Torsten Zesch	Universität Duisburg-Essen, Germany
Heike Zinsmeister	Universität Hamburg, Germany

GSCL 2017 – Sponsors

Gold Sponsors

Silver Sponsors

Proceedings Sponsors

The project "Digitale Kuratierungstechnologien" (DKT) is supported by the German Federal Ministry of Education and Research (BMBF), "Unternehmen Region", instrument Wachstumskern-Potenzial (no. 03WKP45). More information: http://www.digitale-kuratierung.de.

Contents

Online-Media and Online-Content

Miscellaneous

Processing German: Basic Technologies

Reconstruction of Separable Particle Verbs in a Corpus of Spoken German

Dolores Batinić[✉] and Thomas Schmidt

Institut für Deutsche Sprache, Mannheim, Germany
{batinic,thomas.schmidt}@ids-mannheim.de

Abstract. We present a method for detecting and reconstructing separated particle verbs in a corpus of spoken German by following an approach suggested for written language. Our study shows that the method can be applied successfully to spoken language, compares different ways of dealing with structures that are specific to spoken language corpora, analyses some remaining problems, and discusses ways of optimising precision or recall for the method. The outlook sketches some possibilities for further work in related areas.

1 Introduction

German verb particles may occur either attached to their verb stems (compare English: *hand in* sth) or separated from them (compare English: *hand* sth *in*). For instance, consider examples 1 and 2, both taken from the FOLK corpus:

(1) SK: och pascal du muss dein geld nich **raushauen**
(2) JL: das ding **haun** wir **raus**

When searching for occurrences of a separable verb in most currently available online corpora, the user can retrieve directly only those segments in which the verb is attached to the verb particle. In order to retrieve *all* occurrences of a separable verb, the user must query the base verb *hauen* or the verb particle *raus* separately and inspect the context of the retrieved segments. This kind of query may be cumbersome, especially if the corpus interface does not provide a context filter.

For solving the issue of erroneously lemmatised separable verbs, Volk et al. (2016) proposed an algorithm for recomputing verb lemmas that occur in sentences with separated particles, which performs with a high precision on a corpus of written German. Since we work with spoken language, we investigated how their principle of lemma reconstruction performs on the FOLK corpus (Research and Teaching Corpus of Spoken German, (Schmidt 2016)), accessible via the DGD (Database for Spoken German, (Schmidt 2014)). Detecting separable verbs in a corpus of spoken language such as FOLK is challenging because firstly, a segmentation into sentences is not available, and secondly, the verbs may differ from the standard German variants. In order to provide a more efficient corpus

© The Author(s) 2018
G. Rehm and T. Declerck (Eds.): GSCL 2017, LNAI 10713, pp. 3–10, 2018.
https://doi.org/10.1007/978-3-319-73706-5_1

querying in the DGD as well as a reliable analysis of verb lemma counts, we experimented with different adaptations of Volk et al. (2016)'s algorithm on our corpus data.

The motivation for this study was the ongoing work in a project on the lexicon of spoken German (LeGeDe: Lexik des gesprochenen Deutsch) at the Institute for the German Language in Mannheim. Currently the project focuses on the study of perception and motion verbs. Since they happen to be very productive in terms of pair combinations (e.g., *sehen – absehen, ansehen, aussehen; gehen – abgehen, angehen, ausgehen*, etc.), it is of great importance to be able to identify different particle verbs and to reliably calculate their corpus frequencies.

2   Detecting Separable Particle Verbs

To reconstruct the lemma of a separable verb, Volk et al. (2016) attach the verb particle to the lemma of the nearest preceding finite verb. If the reconstruction exists (as confirmed by a lookup in a word list), the previous verb lemma is replaced with the reconstructed lemma.

The same principle for reconstruction of separable verb particles can be applied to the FOLK corpus, since it is PoS-tagged and lemmatised in an analogous manner (i.e., with TreeTagger using STTS (Schmid 1995; Westpfahl and Schmidt 2016). However, a difference which must not be ignored is that FOLK has no proper sentence boundaries. Instead, it is segmented into *contributions*: sequences of words not interrupted by a pause longer than 0.2 s. A schematic view of a contribution written according to simplified GAT2-conventions (Selting et al. 2009) is shown in example 3.

(3) CH: **guck** dir hier mal den profi **an**

In many cases, such as in 3, the contribution corresponds to what would be a sentence in a corpus of written language. However, since the segmentation is schematic and based on a surface feature ("inter-pausal units"), rather than a linguistic analysis, syntactic dependencies do not necessarily end up in one and the same contribution. For our object of study, this means that a particle verb may have the verb stem in one contribution and the verb particle in a following one, as in example 4.

(4) CJ: nun
    (pause length: 0.21 s)
    CJ: **sah** er
    (pause length: 0.54 s)
    CJ: schon viel freundlicher **aus** (.) klar

Since the segmentation relies on a chronological axis, in some cases, the segments of one speaker may get interrupted by another speaker, but still continue afterwards, as in example 5.

(5) LB: **guckt** eusch mal
       XM: (is rischtich)
       LB: die form des signals **an**

An ideal segmentation would reunite the segments having the separable verb and the respective verb particle under the same contribution. Since this is currently not a part of the corpus segmentation, we performed a lemma reconstruction not only contribution-wise, but also by considering the previous contributions of the same speaker. In addition to detecting the separable verbs, we assumed that this approach could be useful for improving the corpus segmentation in the future, since it would connect two syntactically dependent segments.

In the implementation part, we relied on the principle of Volk et al. (2016): we searched for all the occurrences of the verb particles (e.g., *ein*, tagged as PTKVZ) and combined them with the preceding finite verb (e.g., *sehen*, tagged as VVFIN). If the combined verb form (i.e., *einsehen*) existed, we assigned, on a new annotation layer, the reconstructed lemma to the finite form and an indicator pointing to that lemma to the particle. Schematically, our annotation layers had the form as in Table 1.

Table 1. Annotation layers

ID	w1	w2	w3	w4
Transcription	des	sieht	gut	aus
Normalisation	das	sieht	gut	aus
Lemmatisation	das	sehen	gut	aus
Reconstruction	das	**aussehen**	gut	[w2]
STTS tag	PDS	VVFIN	ADJD	PTKVZ

To check the existence of the verb, we used the list of separable verbs collected by Andreas Göbel[1] and extended it by adding reduced verb particle variants common in spoken language, such as *drauf* for *darauf*, *ran* for *heran*, *rein* for *herein*, *rauf* for *herauf*, etc. The resulting verb list contains a total of 7685 separable verbs.

As suggested by Volk et al. (2016), we recombined the verb particles with the lemmas tagged as modal verbs or auxiliaries as well, since they might turn out to be separable verbs after the verb lemma reconstruction: if the particle *vor* (EN: ahead, before) succeeds the auxiliary verb *haben* (EN: to have), the reconstructed particle verb is *vorhaben* (EN: to intend). Concerning coordinated or multiple particles, we reconstructed both (or more) variants: In the segment *machen sie einmal mit der faust auf und zu*, both alternatives *aufmachen* (EN: to open) und *zumachen* (EN: to close) were added to the layer of reconstructed lemmas. We also considered non-standard pronunciations, for example the expressions such

---

[1] http://www.verblisten.de, 01.06.2017.

as*hersch uf*, which is a variant of *hörst du auf* (EN: will you stop that). However, it was beyond our aim to reconstruct the lemmas of highly dialectal verbs, such as the Alemannic *feschthebe* (literally: *festheben* in standard German), which has another base verb lemma in standard German (*festhalten*, EN: to hold tight).

To measure the frequency of the separated verbs crossing the current contribution boundaries, we performed the verb particle reconstruction for each of the following cases:

1. Contribution as boundary (the contribution boundaries are limits within which the reconstruction is performed);
2. Turn as boundary (the reconstruction is performed on the sequence of contributions belonging to one speaker);
3. No boundaries (the reconstruction can skip preceding contributions of another speaker).

For cases 2 and 3, we set a maximal distance of 23 words between the verb and verb particle, since this was the longest distance between a correctly reconstructed verb lemma in the GOLD standard (example 6).

(6) AAC2: äh (.) achtnhalb jahre im verein gespielt (.) und jetzt **spiele** ich nur (.) ähm aus spaßmit meinen freunden aus der stufe h noch ei (.) aus_m
    AAC2: nach_er schule hh in so ner mittwochsliga **mit** (.) in so ner (.) indoorhalle

We first performed the reconstruction on the FOLK GOLD standard (Westpfahl and Schmidt 2016), which contains 145 manually annotated transcript excerpts (99247 tokens). Afterwards, we tested the usability of the methods on the entire FOLK corpus where lemmatisation and PoS tagging have not been checked manually (1.95 Million tokens, tagger accuracy: 95%).

3   Results and Discussion

When considering contribution boundaries as limits for particle verb reconstruction, 597 out of a total 5240 (11%) verbs tagged as finite verbs in the GOLD standard were detected as separable. For the other two approaches that number was slightly higher, amounting to 626 and 627 verbs, respectively. To evaluate the reconstructions on a qualitative level, we examined 100 randomly selected segments from the GOLD standard. We marked as correct all the reconstructions which had a dictionary entry in Duden online[2]. In our evaluation, only the smallest part of the separable verbs actually crossed the contribution borders: it occurred in only one example out of 100 (example 7).

(7) KD: we also ich **geh** jetz ma von dem
    KD: punkt **aus** wo sie dann schon (.) zumindest buchstaben laut zuordnung beherrschen

---

[2] http://www.duden.de/, 01.06.2017.

The precision of the verb particle reconstruction on this excerpt of the GOLD standard was very high (0.99) for all approaches. The only incorrect or missed reconstructions in the evaluation set were either due to the verb particles preceding the verb stem (*mit nach Thailand nehmen*) or to the nested clauses between the verb and the particle (example 8).

(8) LHW1: und dann **gehst** du wieder parallel zu der linie mit der du zum brathähnchen gekommen bist wieder vom brathähnchen **weg**

A closer inspection of the differences between the three approaches – this time based on the entire GOLD standard, rather than an excerpt – revealed that reconstructing the separable verbs within one-speaker turns produced satisfying results: 26 out of 31 verbs which were placed outside the contribution boundaries were correctly identified as separable verbs. Results for the two approaches crossing contribution boundaries were almost identical: the skipping-method additionally produced one correct and one erroneous reconstruction. Almost all incorrect examples were reconstructions of modal and auxiliary verbs and coordinated verb particles. In the evaluation of all reconstructions concerning auxiliary and modal verbs in the GOLD standard, the lemma was correctly reconstructed in 22 out of 30 cases (73.3%). Since the reconstruction of verb particles with modal or auxiliary verbs are uncommon (only 0.9% of all modals and auxiliaries in GOLD standard), it may be advantageous to correct the erroneous reconstructions in a post-processing step. Alternatively, one could reconstruct only verbs such as *vorhaben* (EN: to intend), *anhaben* (EN: to wear) or *loswerden* (EN: to get rid off), whose particles unambiguously belong to the explicit auxiliary or modal stems, and avoid reconstructing verbs such as *rausmüssen* (EN: to have to go out), whose status as separable verbs may be debated: In the examples such as in *ich muss raus* the particle *raus* can also be seen as a part of an unrealised motion verb such as *gehen* (*ich muss raus [gehen]*).

Applying the same methods to the entire FOLK corpus, a total of 7% of all finite verb tokens in the corpus were reconstructed, resulting in 1059 different verbs (types) for the contribution-oriented approach, 1140 types for the turn-oriented approach and 1156 for the skipping approach. We measured the accuracy of the reconstruction by dividing the number of correctly reconstructed verbs (true positives) and correctly non-reconstructed verbs (true negatives) by the total of analysed examples. We evaluated all the examples in which the verb reconstructions were unambiguous and clearly understandable (97 out of a sample of 100). As shown in Table 2, each method achieved an accuracy of 0.9. As might have been expected, the contribution-oriented reconstruction had a higher precision, but lower recall than the other two types of reconstruction.

A closer look at the reconstruction differences revealed that crossing contribution boundaries would be profitable when prioritising recall over precision, otherwise a contribution-oriented approach to reconstruction might be the better option for automatically tagged data. In comparison to turn-oriented reconstruction, skipping contributions produces much more false positives (in a small examination of the differences between the two, we observed 3 correct reconstructions and 17 incorrect ones). A closer inspection of the differences revealed

8   D. Batinić and T. Schmidt

Table 2. Evaluation results

	Precision	Recall	Accuracy
Contribution as limit	0.91	0.95	0.90
Turn as limit/no limits	0.88	0.98	0.90

that several erroneous reconstructions were due to the independently used verb particles being mistaken for coordinations. For a higher precision, reconstructing multiple particles per verb can hence be avoided in future. The most frequently identified separated verbs are shown in Table 3.

Table 3. Frequently reconstructed verb lemmas

Lemma	Before	Reconst.	After	Reconst./after
aussehen	243	524	767	68%
anfangen	543	264	807	33%
ankommen	134	153	287	53%
rauskommen	104	133	237	56%
hingehen	140	117	257	46%
angucken	188	116	304	38%
aufhören	76	113	189	60%
aufpassen	121	110	231	48%
ausgehen	147	107	254	42%
reinkommen	58	92	150	61%

Reduced variants of the particles dominated clearly over the non-reduced variants. Moreover, in most cases there were no occurrences of the non-reduced variants beginning with *heraus*, *daran*, *daraus*, etc. neither before not after the reconstruction, whereas the reconstruction method was productive in such cases (*rausholen*: 35 before, 60 after; *drankommen*: 7 before, 39 after, etc.)

During the examination of verb particle reconstructions we encountered several ambiguous cases in which the correctness of a reconstruction would require further linguistic examination, such as repetitions of the same verb particle (example 9), truncations (10), self-corrections (11) and coordinated particles (12).

(9) BUC1: und (.) °hh jetzt **geh** mal von der linken (.) oberen ecke
BUC2: ja
BUC1: äh (.) so einen zentimeter **raus** praktisch so schräg **raus**
(10) VW: so die frau (.) **lebt** sozusagen
VW: oder beide **leben** ihre emotionale seite halt **aus** die sie im alltag [...] nicht ausleben können
(11) DJ: °h währenddessen **bricht** der vulkan weiter **auf**
DJ: **aus**

(12) US: °h °h diesen werbespot da is_n total betrunkener
der kriegt dann von vo der **läuft** auf der straße so **hin**
und **her** also der is wirklich sturzbetrunken

4 Related Work

Volk et al. (2016) proposed a method for detecting and recombining German separable verbs by locating the verb particles in the sentences and attaching them to the preceding verb stems. They report a precision of 97% when working with correct PoS-tags. Besides recomputing the lemma, Volk and colleagues also integrate a PoS-correction of multi-word adverbs such as *ab und an* or *ab und zu* that are frequently mistagged as verb particles. Bott and im Walde (2015) recompiled the lemmas of separable verbs by relying on a dependency parser, which proved to improve the performance of the prediction of the semantic compositionality of German particle verbs. Nagy and Vincze (2014) introduced a machine learning-based tool VPCTagger for identifying English particle verbs. For theoretical aspects regarding particle verbs see Stiebels (1996), Lüdeling (2001) and Poitou (2003).

5 Conclusion and Outlook

Our study shows that the method proposed by Volk et al. (2016) can be transferred successfully to a spoken language corpus like FOLK. An additional annotation layer can automatically be added in which information useful for frequency counts and corpus queries is represented with sufficient accuracy. Our analyses have also revealed approaches to optimising this procedure for either higher precision or higher recall.

Another highly frequent phenomenon in spoken language, which is structurally similar and could thus be treated in an analogous manner, are pronominal adverbs (see also Kaiser and Schmidt (2016)). Here, too, we observe alternations between combined forms (example 13) and separated forms (example 14).

(13) OB: (.) ich hab kohle **dafür** gekricht
(14) CT: ja auf ihre (.) also das **da** zahln wir nix **für**

Using the same approach with different PoS tags (ADV and APPR) and a suitable list of pronominal adverbs may serve to reconstruct these forms. We plan to test this approach in the future.

References

Bott, S., im Walde, S.S.: Exploiting fine-grained syntactic transfer features to predict the compositionality of German particle verbs. In: Proceedings of the 11th Conference on Computational Semantics (IWCS), London, pp. 34–39 (2015)

Kaiser, J., Schmidt, T.: Einführung in die Benutzung der Ressourcen DGD und FOLK für gesprächsanalytische Zwecke. Handreichung: Einfache Recherche-Anfragen als Übungs-Beispiele. Institut für Deutsche Sprache, Mannheim. (2016) http://nbn-resolving.de/urn:nbn:de:bsz:mh39-55360

Lüdeling, A.: On Particle Verbs and Similar Constructions in German. CSLI, Stanford (2001)

Poitou, J.: Fortbewegungsverben, Verbpartikel, Adverb und Zirkumposition. Cahiers d'études Germaniques **2003**, 69–84 (2003)

Schmid, H.: Improvements in part-of-speech tagging with an application to German. In: Proceedings of the ACL SIGDAT-Workshop, pp. 47–50 (1995)

Schmidt, T.: The database for spoken German - DGD2. In: Proceedings of the Ninth Conference on International Language Resources and Evaluation (LREC 2014), Reykjavik, Iceland, pp. 1451–1457 (2014)

Schmidt, T.: Good practices in the compilation of FOLK, the research and teaching corpus of spoken German. Int. J. Corpus Linguist. **21**(3), 396–418 (2016)

Selting, M., Auer, P., Barth-Weingarten, D., Bergmann, J.R., Bergmann, P., Birkner, K., Couper-Kuhlen, E., Deppermann, A., Gilles, P., Günthner, S., Hartung, M., Kern, F., Mertzlufft, C., Meyer, C., Morek, M., Oberzaucher, F., Peters, J., Quasthoff, U., Schütte, W., Stukenbrock, A., Uhmann, S.: Gesprächsanalytisches Transkriptionssystem 2 (GAT 2). Gesprächsforschung - Online-Zeitschrift zur verbalen Interaktion **10**, 353–402 (2009)

Stiebels, B.: Lexikalische Argumente und Adjunkte: zum semantischen Beitrag von verbalen Präfixen und Partikeln. Studia Grammatica 39. Akademie Verlag, Berlin (1996)

Nagy, I.T., Vincze, V.: VPCTagger: detecting verb-particle constructions with syntax-based methods. In: Proceedings of the 10th Workshop on Multiword Expressions (MWE), Gothenburg, Sweden, pp. 17–25 (2014)

Volk, M., Clematide, S., Graën, J., Ströbel, P.: Bi-particle adverbs, PoS-Tagging and the recognition of German separable prefix verbs. In: Proceedings of the 13th Conference on Natural Language Processing (KONVENS 2016), Bochum, pp. 297–305 (2016)

Westpfahl, S., Schmidt, T.: FOLK-Gold - A GOLD standard for Part-of-Speech-Tagging of spoken German. In: Proceedings of the Tenth International Conference on Language Resources and Evaluation (LREC 2016), Portorož, Slovenia, pp. 1493–1499 (2016)

Detecting Vocal Irony

Felix Burkhardt[1]([✉]), Benjamin Weiss[2], Florian Eyben[3],
Jun Deng[3], and Björn Schuller[3]

[1] Deutsche Telekom AG, Berlin, Germany
felix.burkhardt@telekom.de
[2] Technische Universität Berlin, Berlin, Germany
[3] audEERING GmbH, Gilching, Germany

Abstract. We describe a data collection for vocal expression of ironic utterances and anger based on an Android app that was specifically developed for this study. The main aim of the investigation is to find evidence for a non-verbal expression of irony. A data set of 937 utterances was collected and labeled by six listeners for irony and anger. The automatically recognized textual content was labeled for sentiment. We report on experiments to classify ironic utterances based on sentiment and tone-of-voice. Baseline results show that an ironic voice can be detected automatically solely based on acoustic features in 69.3 UAR (unweighted average recall) and anger with 64.1 UAR. The performance drops by about 4% when it is calculated with a leave-one-speaker-out cross validation.

1 Introduction

Verbal irony occurs when someone says something that is obviously not expressing the real intention or meaning; sometimes it is even the opposite. This is usually achieved by gross exaggeration, understatement or sarcasm. The speakers rely on the receiver's knowledge to decode the real meaning, usually because the context of the semantics does not fit (example: saying "beautiful weather" when it starts to rain) or the utterance contains contrasting polarities, as for example in "slow internet is exactly what I need."

Beneath the semantic contrast, this can be also be achieved by contrasting the "tone of voice" with the sentiment of the words. According to the Relevance Theory (Wilson and Sperber 2012), irony in speech can be considered as an attitude towards the statement, consequently modifying its meaning – for example a verbally positive statement realized with a prosody that indicates a negative attitude. An open question as of yet is whether a purely acoustic form of irony exists, i.e., an "ironic tone of voice" irrespective of the words being uttered. Studies on this have not been very promising, indicating that prosody alone is insufficient to signal irony (Bryant and Fox Tree 2005). Still, whereas a "dry" version does not reveal this ironic attitude in prosody, "dripping" irony does differ from sincere utterances in prosody (Bryant and Fox Tree 2005).

G. Rehm and T. Declerck (Eds.): GSCL 2017, LNAI 10713, pp. 11–22, 2018.
https://doi.org/10.1007/978-3-319-73706-5_2

Single target words exhibit acoustic differences between ironic from literal meaning in female German. The ironic versions were lower in fundamental frequency (F0), had longer and hyper-articulated vowels, and more energy (Scharrer et al. 2011). A similar study for French females reveled an expanded F0 range with higher average F0, syllable lengthening, and a raised F0 contour instead of a falling one as discriminating ironic from sincere versions in target words (González-Fuente et al. 2016). A subsequent re-synthesis experiment on the same target words confirmed the effects for all three features combined, for lengthening only, and pitch contour only. For English, particularly a lower average F0, but also lower F0 range, longer durations and higher intensity are found in ironic conditions (Cheang and Pell 2008; Rockwell 2000). Directly contrasting adjunct sentences, ironic utterances were slower (Bryant 2010).

This incongruence in affective prosody with the valence of the semantics (Bryant and Fox Tree 2002; Woodland and Voyer 2011) can be even observed in fMRI data (Matsui et al. 2016). From our daily experience probably most of us will tend to agree that this discrepancy between content and prosodic attitude can be easily detected – for the "dripping" case of intentionally signaling this ironic attitude at least. Therefore, an automatic irony detector should be feasible to develop based on sufficient training data, but to our knowledge no such data has be collected and investigated with the goal of automatic classification as of yet.

Irony detection is obviously an important step in human-machine communication as it can reverse the meaning of an utterance. One use-case for example would be the automatic sentiment analysis of vocal social media data for market research, another use-case would be to enhance the semantic processing of robots and software agents with speech interfaces.

This paper is structured as follows. Section 2 describes the Android app that was used to collect data on ironic utterances. Section 3 explains the acoustic classification process, whereas Sect. 4 does the same for the textual sentiment categorization. In Sect. 5, the data collection process is described, and in Sect. 6 the way the data was labeled for emotional expression as well as textual sentiment. It is followed by Sect. 7, which reports on some experiments we performed. The paper concludes with a summary and outlook in Sect. 8.

2 The Irony Simulation App

For data collection, an Android app was created which is capable of recording audio, streaming it to audEERING's sensAI API service for analysis of emotional parameters. The app provides various pictures on which the users can comment on to evoke angry or ironic responses. However, these pictures only served as a first stimulation, but users need not to adhere to the pictures. They were free to test the app with any neutral, emotional, ironic, or angry comment they would come up with. Likewise, although the chosen image was logged by the app, it was not used any further.

Fig. 1. Main screen of the irony data collection app, while a result is being shown. (Color figure online)

The main focus of the study was on the detection of irony. The detection of anger was added to compare findings with earlier studies that focused on angry speech detection (Burkhardt et al. 2009).

The app displays the result to user immediately after processing. For the preliminary irony detection module, a text transcription based on the Google Speech Services API and a ternary text sentiment classification was added to audEERING's sensAI API. Further, for the purpose of collecting a spoken irony database, all requests to the API made through this app were saved on the back end in order to build the database from these recordings.

In order to get an estimate on the individual usage behavior, the Android ID of the client is submitted and stored on the server. The owners name, however, is not collected. Of course we cannot control that all utterances stemming from one Android device are made by the same person.

Figure 1 displays the main screen of the irony data collection app, while it is showing a recognition result. The results shown in the App include the text transcript, the level of emotional activation (*Aktivierung*), valence (*Valenz*), and anger (*Ärger*), as well as the textual transcription (first line) and the textual Sentiment (positive/neutral/negative). Binary classification results for irony and anger are shown through the buttons at the top. The green button indicates a negative result (no irony/anger), and a flashing red button would indicate a positive result (irony and/or anger). The irony detection in the data collection app is based on rules that detect a mismatch between positive textual sentiment and negative acoustic valence or a very low acoustic activation alongside a positive sentiment. Binary anger detection is based on a threshold to the anger level, as well as thresholds to activation and valence.

The user can listen to his/her recording again by pressing the play button. Using the pencil icon on the right, the user can open a feedback dialogue, where irony and anger (binary yes/no labels) can be corrected and transmitted to the server. This feedback is optional to the user, however, most trial users were encouraged to make use of this functionality.

3   Acoustic Irony Classification

For transparency and reproducibility, here, we apply the widely used acoustic emotion classification framework to the irony database, which depends on the popular acoustic feature extraction tool openSMILE (Eyben et al. 2013) and Support Vector Machines (*SVMs*). Two audio feature sets, established in the field of computational paralinguistic, were used; the extended Geneva Minimalistic Acoustic Parameter Set (eGEMAPS) and the larger-scale Interspeech 2013 Computational paralinguistic Challenge feature set (ComParE) (Schuller et al. 2013).

eGeMAPS is a *knowledge-driven* data set that exploits the first two statistical moments (*mean* and *coefficient of variation*) to capture the distribution of *low-level descriptors* (LLDs) describing spectral, cepstral, prosodic and voice quality information, creating an 88 dimensional acoustical representation of an utterance. It was specifically designed by a small group of experts to be a basic standard acoustic parameter set for voice analysis tasks including paralinguistic speech analysis. For full details the reader is referred to Eyben et al. (2016). ComParE, on the other hand, is a large-scale brute forced acoustic feature set which contains 6 373 features representing prosodic, spectral, cepstral and voice quality LLDs. A detailed list of all LLDs for ComParE is given in Table 3. For full details on ComParE the reader is referred to Eyben et al. (2013).

For classifiers, we use open-source implementations from Scikit-learn (Pedregosa et al. 2011), where SVMs with linear kernels are considered. We scale all features to zero mean and unit standard deviation by using the parameters derived from the training set. For the baseline, we introduce two types of cross validations: 5-fold cross validation (5-fold CV) and leave-one-speaker-out cross-validation (LOSO-CV). By that, it is desired that the results obtained are more representative.

4   Textual Sentiment Classification

The textual sentiment analysis is based on the GATE framework (Cunningham et al. 2011). The name of the grammar formalism used with GATE is Jape. The first steps involve splitting and tokenization. We added Stanford Part-of-Speech Tagger (Toutanova and Manning 2000) and a lemmatizer based on a German lexicon derived from the Morphy project (Lezius 2010). A German polarity lexicon (Waltinger 2010) is then used to identify tokens that carry positive or negative polarity. Further gazetteers annotate negation tokens (like *"not* great") and strengthener words (like *"super* bad"). Because tokens can be both polarity words and negation or strengthener words, we use a Jape transducer to neutralize them when they occur in conjunction with a polarity token. Another set of Jape rules reverses the polarity for polarity words that are negated (example: "not bad") or used as strengtheners (example "super bad"). Finally, a polarity decision for the whole input sentence is computed by a comparison of the number of positive and negative tokens. The number of polarity tokens with the most frequent polarities, normalized by the number of all tokens, is being used as a confidence measure.

5 Data Collection

We conducted a workshop with lay people to gain experience on how the automatic recognition of sentiment will be perceived by users and also with the aim to collect some user data. Because we were not aware of any "ironic voice" acoustic data collection with the aim of automatic classification, we could not compute a model for irony and the app indicated irony when meeting a discrepancy between textual sentiment and vocally expressed valence, as described in Sect. 2.

During the workshop, 12 test users got introduced to the app and tried it out. There were nine male and three female participants, aged 25 to 48, average age 37 years, with 7.8 standard deviation, all native Germans currently living in Berlin. The testing subjects installed the app on their private Android phones and were instructed to use it for one week. After this time span, the server that the app needs to analyse the audio was shut down. They were compensated for their time and all of them signed agreement that the data will be recorded, stored and can be used and distributed for scientific research.

After the try-out period a set of 937 labeled samples had been collected, see Sect. 6 for details on the label process. The number of different Android device IDs is 21. There is 3910 s (about one hour) audio in total. The maximum, minimum, and average length are 21.84, 0.84, and 4.18 s respectively. The number of recognized words are at most 35 and at least 1 with average value 7.14 and standard deviation 4.75. Using a majority voting on the labelers results in 51.57% of vocally ironic, 41.91% angry and 22.15% both ironic and angry utterances.

6 Data Labeling

Two sets of data labels are available in the database: The first is obtained directly from the users who contributed the speech data through the feedback functionality of the app; the second is obtained by manual rating from six labelers.

For rating, the crowd-sourcing platform iHEARu-Play (Hantke et al. 2015) was used. The platform was developed to allow raters to easily work from anywhere they feel comfortable through a web-based interface. No actual crowd-sourcing features which the platform provides (automatic rater trustability, dealing with raters who only rate parts of the database, etc.) were used here. Six student (psychology, linguistics, theatre) and professional (some of the authors of the paper) raters rated the whole database for angry and ironic "sound of

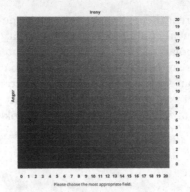

Fig. 2. Two dimensional rating matrix used for human labeling of perceived irony and anger levels. Participants were prompted for their response with these instructions: *Please rate the level of anger and irony that you hear from the voice in the recording. Try to ignore the wording and listen only to the sound of the voice! If you are unsure, or no audio is audible, please choose the point (20, 20 → upper right corner) – Do not choose this point otherwise (use 19;19, 20;19, or 19;20 instead if you really need to rate a very high anger and high irony level).*

voice". They were instructed to ignore the textual content of the utterances as much as possible and judge solely based on how the speech is expressed non-verbally. The judgements were given as a point in a two dimensional matrix with the dimensions "irony" and "anger", with twenty points on each axis for quasi-continuous rating. Figure 2 shows the labeling interface in which the raters had to choose one point which reflected anger and irony levels.

In order to obtain, from the six individual ratings, a single gold standard for automatic classification, the Evaluator Weighted Estimator (EWE) is used (Grimm and Kroschel 2005; Grimm et al. 2007). The EWE is a weighted average of the individual ratings, where the weights are based on the average inter-rater reliability of each rater. All evaluations in this paper are based on the EWE.

To compute EWE the following steps are performed:

1. Normalisation of ratings of each rater to 0 mean, and maximum range (1..5)
2. Normal (arithmetic) averaging of all ratings for each segment
3. Computation of EWE weights (r) as Pearson correlation of each rater's ratings with the average rating from b)
4. Normalisation of EWE weights to sum 1 and min and max. correlations to 0 and 1
5. Computation of final EWE average rating by weighted average using EWE weights

In Tables 1 and 2, the pairwise rater agreements and the agreement of each rater with the EWE (also including this rater) are displayed as cross correlation coefficients. Further, the mean, minimum, and maximum pairwise rater correlations are shown (excluding correlations of raters and EWE).

Table 1. Pairwise cross correlation (pearson correlation coefficient ρ) for the dimension irony. Mean, Min., Max. values exclude the Evaluator Weighted Estimator (mean) (EWE) values (first row).

Rater #		1	2	3	4	5	6	AVG
EWE		0.89	0.68	0.70	0.82	0.87	0.73	0.81
1			0.52	0.59	0.66	0.77	0.67	0.73
2				0.44	0.48	0.45	0.42	0.57
3					0.54	0.59	0.56	0.63
4						0.68	0.58	0.68
5							0.65	0.72
6								0.66
Mean ρ	0.57							
Min ρ	0.42							
Max ρ	0.77							

Table 2. Pairwise cross correlation (pearson correlation coefficient ρ) for the dimension anger. Mean, Min., Max. values exclude the Evaluator Weighted Estimator (mean) (EWE) values (first row).

Rater #		1	2	3	4	5	6	AVG
EWE		0.75	0.78	0.76	0.79	0.73	0.76	0.80
1			0.45	0.59	0.52	0.59	0.58	0.64
2				0.60	0.57	0.56	0.67	0.66
3					0.56	0.59	0.58	0.67
4						0.60	0.57	0.66
5							0.57	0.66
6								0.68
Mean ρ	0.57							
Min ρ	0.45							
Max ρ	0.67							

6.1 Textual Data Labeling

The textual data was labeled for sentiment by one of the authors with one of the three categories "neutral", "positive" or "negative". As stated in Sect. 2, the textual data is the result of the Google Speech API ASR. It was not manually corrected but used including the errors. An evaluation of the performance by computing WER or BLEU has not been done yet.

We found several situations where the decision for a sentiment was rather difficult. For example when two sentiments were given (example: "this is good but this is bad"), a question was raised (example: "is this bad?"), someone else's

Table 3. ComParE acoustic feature set: 65 low-level descriptors (LLD).

4 energy related LLD	Group
RMS energy, zero-crossing rate	Prosodic
Sum of auditory spectrum (loudness)	Prosodic
Sum of RASTA-filtered auditory spectrum	Prosodic
55 spectral LLD	Group
MFCC 1–14	Cepstral
Psychoacoustic sharpness, harmonicity	Spectral
RASTA-filt. aud. spect. bds. 1–26 (0–8 kHz)	Spectral
Spectral energy 250–650 Hz, 1 k–4 kHz	spectral
Spectral flux, centroid, entropy, slope	Spectral
Spectral Roll-Off Pt. 0.25, 0.5, 0.75, 0.9	Spectral
Spectral variance, skewness, kurtosis	spectral
6 voicing related LLD	Group
F_0 (SHS and Viterbi smoothing)	Prosodic
Prob. of voicing	Voice qual.
log. HNR, jitter (local and δ), shimmer (local)	Voice qual

opinion was given (example: "he says it's bad"), contradictory statements were given (example: "I love it when it's bad") or complex statements ("it'd be better to spend the money on crime defense" being rated positive because of the word "better", whereas "this I'd like to see every day" being rated as neutral because no polarity words are used).

7   Experiments

7.1   Irony and Anger Expression

In our following experiments with acoustic irony and anger classification, we unified the EWE values from four raters by using majority voting and then mapping them to two classes. In particular, we use the average value of the unified EWE values to form the binary labels: for acoustic irony classification, if an unified EWE value for one utterance is larger than the average value, then its label is assigned as "irony", otherwise as "non-irony". Similarly, we assign the binary anger labels (i.e., "anger" or "non-anger") for the collected database.

7.2   Sentiment Analysis

We evaluated and tuned the textual sentiment analyzer using the data collection. The set of 915 unique sentences was split into 165 samples for a test set and 750 for training. Out-of-the-box, 109 out of the 165 test samples were correctly classified by the Sentiment analyzer (33.93% error rate). We then went through the training set and corrected wrong classifier decisions by adding rules or editing the polarity lexicon and reached an error rate of 27.27%. With larger data-sets it would make sense to use machine learning for classification.

7.3 Acoustic Emotion Classification Results

First, Table 4 shows the acoustic irony classification results for the LOSO-CV (leave-one-speaker-out cross validation) and 5-fold CV with different complexity parameters of the linear SVM. It can be seen clearly that ComParE outperforms eGEMAPS by a noticeable margin. Specifically, the baseline systems using eGEMAPS achieve promising UARs (Unweighted Average Recall) of 54.6% and 65.6% for the LOSO-CV and 5-fold CV evaluation schemes. In the meantime, the ComParE-based systems result in UARs of 61.4% and 69.6%, which is higher than the one obtained by the corresponding eGEMAPS systems. It is worth noting that the ComParE-based systems are more robust to the choice of the complexity of the linear SVM when compared to the eGEMAPS baseline systems for the acoustic irony classification.

Next, Table 5 presents the experimental results for the acoustic anger classification task, which was previously defined in Sect. 7.1. Based on Table 5, we can see that the baseline systems obtain representative performance in terms of UARs for both the LOSO-CV and 5-fold CV evaluation schemes. It is surprising for the LOSO-CV scheme that the linear SVM system using the eGEMAPS feature set of only 88 features reaches a UAR of 63.5%, which is clearly higher than

Table 4. Results in terms of Unweighted Average Recall (UAR, %) for the binary irony classification task with ComParE and eGEMAPS acoustic feature sets. LOSO-CV stands for the leave-one-speaker-out cross validation while 5-fold CV stands for the 5-fold cross validation. C indicates the complexity parameter for the linear SVM, which corresponds to the penalty parameter of the error term.

Feature	C(SVM)	UAR	
		LOSO-CV	5-fold CV
ComParE	1e−4	**61.4**	69.3
ComParE	1e−3	61.3	**69.6**
ComParE	1e−2	60.6	67.9
ComParE	1e−1	60.3	67.9
ComParE	1.0	60.4	67.9
eGEMAPS	1e−4	48.9	53.4
eGEMAPS	1e−3	53.9	65.5
eGEMAPS	1e−2	**54.6**	**65.6**
eGEMAPS	1e−1	54.0	64.1
eGEMAPS	1.0	52.6	64.8

Table 5. Results in terms of Unweighted Average Recall (UAR, %) for the binary anger classification task with ComParE and eGEMAPS acoustic feature sets.

Feature	C(SVM)	UAR	
		LOSO-CV	5-fold CV
ComParE	1e−4	**59.0**	**67.1**
ComParE	1e−3	57.5	63.7
ComParE	1e−2	58.8	62.5
ComParE	1e−1	58.6	62.5
ComParE	1.0	58.6	62.5
eGEMAPS	1e−4	62.2	53.1
eGEMAPS	1e−3	**63.5**	63.8
eGEMAPS	1e−2	63.0	**64.1**
eGEMAPS	1e−1	61.5	63.3
eGEMAPS	1.0	57.6	61.8

the one obtained by the ComParE feature set of 6 373 features. This suggests that the knowledge-driven feature set of eGEMAPS contain more informative information for the anger classification task in the LOSO situation. For the 5-fold CV scheme, the ComParE baseline system surpasses the eGEMAPS system, just like we found with the acoustic irony classification.

8   Summary and Outlook

We described a data collection for vocal irony and anger expression based on an Android app that was specifically developed for this study. The main aim of the investigation was to find evidence for a non-verbal expression of irony. A data set of 937 utterances was collected and labeled by six listeners for irony and anger (for comparability with earlier studies). The automatically recognized textual content was labeled for sentiment. We conducted experiments to classify ironic utterances based on sentiment and tone-of-voice with machine learning. The results show that irony can be detected automatically solely based on acoustic features in 69.3 UAR (unweighted average recall) and anger with 64.1 UAR. The performance drops by about 4% when it is calculated with a leave-one-speaker-out cross validation.

It is planned to make the collected data available to the research community to foster the investigation of ironic speech.

There are many ideas to work on these topics, including:

- Enhance the textual sentiment classifier by adding more rules.
- When sufficient data was collected, enhance the textual sentiment classifier by machine learning.

- Investigate the influence of ASR errors on the sentiment classification, for example by contrasting it with a manual transcription.
- Differentiate between categories of vocal and non-vocal expressions of irony and investigate with which modalities are best suited to detect them.
- Add more modalities, beneath text and voice, to irony detection, for example facial recognition.
- Investigate in how far prosodic expression of irony is culture dependent, as indicated by (Cheang and Pell 2009), for example by testing the data with non-German speakers from different cultures.
- Validate hypotheses on acoustic expressions of irony by re-synthesis experiments, for example by systematically varying acoustic features of target sentences with speech synthesis and doing a perception test.

References

Bryant, G.A.: Prosodic contrasts in ironic speech. Discourse Process. **47**(7), 545–566 (2010)

Bryant, G.A., Fox Tree, J.E.: Recognizing verbal irony in spontaneous speech. Metaphor Symb. **17**(2), 99–117 (2002)

Bryant, G.A., Fox Tree, J.E.: Is there an ironic tone of voice? Lang. Speech **48**(3), 257–277 (2005)

Burkhardt, F., Polzehl, T., Stegmann, J., Metze, F., Huber, R.: Detecting real life anger. In: Proceedings ICASSP, Taipei, Taiwan, April 2009

Cheang, H.S., Pell, M.D.: The sound of sarcasm. Speech Commun. **50**(5), 366–381 (2008)

Cheang, H.S., Pell, M.D.: Acoustic markers of sarcasm in cantonese and english. J. Acoust. Soc. Am. **126**(3), 1394–1405 (2009)

Cunningham, H., Maynard, D., Bontcheva, K., Tablan, V., Aswani, N., Roberts, I., Gorrell, G., Funk, A., Roberts, A., Damljanovic, D., Heitz, T., Greenwood, M.A., Saggion, H., Petrak, J., Li, Y., Peters, W.: Text Processing with GATE (Version 6) (2011). ISBN 978-0956599315. http://tinyurl.com/gatebook

Eyben, F., Weninger, F., Gross, F., Schuller, B.: Recent developments in openSMILE, the munich open-source multimedia feature extractor. In: Proceedings of the ACM Multimedia 2013, Barcelona, Spain, pp. 835–838. ACM (2013)

Eyben, F., Scherer, K.R., Schuller, B.W., Sundberg, J., André, E., Busso, C., Devillers, L.Y., Epps, J., Laukka, P., Narayanan, S.S., Truong, K.P.: The geneva minimalistic acoustic parameter set (GeMAPS) for voice research and affective computing. IEEE Trans. Affect. Comput. **7**(2), 190–202 (2016)

González-Fuente, S., Pilar, P., Noveck, I.: A fine-grained analysis of the acoustic cues involved in verbal irony recognition in French. In: Proceedings of the Speech Prosody, pp. 1–5 (2016)

Grimm, M., Kroschel, K.: Evaluation of natural emotions using self assessment manikins. In: Proceedings of the IEEE Workshop on Automatic Speech Recognition and Understanding (ASRU) 2005, Cancun, Maxico, pp. 381–385. IEEE, November 2005

Grimm, M., Mower, E., Kroschel, K., Narayanan, S.: Primitives based estimation and evaluation of emotions in speech. Speech Commun. **49**, 787–800 (2007)

Hantke, S., Eyben, F., Appel, T., Schuller, B.: iHEARu-PLAY: introducing a game for crowdsourced data collection for affective computing. In: Proceedings of the 1st International Workshop on Automatic Sentiment Analysis in the Wild (WASA 2015) Held in Conjunction with the 6th Biannual Conference on Affective Computing and Intelligent Interaction (ACII 2015), Xi'an, P.R. China, pp. 891–897. AAAC, IEEE, September 2015

Lezius, W.: Morphy - German morphology, part-of-speech tagging and applications. In: Proceedings of the 9th EURALEX International Congress, pp. 619–623 (2000)

Matsui, T., Nakamura, T., Utsumi, A., Sasaki, A.T., Koike, T., Yoshida, Y., Harada, T., Tanabe, H.C., Sadato, N.: The role of prosody and context in sarcasm comprehension: behavioral and fMRI evidence. Neuropsychologia **87**, 74–84 (2016)

Pedregosa, F., Varoquaux, G., Gramfort, A., Michel, V., Thirion, B., Grisel, O., Blondel, M., Prettenhofer, P., Weiss, R., Dubourg, V., Vanderplas, J., Passos, A., Cournapeau, D., Brucher, M., Perrot, M., Duchesnay, É.: Scikit-learn: machine learning in python. J. Mach. Learn. Res. **12**, 2825–2830 (2011)

Rockwell, P.: Lower, slower, louder: vocal cues of sarcasm. J. Psycholinguist. Res. **29**(5), 483–495 (2000)

Scharrer, L., Christmann, U., Knoll, M.: Voice modulations in German ironic speech. Lang. Speech **54**(4), 435–465 (2011)

Schuller, B., Steidl, S., Batliner, A., Vinciarelli, A., Scherer, K., Ringeval, F., Chetouani, M., et al.: The INTERSPEECH 2013 computational paralinguistics challenge: social signals, conflict, emotion, autism. In: Proceedings of the INTERSPEECH, Lyon, France, pp. 148–152. ISCA (2013)

Toutanova, K., Manning, C.D.: Enriching the knowledge sources used in a maximum entropy part-of-speech tagger. In: Proceedings of the Joint SIGDAT Conference on Empirical Methods in Natural Language Processing and Very Large Corpora (EMNLP/VLC-2000), pp. 63–70 (2000)

Waltinger, U.: GermanPolarityClues: a lexical resource for German sentiment analysis. In: Proceedings of the Seventh conference on International Language Resources and Evaluation (LREC 10) (2010)

Wilson, D., Sperber, D.: Explaining irony. In: Meaning and Relevance, pp. 123–146. Cambridge University Press, Cambridge (2012)

Woodland, J., Voyer, D.: Context and intonation in the perception of sarcasm. Metaphor and Symb. **26**(3), 227–239 (2011)

The Devil is in the Details: Parsing Unknown German Words

Daniel Dakota[✉]

Department of Linguistics, Indiana University, Bloomington, USA
ddakota@indiana.edu

Abstract. The statistical parsing of morphologically rich languages is hindered by the inability of parsers to collect solid statistics because of the large number of word types in such languages. There are however two separate but connected problems, reducing data sparsity of known words and handling rare and unknown words. Methods for tackling one problem may inadvertently negatively impact methods to handle the other. We perform a tightly controlled set of experiments to reduce data sparsity through class-based representations in combination with unknown word signatures with two PCFG-LA parsers that handle rare and unknown words differently on the German TiGer treebank. We demonstrate that methods that have improved results for other languages do not transfer directly to German, and that we can obtain better results using a simplistic model rather than a more generalized model for rare and unknown word handling.

1 Introduction

Parsing morphologically rich languages (MRLs) has proven to be a challenge for statistical constituent parsing. The relative success for English has not been achieved on other languages, particularly MRLs as the computational methods and algorithms that yield good results are not directly transferable to other languages, which have been shown to be intrinsically harder to parse (Nivre et al. 2007). This can be attributed to the various linguistic properties these languages possess (e.g. freer word order), which present difficulties for capturing their more complex syntactic behaviors. Such properties are attributed to a higher degree of inflectional morphology, resulting in increased data sparsity from a substantial proportion of word types occurring rarely in a text (Tsarfaty et al. 2010).

German contains characteristics of more rigid word order languages like English, such as verb placement, but also possesses many phenomena that are present in MRLs, such as generally freer word order, resulting in being coined as a morphologically rich-less configurational language (MR&LC), a position between configurational and non-configurational languages (Fraser et al. 2013). The language also possesses problematic phenomena for NLP, such as case syncretism, which require information between morphology and syntax to more accurately disambiguate constituents.

G. Rehm and T. Declerck (Eds.): GSCL 2017, LNAI 10713, pp. 23–39, 2018.
https://doi.org/10.1007/978-3-319-73706-5_3

In order to improve statistical parsing in general, but especially for MRLs, two problems must be addressed: the need to reduce data sparsity and the treatment of unknown words. Many tools, such as POS taggers and parsers, have sophisticated internal mechanisms to handle unknown words and by default often perform better than simplistic probability models. However, the weakness of sophisticated models is they can over-generalize, biasing themselves against the very words for which they are trying to compensate. A simplistic unknown word handling model, which is not affected in this way, can benefit greatly from both the reduction of data sparsity and simplistic treatment of unknown words, surpassing results from more sophisticated models. We examine two separate but connected problems, the interaction between parser-internal probability models for handling unknown and rare words and performance, while also simultaneously reducing data sparsity issues using Brown clustering and word signatures for rare and unknown words.

The paper is structured as follows. We discuss previous literature in Sect. 2 followed by our experimental methodology in Sect. 3. Results and discussion are presented in Sects. 4 and 5 respectively, before concluding in Sect. 6.

2   Related Work

2.1   Handling Rare and Unknown Words

Handling rare and unknown words are two separate but equal components that are intrinsic to many NLP applications such as lemmatizers, POS taggers, and parsers. The task becomes more difficult when processing MRLs, due to the exponential increase of word forms as they have higher ratios of word forms to lemmas (Tsarfaty et al. 2010). The chosen methodology however has different practical applications for different NLP tasks. A POS tagger may only need to be concerned with the lexical level within a trigram, whereas a parser may be concerned with an unlexicalized trigram and the constraints of its own grammar. Thus the probability models and goals of the tools are not the same, and an effective method used for one task may not be ideal for another.

A reasonable treatment of rare words that occur below a given threshold is to handle them identically to unknown words due to the inability to obtain reliable distributional statistics. The most simple approach is to reserve a proportion of the probability mass, assigning each word equal weight and mapping them to an UNK symbol. This *simple lexicon* is universal in its application, but suffers from an oversimplification of the problem, and its inability to make more informed decisions. Specifically, each unknown word will be given equal weight when intuitively we know that certain words are more likely to occur in a sequence. These probabilities are obtained from a training set with the majority tag becoming the default (see Attia et al. (2010) for a comprehensive parsing example), which, strictly speaking, determines the the tool's performance on any subsequent application.

More sophisticated approaches try to allow for generalization while taking into account that not all words are equally likely. The PCFG-LA Berkeley parser

(Petrov and Klein 2007a,b) uses rare words to estimate unknown words by obtaining statistics on the rare words with latent tags and uses linear smoothing to redistribute the emission probabilities across the rare words (Huang and Harper 2009, 2011). Although this allows for good generalizations in a PCFG-LA setting, this has been shown to cause rare words to suffer more from over-fitting than frequent words (Huang and Harper 2009) and to not effectively handle out-of-vocabulary (OOV) words as it can only generate probabilities of words seen in the training data (Huang et al. 2013). The parser also has what is referenced as a *sophisticated model*, which uses a more linguistically informed approach to handle OOV words by exploiting word formation characteristics, such as affixes and capitalization, but its approach has been shown to be biased towards an English lexicon (Hall et al. 2014). The development of language specific signatures can considerably improve performance (Huang and Harper 2009; Attia et al. 2010), but is often ignored in practice.

2.2 Word Clustering

A by-product of a more robust morphological system in a language is an increase in word forms, resulting in an increase of data sparsity. Various forms of clustering have been utilized to reduce sparsity issues and increase class-based representations to improve performance through better probability emissions.

Brown clustering (Brown et al. 1992) is a unsupervised hard clustering algorithm that obtains a pre-specified number of clusters (C). The algorithm assigns the C most frequent tokens to their own cluster. The $C+1$ most frequent token is assigned to one of the pre-specified C clusters by creating a new cluster and merging the $C+1$ cluster with the cluster that minimizes the loss in likelihood of the corpus based on a bigram model determined from the clusters. This is repeated for every each $(C+N)th$ individual word types within the corpus, resulting in a binary hierarchical structure with each cluster encoded with a bit string. Words can be replaced by their bit string, thus choosing a short bitstring can drastically reduce the number of words in a corpus, allowing for a flexible granularity between POS tags and words. The distributional nature of the algorithm lends itself to the problem of clustering words that behave similarly syntactically by grouping words based on their most likely distribution, adding a semantic nuance to the clustering.

On what linguistic information: words, lemmas, or inflected forms; to perform clustering for MRLs is not obvious. Various linguistic information on which Brown clustering has been performed has yielded different results for different languages. This is further compounded by how cluster information can be incorporated into different parsers and the impact this has on each parser's performance.

Koo et al. (2008) demonstrated that cluster-based features for both English and Czech outperformed their respective baselines for dependency parsing. Ghayoomi (2012) and Ghayoomi et al. (2014) created clusters using word and POS information to resolve homograph issues in Persian and Bulgarian respectively, significantly improving results for lexicalized word-based parsing.

Candito and Crabbé (2009) clustered on *desinflected* words, removing unnecessary inflection markers using an external lexicon, subsequently combining this form with additional features. This improved results for unlexicalzed PCFG-LA parsing for both medium and higher frequency words (Candito and Seddah 2010), but was comparable to clustering the lemma with its predicted POS tag.

In contrast to Candito et al. (2010) who did not achieve substantial improvements for French dependency parsing using clusters, Goenaga et al. (2014) created varying granularities of clusters using words (for Swedish) and lemmas plus morphological information (for Basque) to obtain noticeable improvements for dependency parsing. Versley (2014) noted that cluster-based features improved discontinuous constituent parsing results for German considerably, but results are influenced by cluster granularities.

3 Methodology

3.1 Clustering Data

The data used for generating Brown clustering is a German Wikipedia dump consisting of approximately 175 million words (Versley and Panchenko 2012). The data includes POS and morphological information representative of the annotation schemas of TiGer. A single sequence of POS tags and morphological features was assigned using the MATE toolchain (Björkelund et al. 2010) with a model trained using cross-validation on the training set via a 10-fold jackknifing method assigning information regarding lemmas, POS tags, and morphology. We added the TiGer corpus into the Wikipedia data and retained punctuation, which may provide contextual clues for certain words for clustering purposes. We clustered on raw words, lemmas, and a combination of lemma and part of speech tags (lemma_POS) to obtain 1000 clusters for tokens occurring with a minimum frequency of 100.

3.2 Methods

For training and development, the TiGer syntactic treebank 2.2 (Brants et al. 2004) was utilized, specifically the 5k train and dev set from the SPMRL 2014 shared task data version (Seddah et al. 2014). Importantly, punctuation and other unattached elements are attached to the tree following Maier et al. (2012), resolving crossing-branches (for a full description of the data preprocessing, see Seddah et al. (2013b)).

Parsing experiments were performed using the Berkeley parser (Petrov and Klein 2007a,b) and the Lorg parser (Attia et al. 2010) which is a reimplementation of the Berkeley parser. The parsers learn latent annotations and probabilities (Matsuzaki et al. 2005; Petrov et al. 2006) in a series of split/merge cycles that evaluate the impact of these new annotations and merge back those deemed least useful, performing smoothing after each cycle, while calculating the EM after each step.[1]

The Lorg parser uses a *simple lexicon* unless a specific language signature file is specified.[2] In principle this is equivalent to the Berkeley setting of *simple lexicon* option, a point that will be further investigated in Sect. 4. The default unknown threshold for Lorg is five while the default rare word threshold for Berkeley it is 20. We experimented with German signatures for German unknown words and clusters to test the impact on results.

3.3 Evaluation

The SPMRL 2013 shared task scorer (Seddah et al. 2013b) was used for evaluation to report F-scores and POS accuracy. This script is a reimplementation of EVALB (Sekine and Collins 1997), but allows for additional options, such as completely penalizing unparsed sentences, which we include. We do not score grammatical functions and remove virtual roots with a parameter file, but do score for punctuation. We report results for both providing the parser with gold POS tags and parser-internal tagging on the development set[3] reporting the average over four grammars using four different random seeds $(1, 2, 3, 4)$ as Petrov (2010) noted that EM training within a PCFG-LA framework is susceptible to significant performance differences.

4   Results

4.1   Rare and Unknown Word Thresholds

Figures 1a to c show results for different settings of the unknown threshold for Lorg and the rare word threshold for Berkeley. The influence of the unknown threshold on Lorg's performance is negligible when the parser is given tags, but is significant for parser-internal tagging, with performance dropping by around 10% absolute. This is expected considering how easily influenced the *simplex lexicon* is by word frequencies. The small data sets may have an impact, but preliminary experiments with the full data sets show a similar trend, but less pronounced. The impact the rare word threshold have on Berkeley (see Fig. 1b) using the *sophisticated lexicon* however is not as pronounced for both gold tags and parser-internal tagging. The internal smoothing algorithm seemingly allows it to be less influenced by a change in its rare word thresholds, even with a small data set, as more words are simply subsumed, keeping the grammar rather intact.

---

[1] We trained without grammatical functions, due to the time it took in preliminary experiments to parse TiGer with grammatical functions, and use a split/merge cycle of 5.

[2] This currently only exists for English, French, and Arabic.

[3] The test set is left for final evaluation after further experimentation, although we note that the TiGer test set has been shown to be substantially harder to parse than the dev set (see Maier et al. 2014).

It is worth nothing however that the optimal setting is around 5 and not the default setting of 20. In order to examine the impact smoothing has on Berkeley, we performed experiments using the parser's *simple lexicon* option, presented in Fig. 1c, which is said to be the same as Lorg's *simple lexicon* model. These results stand in contrast to not only the results with the Berkeley's *sophisticated lexicon* smoothing of rare words, but the *simple lexicon* model of Lorg. Although the curves in Figs. 1b and c are similar, the actual performance is better using Berkeley's *sophisticated lexicon* approach, but these results can be partially attributed to the number of unparsed sentences (in the 100 s in some cases) for which the parser is penalized, as it is unable to find rules within its grammars for the given inputs. There is a substantial increase in F-score from a threshold of 1 to 5, but minimal increases there afterwards, with the best performance at a threshold of 15. The stark differences between the *simple lexicon* model implemented by Berkeley and Lorg suggests that there are undocumented implementation differences which are not strictly identical.

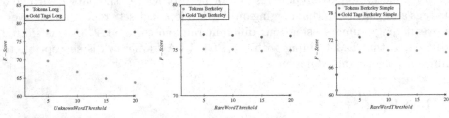

a UNK Thresholds Lorg Simple Lexicon

b Rare Word Thresholds Berkeley Sophisticated Lexicon

c Rare Word Thresholds Berkeley Simple Lexicon

Fig. 1. Rare and unknown word thresholds

In order to examine on what linguistic representations Brown clustering can be performed that has yielded improvements for other languages, we perform experiments on German by replacing all terminals with their POS tags, their lemmas, and lemmas and pos_information, with results presented in Table 1. Only results for the best performing unknown threshold (UNK TH.) for each parser is given, as well as for the lexicon reduction (Lex. Red.). Lexicon reduction is defined as the proportional decrease in the vocabulary size of the word types from the original Tiger dev set to the dev set replaced with clusters and UNK types.

For both lemmas and lemma_POS, all terminals with the following tags were replaced with their tags respectively: CARD, FM, and XY. Punctuation was left in its original form. When replacing terminals with POS tags, there is a drop in the F-score between gold tags and parser-internal tag of between 4–6% absolute for Lorg while this drops to between 1–2.5% for Berkeley. Every Lorg with gold tags outperforms its Berkeley counterpart, which is noteworthy given

Table 1. Results for orig, lemma, POS, and lemma_pos blue = gold POS tags | red = parser-internal tags

Parser	Terminal type	Parsed	UNK TH	F-score	POS Acc.	Lex. Red.
Lorg	orig	tokens	1	71.80	90.81	N/A
	orig	tagged	5	77.94	99.54	N/A
	POS	tokens	15	74.64	99.54	**99.61**
	POS	tagged	15	74.65	99.54	**99.61**
	lemma	tokens	1	71.54	90.87	27.83
	lemma	tagged	5	77.25	99.53	27.83
	lemma_pos	tokens	1	73.15	93.70	18.95
	lemma_pos	tagged	5	77.30	99.54	18.95
Berkeley	orig	tokens	5	75.10	94.04	N/A
	orig	tagged	5	76.69	99.87	N/A
	POS	tokens	15/20	74.20	98.89	**99.61**
	POS	tagged	15/20	74.17	99.92	**99.61**
	lemma	tokens	5	73.56	92.89	27.83
	lemma	tagged	5	75.91	99.83	27.83
	lemma_pos	tokens	10	75.21	95.97	18.95
	lemma_pos	tagged	10	76.01	99.93	18.95

that Lorg consistently has a higher number of unparsed sentences for which it is penalized, while Berkeley outperforms Lorg for parser-internal tagging, except for POS terminals. This suggests that the default handling of rare and unknown words is influential on the parsers subsequent performance on any downstream application without further enhancements, as Berkeley outperforms Lorg in its base form. Furthermore, a threshold of 1 on Lorg consistently achieving the best results should not be surprising as Attia et al. (2010) explicitly note that lower thresholds for Lorg perform best, thus the default thresholds are not necessarily ideal for a given language. This is supported by Seddah et al. (2013a), who noted that a threshold of 1, or true unknown words, resulted in the best performance for French out-of-domain parsing.

For Berkeley, the original treebank outperforms all other variations with gold POS tags, but for Lorg, replacing the terminals with their POS actually achieves the best performance for parser-internal tagging with lemma_pos performing second best overall. The results regarding replacing POS tags confirm the findings of Benoît and Candito (2008). Given that latent variables are obtained by splitting a terminal into two categories, it would seem reasonable that variation in terminals is needed for better approximation of latent categories, as such differences percolate up the tree. However, it is interesting to note that terminals consisting of POS tags still outperform replacing terminals with lemmas for parser-internal tagging. Replacing terminals with lemmas likely results in increased ambiguity of the syntactic nature of terminals.

4.2   Suffix Results

Not all words in the treebank have a cluster ID. In such cases, words can be considered rare or even unknown, even though they may appear in both the training and dev set, but are infrequent. In order to group infrequent words into more generalized categories, each non-clustered word is replaced with a UNK token, with various suffix lengths. Here a suffix is not inherently linguistically oriented, but strictly character length. Table 2 shows the impact that various suffix lengths of unknown words have on performance on Lorg.[4] The experiment *raw+orig* replaces terminals with cluster IDs and leaves the original terminal untouched if no cluster can be found. For all other experiments, words with no assignable cluster were changed to UNK_suffix$N$ where N is the length of the suffix on the UNK token (e.g. UNK_suffix2 for the word *spielen* "to play" would be UNK_en). The parser with gold POS tags shows little variation in performance on the suffix length. For parser-internal tags, there is slightly more variation but not substantial.

Table 2. Suffix length for UNK words for Lorg

Token type	Parsed	UNK TH.	F-score	POS Acc.	Lex. Red.
raw+orig	tokens	1	75.90	93.45	59.24
raw+orig	tagged	1	78.16	99.52	59.24
raw+unk_suffix0	tokens	1	75.88	93.26	**93.86**
raw+unk_suffix0	tagged	1	78.26	99.45	**93.86**
raw+unk_suffix1	tokens	5	76.14	94.05	93.45
raw+unk_suffix1	tagged	5	78.05	99.53	93.45
raw+unk_suffix2	tokens	5	76.27	94.23	91.09
raw+unk_suffix2	tagged	10	78.20	99.40	91.09
raw+unk_suffix3	tokens	1	76.05	93.86	86.61
raw+unk_suffix3	tagged	5	78.10	99.40	86.61
raw+unk_suffix4	tokens	1	76.03	93.92	80.63
raw+unk_suffix4	tagged	5	78.34	99.49	80.63

Although the best suffix length is not clear, we choose a suffix of length 2 for our additional experiments for three reasons: (1) it achieves the best results on average for parser-internal tagging; (2) it adequately balances between lexicon reduction and additional information as the German alphabet consists of 30 letters,[5] thus a suffix of length two will have at most $30^2 = 900$ possible combinations where a suffix of length 4 will have $30^4 = 810000$ possible combinations;

---

[4] Experiments with Berkeley showed less variation.

[5] We note that not all possible letter sequences are likely or plausible (e.g. ßß).

(3) a suffix of length 2 has more linguistic motivation as most inflectional morphology in German is identifiable within 2 characters thus categorization of unknown words in terms of POS type is feasible, though not absolute.

4.3 Cluster and Signature Results

In order to examine the interaction between different signatures, cluster-based features, and lexicon reduction, we performed experiments with various additional modifications of unknown words as well as open and closed classes to better understand the interaction between such treebank representations and parsing models, presented in Tables 3 and 4. If a token had no corresponding cluster it was replaced with a UNK representation with additional information attached, with capitalization (C) indicated on all tokens (clustered and UNK). We also experimented with not replacing closed class words with their corresponding cluster ID, and instead leaving them in place ($noCC$). Once again, we see little difference in F-scores when providing the parser tags, but we see more range with parser-internal tagging.

Table 3. Results for Lorg on raw words and lemma_pos clusters

Token type	Parsed	UNK TH.	F-score	POS Acc.	Lex. Red.
Craw	tokens	1	76.47	94.22	**93.38**
Craw	tagged	5	78.34	99.52	**93.38**
raw_suffix2	tokens	5	76.27	94.23	91.09
raw_suffix2	tagged	10	78.10	99.40	91.09
Craw_suffix2	tokens	1	76.50	94.57	89.98
Craw_suffix2	tagged	1	78.17	99.40	89.98
raw_noCC	tokens	1	76.00	93.68	92.73
raw_noCC	tagged	1	78.10	99.54	92.73
Craw_suffix2_noCC	tokens	1	76.57	94.93	88.86
Craw_suffix2_noCC	tagged	5	78.20	99.54	88.86
Clemma_pos	tokens	1	76.86	96.54	93.32
Clemma_pos	tagged	1	77.44	99.51	93.32
lemma_pos_suffix2	tokens	1	76.78	96.69	91.63
lemma_pos_suffix2	tagged	1	77.67	99.52	91.63
Clemma_pos_suffix2	tokens	5	76.77	96.63	90.54
Clemma_pos_suffix2	tagged	5	77.46	99.54	90.54
lemma_pos_noCC	tokens	1	73.67	94.08	**94.04**
lemma_pos_noCC	tagged	10	77.48	99.53	**94.04**
Clemma_pos_suffix2_noCC	tokens	1	76.08	95.61	90.53
Clemma_pos_suffix2_noCC	tagged	5	77.45	99.53	90.53

Results for Lorg indicate a distinct split. When *noCC* is not included, lemma_pos clusters obtain consistently higher performance, but when *noCC* is included, raw words perform consistently better. One reason may be that there is still too much ambiguity present with a lemma_pos combination, particularly with articles. However, we are still able to increase results for parser-internal tagging by over 5% absolute and more than .3% with gold tags. It is worth noting that the best achieved score is using gold tags with a suffix of length 4 (see Table 2) or simply marking capitalization on raw clusters and unknown words (see Table 3).

Table 4. Results for Berkeley raw words and lemma_pos clusters

Token type	Parsed	UNK TH.	F-score	POS Acc.	Lex. Red.
Craw	tokens	5	75.59	93.72	**93.38**
Craw	tagged	5	76.89	99.76	**93.38**
raw_suffix2	tokens	5	75.28	93.82	91.09
raw_suffix2	tagged	10	76.50	99.84	91.09
Craw_suffix2	tokens	5	75.66	94.27	89.98
Craw_suffix2	tagged	5	76.65	99.76	89.98
raw_noCC	tokens	1	75.23	93.29	92.73
raw_noCC	tagged	1	76.95	99.36	92.73
Craw_suffix2_noCC	tokens	1	75.73	94.68	88.86
Craw_suffix2_noCC	tagged	10	76.60	99.87	88.86
Clemma_pos	tokens	5	75.76	96.27	93.32
Clemma_pos	tagged	5	75.90	99.87	93.32
lemma_pos_suffix2	tokens	5	75.64	96.46	91.63
lemma_pos_suffix2	tagged	5	75.93	99.85	91.63
Clemma_pos_suffix2	tokens	10	75.82	96.69	90.54
Clemma_pos_suffix2	tagged	10	75.93	99.88	90.54
lemma_pos_noCC	tokens	1	72.49	93.33	**94.04**
lemma_pos_noCC	tagged	1	75.81	93.32	**94.04**
Clemma_pos_suffix2_noCC	tokens	1	75.00	95.23	90.53
Clemma_pos_suffix2_noCC	tagged	1	75.91	99.83	90.53

For Berkeley there are some similar trends (see Table 4), including the steep decline in lemma_pos performance when *noCC* is included. Although we are able to improve results over the Berkeley baselines, the increase in performance is around .3% absolute for gold tags and .6% for parser-internal tagging, although there is significantly less variation between settings.

5   Discussion

There is no direct correlation between lexicon reduction and parser performance. Clearly, reducing the lexicon helps performance, but it is not the case that the largest reduction results in the best performance. As discussed in Sect. 2, previous research has yielded strategies that have improved performance in other languages, such as lemmatization, but these do not benefit German to the same extent. This suggests that for German, simply reducing the lexicon is not enough, rather certain linguistic information, particularly at the morphological level, may need to be retained for certain word classes to help resolve errors.

A break-down of the most frequent UNK tokens is presented in Tables 5 and 6 extracted from the Craw_suffix2_noCC data from the train and dev set respectively. For some suffixes, NNs are either the only tag or represent almost all

Table 5. Top 10 UNK_ in raw train

UNK type	Count	Top 3 POS categories		
CUNK_en	897	NN (836)	NE (36)	ADJA (15)
UNK_en	624	ADJA (279)	VVINF (134)	VVFIN (89)
CUNK_er	429	NN (332)	NE (72)	ADJA (22)
CUNK_ng	255	NN (231)	NE (23)	ADJA (1)
CUNK_te	127	NN (115)	ADJA (8)	NE (3)
CUNK_es	112	NN (86)	NE (18)	ADJA (7)
CUNK_rn	110	NN (110)		
UNK_er	108	ADJA (79)	ADJD (18)	NN (7)
CUNK_in	106	NN (69)	NE (37)	
CUNK_cl	103	NN (74)	NE (27)	PITA (1)

Table 6. Top 10 UNK_ in raw Dev

UNK type	Count	Top 3 POS categories		
CUNK_en	884	NN (795)	NE (32)	VVPP (6)
CUNK_er	515	NN (351)	NE (123)	ADJA (34)
UNK_en	462	ADJA (185)	VVINF (122)	VVFIN (82)
CUNK_ng	265	NN (253)	NE (10)	FM/ADJD (1)
CUNK_te	174	NN (166)	NE (4)	ADJA (4)
CUNK_rn	108	NN (103)	NE (3)	ADV (2)
CUNK_ft	101	NN (95)	NE (6)	
UNK_er	94	ADJA (68)	ADJD (17)	NN (6)
CUNK_es	91	NN (74)	NE (11)	ADJA (6)
UNK_te	89	VVFIN (49)	ADJA (38)	ADV/NN (1)

words in the signature. This can most likely be attributed to German orthography, where all nouns, both common and proper, are capitalized. From a syntactic perspective, they behave similarly, even though they may have different POS tags with NN being a common noun and NE being a proper noun. Results indicate this is perhaps the single most important signature, especially given German's notorious ability to generate new compounds words, many of which will seldom be seen.

The consistency between the types of UNK found between the two sets is indicative of why the suffix information is pertinent, as, although none of the words have a corresponding cluster ID, their POS tag and suffix information allow more unknown words to be grouped together for better probabilistic emissions. From a *simple lexicon* perspective, such a grouping of words should allow for better probabilistic modeling due to an increase in frequency.

However, the distinction between adjectives and verbs is a point that could use more refined signature differences, which is most evident with the *UNK_en* signature which handle words ending in *en*. Linguistically the intermingling makes sense as infinitive verbs will end in *-en*[6] while strong adjective endings will also have the same ending. Obtaining morphological characteristics of this UNK type, either case or person information, may resolve this overlap and improve performance as adjective and verbs exhibit syntactically different behaviors. However, past participles can behave similarly to adjectives when used as such, which may also influence the coalescence in this unknown type.

Further exploration of the POS tags and larger groups of the UNK words will allow for a better understanding of how the parsers choose to tag these words and whether they align consistently with provided tags as well as the linguistic characteristics of the true word.

5.1 External POS Tagger

We also examined the interaction between using an external POS tagger trained on the same data set, but with its own rare and unknown word probabilistic model on parsing performance. We trained the TnT tagger (Brants 2000) on the Craw_suffix2_noCC and Clemma_pos training sets and tagged the development sets respectively. TnT is a language-independent HMM tagger that employs multiple smoothing techniques using linear interpolation and handles unknown words using suffix information. The predicted tags were used as input for both Lorg and Berkeley, results of which are presented in Table 7. Using the TnT tags with the Berkeley parser are extremely similar to results with Berkeley-internal tagging, consistent with the findings of Maier et al. (2014). However, this may be attributed to the fact that both use smoothing within their probabilistic models and simply converge to a similar outcome. However, the results for Lorg are worse than those seen in Table 3. This is good evidence that the smoothing techniques used to generate tags by TnT directly conflict with the preferred tags generated by *simple lexicon* grammar model of Lorg and is ultimately detrimental to its

---

[6] or "-n" in many cases.

performance. This motivates that a closer examination between the interaction of different methods of both unknown word handling among not just among parsers, but also this interaction between parsers and POS taggers in a pipeline approach. Different tools in the pipeline handle unknown words differently and the chosen methods will influence the interactions between tools in the pipeline, impacting performance.

Table 7. TnT results

Token type	System	F-Score	POS Acc.
Craw_suffix2_noCC	TnT	n/a	94.43
	Lorg w/TnT Tags	74.62	94.28
	Berkeley w/TnT Tags	**75.70**	**94.65**
Clemma_pos	TnT	n/a	**96.66**
	Lorg w/TnT Tags	**76.03**	96.26
	Berkeley w/TnT Tags	75.56	96.26

5.2   Number of Clusters

In order to examine how much the impact on the number of clusters has on the performance of the *simple lexicon*, we performed a set of experiments with Lorg where we used an unknown threshold of 1 for both Craw_suffix2_noCC and Clemma_pos on parser-internal tagging, presented in Table 8. We chose our initial clustering parameters based on what has been a standard approach, but determining the optimal clustering size is not intuitive and requires extensive experimentation (see Derczynski et al. (2015)), as which clusters are splitting and which are combined when the number of clusters size is changed cannot be determined beforehand. The results indicate little variation between the cluster sizes, with 800 being optimal for the raw clusters and 1000 for the lemma_pos clusters. Interestingly, as the cluster sizes increase, the POS accuracy also increases, although the parsing performance does not. Changing the number of clusters will not increase the overall coverage, but simply alter the learned probabilities of the words already covered. Experiments by Dakota (2016) noted that although a minimum frequency of 100 may cover almost 90% of the tokens, it only covers roughly 30% of the actual token types in the TüBa-D/Z treebank (Telljohann et al. 2015). Reducing the minimum frequency to 3 ultimately yielded the best results for the creation of data-driven POS tags. Changing the minimum frequency a word must appear to be clustered will thus require optimal cluster sizes to be determined anew. Furthermore, when not replacing closed class words (*noCC*), a more in-depth evaluation is needed to see which cluster IDs (and by extension which types of words) are most prevalent and which are not. This will allow a better understanding of which types of words are being covered and excluded, but will naturally be influenced by any adjustment to the minimum frequency during the clustering process.

Table 8. Different cluster sizes

Token type	Cluster size	F-score	POS Acc.
Craw_suffix2_noCC	500	76.48	94.06
	800	76.65	94.64
	1000	76.57	94.93
	1500	76.60	95.12
	2000	76.45	95.22
Clemma_pos	500	76.67	95.73
	800	76.78	96.37
	1000	76.86	96.54
	1500	76.81	96.72
	2000	76.66	96.87

6   Conclusion and Future Work

We have shown that there is an intricate interaction between reducing data sparsity and the handling of unknown words. Better understanding this interaction allowed us to increase parser performance over our baselines, with best results obtained by using Brown clusters created from a combination capitalization and lemma_pos information. Although smoothing helps create better generalized models, it biases itself against the handling of rare and unknown words, which is in line with previous work examining such interactions within a PCFG-LA framework (Huang and Harper 2009, 2011). This technique has somewhat unexpected effects as although it helps with data sparsity, it results in lower performance. We were able to achieve maximum results when using a *simple lexicon* model for unknown word handling, as the simplistic division of the probability mass allowed us to better exploit the clustering of data through cluster IDs and word signatures without the bias against seldom seen word types.

There are a number of interacting variables that occur in the process of reducing data sparsity, each requiring an extensive in-depth evaluation to better understand how a modification or implementation to solve one aspect directly positively or negatively impacts another aspect. Future work will examine what linguistic information can be exploited on different word classes as well as exploring cluster granularity. There is a balance between the reduction of data sparsity and the need to create generalized enough models, the interaction of which is an area worth further exploration, particularly for MRLs; which consistently present such challenges. We will also examine whether the minimum frequencies during the clustering process can help reduce the number of unknown words further while adjusting the cluster numbers, to compensate for too much of an increase.

Acknowledgments. I would like to thank Djamé Seddah for his gracious help and input with the experiment design and implementation. I would also like to thank Sandra Kübler and Andrew Lamont for their comments, as well as the anonymous reviewers.

The research reported here was supported by the Chateaubriand Fellowship of the Office for Science & Technology of the Embassy of France in the United States and by the Labex EFL.

References

Attia, M., Foster, J., Hogan, D., Roux, J.L., Tounsi, L., van Genabith, J.: Handling unknown words in statistical latent-variable parsing models for Arabic, English and French. In: Proceedings of SPRML 2010 (2010)

Crabbé, B., Candito, M.: Expériences danalyse syntaxique statistique du français. In: Actes de la 15éme Conférence sur le Traitement Automatique des Langues Naturelles (TALN08), Avignon, France, pp. 45–54, June 2008

Björkelund, A., Bohnet, B., Hafdell, L., Nugues, P.: A high-performance syntactic and semantic dependency parser. In: Proceedings of the 23rd International Conference on Computational Linguistics: Demonstrations, Beijing, China, pp. 33–36, August 2010

Brants, S., Dipper, S., Eisenberg, P., Hansen, S., König, E., Lezius, W., Rohrer, C., Smith, G., Uszkoreit, H.: TIGER: linguistic interpretation of a German corpus. J. Lang. Comput. 2004(2), 597–620 (2004)

Brants, T.: TnT: a statistical part-of-speech tagger. In: Proceedings of the Sixth Conference on Applied Natural Language Processing, ANLC 2000, Seattle, Washington, pp. 224–231, April 2000

Brown, P., Della, V., Desouza, P., Lai, J., Mercer, R.: Class-based n-gram models of natural language. Comput. Linguist. 19(4), 467–479 (1992)

Candito, M., Crabbé, B.: Improving generative statistical parsing with semi-supervised word clustering. In: Proceedings of the 11th International Conference on Parsing Technologies, IWPT 2009, Paris, France, pp. 138–141 (2009)

Candito, M., Seddah, D.: Parsing word clusters. In: Proceedings of the NAACL HLT 2010 First Workshop on Statistical Parsing of Morphologically-Rich Languages, SPMRL 2010, Los Angeles, California, pp. 76–84 (2010)

Candito, M., Nivre, J., Denis, P., Anguiano, E.H.: Benchmarking of statistical dependency parsers for french. In: Proceedings of the 23rd International Conference on Computational Linguistics: Posters, COLING 2010, Beijing, China, pp. 108–116 (2010)

Dakota, D.: Brown clustering for unlexicalized parsing. In: Proceedings of the 13th Conference on Natural Language Processing (KONVENS 2016), Bochum, Germany, pp. 68–77, September 2016

Derczynski, L., Chester, S., Bøgh, K.: Tune your brown clustering, please. In: Proceedings of Recent Advances of Natural Language Processing (RANLP) 2015, Hissar, Bulgaria, pp. 110–117, September 2015

Fraser, A., Schmid, H., Farkas, R., Wang, R., Schütze, H.: Knowledge sources for constituent parsing of German, a morphologically rich and less-configurational language. Comput. Linguist. 39(1), 57–85 (2013)

Ghayoomi, M.: Word clustering for Persian statistical parsing. In: Isahara, H., Kanzaki, K. (eds.) JapTAL 2012. LNCS (LNAI), vol. 7614, pp. 126–137. Springer, Heidelberg (2012). https://doi.org/10.1007/978-3-642-33983-7_13

Ghayoomi, M., Simov, K., Osenova, P.: Constituency parsing of Bulgarian: word-vs class-based parsing. In: Proceedings of the Ninth International Conference on Language Resources and Evaluation (LREC 2014), Reykjavik, Iceland, pp. 4056–4060, May 2014

Goenaga, I., Gojenola, K., Ezeiza, N.: Combining clustering approaches for semi-supervised parsing: the BASQUE TEAM system in the SPRML'2014 shared task. In: First Joined Workshop of Statistical Parsing of Morphologically Rich Language and Syntactic Analysis of Non-Canonical Languages, Dublin, Ireland (2014)

Hall, D., Durrett, G., Klein, D.: Less grammar, more features. In: Proceedings of the 52nd Annual Meeting of the Association for Computational Linguistics, Baltimore, Maryland, pp. 228–237, June 2014

Huang, Q., Wong, D.F., Chao, L.S., Zeng, X., He, L.: Augmented parsing of unknown word by graph-based semi-supervised learning. In: 27th Pacific Asia Conference on Language, Information, and Computation, Wenshan, Taipei, pp. 474–482, November 2013

Huang, Z., Harper, M.: Self-training PCFG grammars with latent annotations across languages. In: Proceedings of the 2009 Conference on Empirical Methods in Natural Language Processing, Singapore, pp. 832–841, August 2009

Huang, Z., Harper, M.: Feature-rich log-linear lexical model for latent variable PCFG grammars. In: Proceedings of the 5th International Joint Conference on Natural Language Processing, Chiang Mai, Thailand, pp. 219–227, November 2011

Koo, T., Carreras, X., Collins, M.: Simple semi-supervised dependency parsing. In: Proceedings of ACL-08: HLT, Columbus, Ohio, pp. 595–603 (2008)

Maier, W., Kaeshammer, M., Kallmeyer, L.: Data-driven PLCFRS parsing revisited: restricting the fan-out to two. In: Proceedings of the Eleventh International Conference on Tree Adjoining Grammars and Related Formalisms (TAG+11), Paris, France, pp. 126–134, September 2012

Maier, W., Kübler, S., Dakota, D., Whyatt, D.: Parsing German: how much morphology do we need? In: Proceedings of the First Joined Workshop of Statistical Parsing of Morphologically Rich Language and Syntactic Analysis of Non-Canonical Languages (SPMRL-SANCL 2014), Dublin, Ireland, pp. 1–14, August 2014

Matsuzaki, T., Miyao, Y., Tsujii, J.: Probabilistic CFG with latent annotations. In: Proceedings of the 43rd Annual Meeting on Association for Computational Linguistics, ACL 2005, Ann Arbor, Michigan, pp. 75–82, June 2005

Nivre, J., Hall, J., Kübler, S., McDonald, R., Nilsson, J., Riedel, S., Yuret, D.: The CoNLL 2007 shared task on dependency parsing. In: Proceedings of the CoNLL Shared Task Session of EMNLP-CoNLL 2007, Prague, Czech Republic, pp. 915–932, June 2007

Petrov, S.: Products of random latent variable grammars. In: Human Language Technologies: The 2010 Annual Conference of the North American Chapter of the Association for Computational Linguistics, Los Angeles, CA, USA, pp. 19–27, June 2010

Petrov, S., Klein, D.: Improved inference for unlexicalized parsing. In: Proceedings of Human Language Technologies 2007: The Conference of the North American Chapter of the Association for Computational Linguistics, Rochester, NY, pp. 404–411, April 2007a

Petrov, S., Klein, D.: Learning and inference for hierarchically split PCFGs. In: Proceedings of the National Conference on Artificial Intelligence, Vancouver, Canada, pp. 1663–1666, July 2007b

Petrov, S., Barrett, L., Thibaux, R., Klein, D.: Learning accurate, compact, and interpretable tree annotation. In: Proceedings of the 21st International Conference on Computational Linguistics and the 44th Annual Meeting of the Association for Computational Linguistics, Sydney, Australia, pp. 433–440, July 2006

Seddah, D., Candito, M., Anguiano, E.H.: A word clustering approach to domain adaptation: robust parsing of source and target domains. J. Logic Comput. **24**(2), 395–411 (2013a)

Seddah, D., Tsarfaty, R., Kübler, S., Candito, M., Choi, J.D., Farkas, R., Foster, J., Goenaga, I., Galletebeitia, K.G., Goldberg, Y., Green, S., Habash, N., Kuhlmann, M., Maier, W., Nivre, J., Przepiórkowski, A., Roth, R., Seeker, W., Versley, Y., Vincze, V., Woliński, M., Wróblewska, A., de la Clergerie, E.V.: Overview of the SPMRL 2013 shared task: a cross-framework evaluation of parsing morphologically rich languages. In: Proceedings of the Fourth Workshop on Statistical Parsing of Morphologically-Rich Languages, Seattle, Washington, USA, pp. 146–182, October 2013b

Seddah, D., Kübler, S., Tsarfaty, R.: Introducing the SPMRL 2014 shared task on parsing morphologically-rich languages. In: Proceedings of the First Joint Workshop on Statistical Parsing of Morphologically Rich Languages and Syntactic Analysis of Non-Canonical Languages, Dublin, Ireland, pp. 103–109. Dublin City University, August 2014. http://www.aclweb.org/anthology/W14-6111

Sekine, S., Collins, M.: EVALB bracket scoring program (1997). http://nlp.cs.nyu.edu/evalb/

Telljohann, H., Hinrichs, E.W., Kübler, S., Zinsmeister, H., Beck, K.: Stylebook for the Tübingen Treebank of Written German (TüBa-D/Z). Seminar für Sprachwissenschaft. Universität Tübingen, Germany (2015)

Tsarfaty, R., Seddah, D., Goldberg, Y., Kübler, S., Candito, M., Foster, J., Versley, Y., Rehbein, I., Tounsi, L.: Statistical parsing of morphologically rich languages (SPMRL): what, how and whither. In: Proceedings of the NAACL HLT 2010 First Workshop on Statistical Parsing of Morphologically-Rich Languages, SPMRL 2010, Los Angeles, California, pp. 1–12 (2010)

Versley, Y.: Incorporating semi-supervised features into discontinuous easy-first constituent parsing. In: First Joined Workshop of Statistical Parsing of Morphologically Rich Language and Syntactic Analysis of Non-Canonical Languages, Dublin, Ireland (2014)

Versley, Y., Panchenko, Y.: Not just bigger: towards better-quality web corpora. In: Seventh Web as Corpus Workshop (WAC7), Lyon, France, pp. 44–52 (2012)

Exploring Ensemble Dependency Parsing to Reduce Manual Annotation Workload

Jessica Sohl[(✉)] and Heike Zinsmeister

Institute for German Language and Literature, Universität Hamburg,
Hamburg, Germany
jessica.katharina.sohl@studium.uni-hamburg.de,
heike.zinsmeister@uni-hamburg.de

Abstract. In this paper we present an evaluation of combining automatic and manual dependency annotation to reduce manual workload. More precisely, an ensemble of three parsers is used to annotate sentences of German textbook texts automatically. By including a constrained-based system in the cluster in addition to machine learning approaches, this approach deviates from the original ensemble idea and results in a highly reliable ensemble majority vote. Additionally, our explorative use of dependency parsing identifies error-prone analyses of different systems and helps us to predict items that do not need to be manually checked. Our approach is not innovative as such but we explore in detail its benefits for the annotation task. The manual workload can be reduced by highlighting the reliability of items, for example, in terms of a 'traffic-light system' that signals the reliability of the automatic annotation.

1 Introduction

Corpus-based linguistic analyses that rely on annotated data require high-quality annotations to be accepted by the community. Working with reference corpora is not useful in many cases because their data is very limited and not suitable for many research questions. Simultaneously, creating manual annotation for new data is very time-consuming, so it is necessary to make use of automated means. However, it is often not feasible for corpus-linguistic projects to create their own annotation tools. They have to rely on off-the-shelf programs. Fortunately, infrastructure efforts such as CLARIN[1] or META-NET[2] have made existing tools much easier accessible for reuse by the community.

One of the issues of working with off-the-shelf tools is that they are developed for or trained on particular texts, which are not necessarily of the same text type as the data of interest. This means that using off-the-shelf tools often coincides with applying the tools to out-of-domain data.

In this paper, we investigate the approach of applying a set of syntactic dependency parsers that are trained on a large newswire corpus to a corpus of

---

[1] CLARIN-D: https://www.clarin-d.de/en/.
[2] META-NET: http://www.meta-net.eu.

© The Author(s) 2018
G. Rehm and T. Declerck (Eds.): GSCL 2017, LNAI 10713, pp. 40–47, 2018.
https://doi.org/10.1007/978-3-319-73706-5_4

'non-standard' texts to support manual annotation. The idea of such ensemble parsing is introduced in Sect. 2. After briefly discussing related work, we describe the setting of our study (Sect. 3): the set of parsers that constitute our parser ensemble; the training domain, which refers to the actual training data in the case of statistical parsers and to the data the constrained-based parser was incrementally tested and improved on, and finally, the test corpus, which consists of data from our target domain. In Sect. 4, we first present quantitative results (Sect. 4.1): We establish the accuracy of the parsers individually on the 'training domain'; we test the parsers individually on the target domain; and, finally, establish the best combination of three parsers in an ensemble setting. Second, in addition to these quantitative results, we analyze which kind of items the ensemble fails to parse correctly (Sect. 4.2). A detailed qualitative analysis helps to estimate the extent to which the parser ensemble can support manual annotation which is discussed in Sect. 5. The choice of parsers is motivated by taking the perspective of a corpus linguistics or digital humanities project that has only limited means for parser optimization itself but has to rely on well described ready-to-use tools.[3]

2 Ensemble Parsing

The concept of ensemble parsing has been thoroughly discussed by Van Halteren et al. (2001) for part-of-speech tagging. The crucial point is that a cluster of taggers is employed instead of a single tagger. There are several methods of combining the output of a tagger ensemble. In this paper we follow the 'multi-strategy approach' (Van Halteren et al. 2001, p. 201), in which tagger models are employed that result from training different learning algorithms on the same data. The key idea is that different taggers create their analyses in different ways such that their errors are uncorrelated. Van Halteren et al. (2001) suggest that a reasonable weighted combination of the tagger choices can obtain better results than the individual taggers do. Many studies applied the multi-strategy approach in a successful way also to dependency parsing (Brill and Wu 1998, Søegaard 2010, Rehbein et al. 2014, a. o.).

In this paper, we deviate from the original approach and include one constrained-based parser in addition to two statistically trained parsers and investigate to what extent this ensemble can support manual annotation of textbook texts.

3 Setting

We train both statistical parsers on a large reference corpus that was manually annotated and also used as a test-bed for developing our constrained-based parser. This ensures that all members of the ensemble are based on the same linguistic analyses (Fig. 1).

---

[3] We presented this work as an unpublished poster at the DGFS-CL poster session in 2017, http://dfgs2017.uni-saarland.de/wordpress/abstracts/clposter/cl_6_zins.pdf.

1	Das	das	ART	ART	_	2	SUBJ	_	_
2	ist	sein	V	VVFIN	_	0	S	_	_
3	ein	ein	ART	ART	_	4	DET	_	_
4	Beispielsatz	Beispielsatz	N	NN	_	2	PRED	_	_
5	.	.	$.	$.	_	0	ROOT	_	_

Fig. 1. Dependency parse in CoNLL format

3.1   Parser Ensemble

Our ensemble consists of three different parsers. The MALT parser (Nivre et al. 2006) creates its dependency trees by means of transition-based hypotheses.[4] The MATE parser (Björkelund et al. 2010) is partly related but takes second order maximum spanning trees into account for creating its trees.[5] Finally, the JWCDG parser (The CDG Team 1997-15)[6] consists of (manually) weighted hand-written rules which were developed on the basis of Hamburg Dependency Treebank (HDT), see subsection 3.2.[7] For the ensemble, we took into account different combinations of parser outputs. In Sect. 4.1, we will present results for the two highest-scoring ensembles evaluated on the gold standard:

- Ensemble 1 (ENS-1): Majority vote of all three parsers agreeing on the annotation (Match-3) or at least two out of three parsers agreeing (Match-2); MATE as the best individual parser serves as the default when all parsers differ from each other.
- Ensemble 2 (ENS-2): Majority vote of all three parsers agreeing on the annotation (Match-3); MATE serves as the default otherwise, except MATE assigns one of the labels *S* or *OBJA* then the annotation of JWCDG is used instead.

Note that both ensembles rely heavily on the MATE parser: ENS-1 takes the output of MATE except for instances in which the other two parsers agree on a different label. ENS-2 accepts the annotation of MATE except for two labels which MATE generally overgenerates. In such instances, the ensemble takes the annotation of JWCDG independent of whether there is a majority vote or not.

3.2   Training Domain

Our training corpus is the Hamburg Dependency Treebank (HDT).[8] In particular, we used part A of the HDT (Foth et al. 2014) which contains 10,199 sentences produced by manual annotation and subsequent cross-checking for consistency with DECCA (Dickinson and Meurers 2003). The texts of the HDT

---

[4] We trained Maltparser v1.9.0 with default settings which results in a *non-optimized version* that does not do justice to the parser system as such.

[5] We used MATE transition-1.24 for training.

[6] The CDG Team (2997-2915): https://gitlab.com/nats/jwcdg; Version: 1.0.

[7] We had to dismiss the Turbo parser from our ensemble due to compilation problems.

[8] HDT: https://nats-www.informatik.uni-hamburg.de/HDT.

are crawled from the website *heise online*, a German-language technology news service mostly covering IT, telecommunications and technology.[9]

We divided HDT into ten equally sized bins and performed a 10-fold cross-validation of the statistical parsers, MALT and MATE, to estimate their in-domain performances. The final versions of the parsers were trained on the full corpus.

3.3 Test Domain and Gold Standard

Our test domain is textbook texts as used in books for German secondary schools. In particular, we used texts from an unpublished textbook corpus: 144 sentences from three different geography textbooks which correspond to one double page per book. We refer to double pages here because they commonly represent one informational unit in such textbooks. In the evaluation, we average the performances on the three double pages.

We developed a gold standard on the test corpus. To this end, two annotators annotated the data independently from scratch using the tagset of the HDT (see Sect. 3.2). The manual annotation resulted in an inter-annotator agreement (IAA) of unlabeled attachment score (UAS) of 0.95 ($\pm$0.01) and labeled attachment score (LAS) of 0.93 ($\pm$0.01) according to MaltEval (Nilsson and Nivre 2008). We also computed a chance-corrected IAA score for dependency annotation and obtained $\alpha = 0.93$ ($\pm$0.02) agreement (Skjærholt 2014).

4 Results

We present quantitative results for the individual parsers both on the training domain and on the test corpus. We also present quantitative results for two different ensemble settings. In the second part of this section, we take a closer look at the parsing failures and analyze the linguistic structures qualitatively that turned out to be problematic for the parsers.

4.1 Quantitative Results

The quantitative results on parsing accuracy are summarized in Fig. 2.

The x-axis represents our three different data sets: the training data from the HDT ("10-fold cross"), the test corpus ("Gold"), and finally the subset of gold instances on which all three parsers of the ensemble agreed upon ("Match-3"). The x-axis is furthermore divided into two different evaluation scores (see the top header): the labeled attachment score (LAS) to the left-hand side, which provides the percentage of tokens for which the system has predicted both the correct head and the correct dependency relation; the unlabeled attachment score (UAS) to the right-hand side, which is the more relaxed score that only checks for the correct head. The different parsers and ensembles are depicted by

---

[9] Heise online: heise.de.

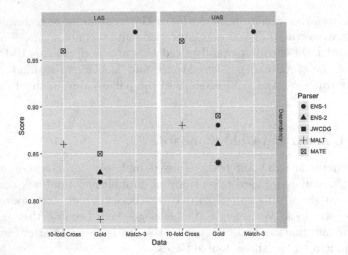

Fig. 2. Parsing accuracy (LAS and UAS) of the parsers (JWCDG, MALT, MATE) and two parser ensembles (ENS-1, ENS-2) on different data sets: training data (10-fold cross-validation), gold standard, and Match-3 items of the gold standard.

different shapes (for details see Sect. 3.1). It is expected that the LAS scores are generally lower than the UAS scores which holds true for all but the Match-3 data which we will discuss further below.

10-fold cross-validation. The evaluation on the HDT corpus of about 10,000 sentences shows that MATE outperforms the non-optimized version of MALT in a range of about 10% points (LAS: 0.86 MALT vs. 0.96 MATE; UAS: 0.88 MALT vs. 0.97 MATE).

Gold standard evaluation. We get a similar tendency on our 144 sentence test corpus (1,697 tokens; on average 566 tokens per double page) even if the difference is not as pronounced and the performance of both parsers drops substantially in comparison to the in-domain cross-validation results (LAS: 0.78 MALT vs. 0.84 MATE; UAS: 0.84 MALT vs. 0.88 MATE on average). The difference between the parsers is still significant (according to a one-tailed t-test for UAS: $t = 3.05$, $df = 2$, $p = 0.04639$; for LAS: $t = 3.1$, $df = 2$, $p = 0.02995$). The constrained-based JWCDG parser has similar performance to MALT and is also outperformed by MATE. The ensemble settings ENS-1 and ENS-2 (cf. Sect. 3.1) outperform JWCDG and MALT but do not quite reach the accuracy of MATE. Interestingly, ENS-1 is better than ENS-2 in assigning the overall dependency structures (UAS) whereas ENS-2 is more reliable in assigning the labels correctly (LAS).

Match-3. The final set is the subset of gold standard instances on which all three parsers of the ensemble agreed in head and label assignment.[10] This subset

---

[10] Figure 2 depicts ENS-1 only but the results hold for both ensemble settings.

consists of 1,128 tokens overall, on average 276 tokens per double page, i. e. about 71% (±0.10) of the tokens are a complete match of the three parsers. The ensembles performed very well on these instances (LAS and UAS both equal 0.98 on average, LAS having a greater variance than UAS). For practical issues it is relevant to look for complete sentences in this set. We observe that Match-3 contains 22 complete sentences, i. e. about 15% of the sentences per double page. All in all 21 out of 22 completely agreed-on sentences are correct.

4.2   Qualitative Results

Some of the parser failures can be related to general challenges in dependency parsing such as the decision of a prepositional phrase functioning as an object (OBJP) or an adverbial (PP) for a given verb (which strongly depends on the training data) and also attachment ambiguities (which require semantic decisions). Table 1 summarizes the major weaknesses of the individual parsers which we employed in creating the parser ensembles (cf. Sect. 3.1).

In addition to these parser-specific errors, we observed domain-specific challenges. The text in textbooks is presented in particular ways. For example, it contains a high amount of lists and exercises that are characterized by incomplete sentences which include list items and nominal structures as in Example (1).

(1) M4 Auswirkungen des Klimawandels am Beispiel "Starkregen"
    'M(aterial)4 Impact of climate change on the example of "severe rain"'

There are also non-finite verbal structures featuring the verb in its canonical VP-final position, cf. *kennen* 'to know' in Example (2).

(2) check-it:
    Merkmale einer thematischen Karte – hier Bodennutzung – kennen
    'check it: - knowing the characteristics of a thematic map – here soil use.'

Table 1. Labels overgenerated by the parsers, potential trigger structures and overall number of errors per parser (APP = generalized apposition, S = root of sentence/fragment, OBJA = accusative object).

Parser	False label	Comment
JWCDG	APP (26%)	Default attachment
	S (15%)	Fragments
		#errors: 262
MALT	S (64%)	Incomplete sentences
		#errors: 335
MATE	S (30%)	Fragments
	OBJA (26%)	Confusion of SUBJ/ OBJA
		#errors: 225

Another issue that is claimed to be characteristic of German scholarly language in text books is complex syntax (e. g., (Griehaber 2013)). Our corpus contains some complex coordinations that are hard to parse even for humans. Example (3) is one of them.

(3) Als praktisch sicher gilt, dass es über den meisten Landflächen wärmere und weniger kalte Tage und Nächte sowie wärmere und häufiger heiße Tage und Nächte geben wird.
'It is virtually certain that there will be warmer and less cold days and nights, as well as warmer and more frequently hot days and nights over most areas.'

We expect that some of the domain-specific structures could be parsed in a more reliable way if the training corpus included data also from the target domain.

5 Conclusion

Our application of ensemble dependency parsing is highly reliable in terms of its ensemble majority vote. However, the ensembles do not outperform the best individual parser. Nevertheless, we can make use of the ensemble to support manual correction. This again means we can very well skip certain labels (e. g., AUX(iliary), DET(erminer), G(enitive)MOD(ifier)) and also complete sentence matches. In addition, we can support manual annotation by highlighting error-prone labels that are easily confused such as OBJP and PP and also areas of the text that are sensitive to errors, e. g., lists and exercises.

The results could be further improved by applying domain adaptation methods such as re-training the statistical parsers and including the gold standard in the training data. More sophisticated methods such as optimizing the parsers' features or combining the parsers with other dependency parsers (e. g., Nivre and McDonald (2008); Köhn and Menzel (2013)) are out of the scope of this project.

Acknowledgments. Jessica Sohl's work was supported by a scholarship of the Hamburg Humanities Graduate School. We would like to thank Sarah Tischer and the anonymous reviewers for helpful comments, and Piklu Gupta for improving our English. All remaining errors remain ours.

References

Björkelund, A., Bohnet, B., Hafdell, L., Nugues, P.: A high-performance syntactic and semantic dependency parser. In: Proceedings of the 23th International Conference on Computational Linguistics (Coling 2010): Demonstrations, pp. 33–36 (2010)

Brill, E., Wu, J.: Classifier combination for improved lexical disambiguation. In: Proceedings of the 36th Annual Meeting of the Association for Computational Linguistics and 17th International Conference on Computational Linguistics, vol. 1, pp. 191–195 (1998)

Dickinson, M., Meurers, W.D.: Detecting Errors in Part-of-Speech annotation. In: Proceedings of the 10th Conference of the European Chapter of the Association for Computational Linguistics (EACL), pp. 107–114 (2003)

Foth, K., Köhn, A., Beuck, N., Menzel, W.: Because size does matter: the Hamburg dependency treebank. In: Proceedings of the Language Resources and Evaluation Conference (LREC 2014), pp. 2326–2333 (2014)

Griehaber, W.: Die Rolle der Sprache bei der Vermittlung fachlicher Inhalte. In: Rhner, C., Hvelbrinks, B. (eds.) Fachbezogene Sprachfrderung in Deutsch als Zweitsprache. Theoretische Konzepte und empirische Befunde zum Erwerb bildungssprachlicher Kompetenzen, pp. 58–74. Juventa, Weinheim (2013)

Köhn, A., Menzel, W.: Incremental and predictive dependency parsing under real-time conditions. In: Proceedings of the International Conference Recent Advances in Natural Language Processing (RANLP 2013), pp. 373–381 (2013)

Nilsson, J., Nivre, J.: Malteval: an evaluation and visualization tool for dependency parsing. In: Proceedings of the Sixth International Conference on Language Resources and Evaluation (LREC 2008), pp. 161–166 (2008)

Nivre, J., McDonald, R.: Integrating graph-based and transition-based dependency parsers. In: Proceedings of the 46th Annual Meeting of the Association for Computational Linguistics: Human Language Technologies (ACL-08: HLT), pp. 950–958 (2008)

Nivre, J., Hall, J., Nilsson, J.: MaltParser: a data-driven parser-generator for dependency parsing. In: Proceedings of the Fifth International Conference on Language Resources and Evaluation (LREC 2006), pp. 2216–2219 (2006)

Rehbein, I., Schalowski, S., Wiese, H.: The KiezDeutsch Korpus (KiDKo) release 1.0. In: Proceedings of the Ninth International Conference on Language Resources and Evaluation (LREC 2014), pp. 3927–3934 (2014)

Skjœrholt, A.: A chance-corrected measure of inter-annotator agreement for syntax. In: Proceedings of the 52nd Annual Meeting of the Association for Computational Linguistics, vol. 1: Long Papers, pp. 934–944 (2014)

Søgaard, A.: Simple semi-supervised training of part-of-speech taggers. In: Proceedings of the ACL 2010 Conference Short Papers, pp. 205–208 (2010)

Van Halteren, H., Zavrel, J., Daelemans, W.: Improving accuracy in word class tagging through the combination of machine learning systems. Comput. Linguist. **27**(2), 199–229 (2001)

Different German and English Coreference Resolution Models for Multi-domain Content Curation Scenarios

Ankit Srivastava, Sabine Weber, Peter Bourgonje[(✉)], and Georg Rehm[(✉)]

Language Technology Lab, DFKI GmbH, Alt-Moabit 91c, 10559 Berlin, Germany
{peter.bourgonje,georg.rehm}@dfki.de

Abstract. Coreference Resolution is the process of identifying all words and phrases in a text that refer to the same entity. It has proven to be a useful intermediary step for a number of natural language processing applications. In this paper, we describe three implementations for performing coreference resolution: rule-based, statistical, and projection-based (from English to German). After a comparative evaluation on benchmark datasets, we conclude with an application of these systems on German and English texts from different scenarios in digital curation such as an archive of personal letters, excerpts from a museum exhibition, and regional news articles.

1 Introduction to Coreference Resolution

Coreference resolution, the task of determining the mentions in a text, dialogue or utterance that refer to the same discourse entity, has been at the core of Natural Language Understanding since the 1960s. Owing in large part to publicly available annotated corpora, such as the Message Understanding Conferences (MUC) (Grishman and Sundheim 1996), Automatic Content Extraction (ACE) (Doddington et al. 2004), and OntoNotes[1], significant progress has been made in the development of corpus-based approaches to coreference resolution. Using coreference information has been shown to be useful in tasks such as question answering (Hartrumpf et al. 2008), summarisation (Bergler et al. 2003), machine translation (Miculicich Werlen and Popescu-Belis 2017), and information extraction (Zelenko et al. 2004).

Figure 1 shows a text consisting of three sentences and demonstrates the occurrence of two nouns and the mentions referring to them; *Prof. Hayes, Hayes, he* (shaded in yellow) and *I, me, Eric* (shaded in blue). The purpose of a coreference resolution system is to identify such chains of words and phrases referring to the same entity, often starting with (proper) noun phrases and referring pronouns.

The curation of digital information, has, in recent years, emerged as a fundamental area of activity for the group of professionals often referred to as knowledge workers. These knowledge workers are given the task to conduct research in

---

[1] https://catalog.ldc.upenn.edu/LDC2013T19.

© The Author(s) 2018
G. Rehm and T. Declerck (Eds.): GSCL 2017, LNAI 10713, pp. 48–61, 2018.
https://doi.org/10.1007/978-3-319-73706-5_5

a particular domain in a very limited time frame. The output of their work is used by newspaper agencies to create articles; museums to construct new exhibitions on a specific topic; TV stations to generate news items. Owing to the diversity of tasks and domains they have to work in, knowledge workers face the challenge to explore potentially large multimedia document collections and quickly grasp key concepts and important events in the domain they are working in. In an effort to help them, we can automate some processes in digital curation such as the identification of named entities and events. This is the primary use case for our paper as coreference resolution plays a significant role in disambiguation as well as harnessing a larger number of entities and events. For example, as seen in Fig. 1, after linking *He* and *Hayes* with *Prof. Hayes*, the knowledge worker gets more information to work with.

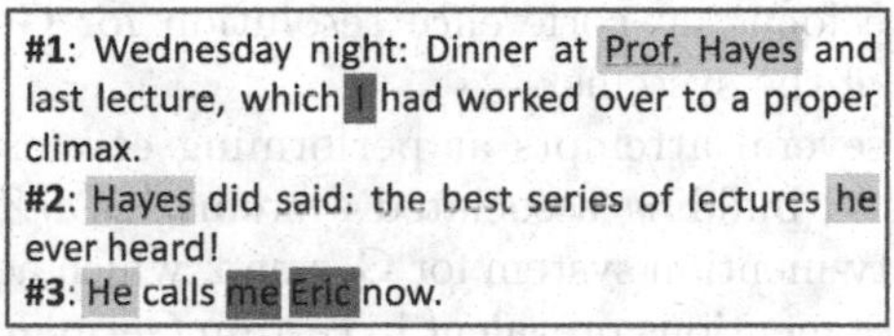

> **#1**: Wednesday night: Dinner at Prof. Hayes and last lecture, which I had worked over to a proper climax.
> **#2**: Hayes did said: the best series of lectures he ever heard!
> **#3**: He calls me Eric now.

Fig. 1. Example of coreference occurrence in English text. Source: Mendelsohn letters dataset (Bienert and de Wit 2014) (Color figure online)

While many coreference systems exist for English (Raghunathan et al. 2010; Kummerfeld and Klein 2013; Clark and Manning 2015, 2016), a freely available[2] competitive tool for German is still missing. In this paper, we describe our forays into developing a German coreference resolution system. We attempt to adapt the Stanford CoreNLP (Manning et al. 2014) Deterministic (rule-based) Coreference Resolution approach (Raghunathan et al. 2010; Lee et al. 2013) as well as the Stanford CoreNLP Mention Ranking (statistical) model (Clark and Manning 2015) to German. We also experiment with projection-based implementation, i.e., using Machine Translation and English coreference models to achieve German coreference resolution.

The main goals of this paper are:

- To evaluate pre-existing English and German coreference resolution systems
- To investigate the effectiveness of performing coreference resolution on a variety of out-of-domain texts in both English and German (outlined in Sect. 4) from digital curation scenarios.

After a brief overview of previous approaches to coreference resolution in English and German (Sect. 2), we describe implementations of three approaches

---

[2] Coreference resolution is language resource dependent and therefore by "freely available" we imply a toolkit which in its entirety (models, dependencies) is available for commercial as well as research purposes.

to German coreference resolution (Sect. 3): the deterministic sieve-based approach, a machine learning-based system, and a English-German crosslingual projection-based system. This is followed by a discussion on applications of coreference (Sect. 4) and concluding notes on the current state of our coreference resolution systems for digital curation scenarios (Sect. 5).

2 Summary of Approaches to Coreference Resolution

A number of paradigms (rule-based, knowledge-rich, supervised and unsupervised learning) have been applied in the design of coreference resolution systems for several languages with regard to whole documents, i.e., to link all mentions or references of an entity within an entire document. While there have been several works giving a comprehensive overview of such approaches (Zheng et al. 2011; Stede 2011), we focus on conference resolution for German and English and summarise some of the systems.

There have been several attempts at performing coreference resolution for German documents and building associated systems.[3] *CorZu* (Tuggener 2016) is an incremental entity-mention system for German, which addresses issues such as underspecification of mentions prevalent in certain German pronouns. While it is freely available under the GNU General Public License, it depends on external software and their respective data formats such as a dependency parser, tagger, and morphological analyser, making it difficult to reimplement it.

BART, the Beautiful/Baltimore Anaphora Resolution Toolkit (Versley et al. 2008), is a modular toolkit for coreference resolution which brings together several preprocesing and syntactic features and maps it to a machine learning problem. While it is available for download as well as a Web Service, there are external dependencies such as the Charniak Reranking Parser. Definite noun matching is resolved via head string matching, achieving an F-score of 73% (Versley 2010). It has been successfully tested on the German TüBa-D/Z treebank (Telljohann et al. 2004) with a claimed F-score of 80.2% (Broscheit et al. 2010).

Definite noun matching cannot be solved via string matching in the domain of newspaper articles. Approximately 50% of the definite coreferent Noun Phrases (NPs) can be resolved using head string matching (Versley 2010). Versley (2010) also used hypernym look-up and various other features to achieve an F-score of 73% for definite anaphoric NPs. Broscheit et al. (2010) claim to get an F1 score of 80.2 on version 4 of the TüBa D/Z coreference corpus using BART.

The goal of the SemEval 2010 Shared Task 1 (Recasens et al. 2010) was to evaluate and compare automatic coreference resolution systems for six different languages, among them German, in four evaluation settings and using four different metrics. The training set contained 331,614 different tokens taken from the TüBa-D/Z data set (Telljohann et al. 2004). Only two of the four competing systems achieved F-scores over 40%, one of them being the BART system

---

[3] For an overview of the development of German coreference systems, see Tuggener (2016).

mentioned above. We use the same dataset and evaluation data to train our statistical system in Sect. 3.1.

Departing from the norm of building mention pairs, one system implemented a mention-entity approach and produced an F-score of 61.49% (Klenner et al. 2010).

The *HotCoref* system for German (Roesiger and Riester 2015) focused on the role of prosody for coreference resolution and used the DIRNDL corpus (Björkelund et al. 2014) for evaluation, achieving F-scores of 53.63% on TüBa-D/Z (version 9) and 60.35% on the SemEval shared task data.

Another system (Krug et al. 2015) adapted the Stanford sieve approach (Lee et al. 2013) for coreference resolution in the domain of German historic novels and evaluated it against a hand annotated corpus of 48 novel fragments with approximately 19,000 character references in total. An F1 score of 85.5 was achieved. We also adapt the Stanford Sieve approach in Sect. 3, with the aim of developing an open-domain German coreference resolution system.

In case of the English coreference resolution, we employ the Stanford CoreNLP implementations. There is a large body of work for coreference resolution in English. While the sieve-based approach (Raghunathan et al. 2010) is a prime example of rule-based coreference resolution, other approaches such as the Mention-Rank model (Clark and Manning 2015) and Neural model (Clark and Manning 2016) have been shown to outperform it.

3 Three Implementations

In this section, we describe the three models of coreference resolution.

- Rule-based (Multi-Sieve Approach): English, German
- Statistical (Mention Ranking Model): English, German
- Projection-based (Crosslingual): coreference for German using English models.

3.1 Rule-Based Approach

For the English version, we employ the deterministic multi-pass sieve-based (open-source) Stanford CoreNLP system (Manning et al. 2014). For the German version, we develop an in-house system and name it **CoRefGer-rule**[4].

The Stanford Sieve approach is based on the idea of an annotation pipeline with coreference resolution being one of the last steps. The processing steps include sentence splitting, tokenisation, constituency and dependency parsing, and extraction of morphological data. In our system **CoRefGer-rule**, we also perform Named Entity Recognition.

What is typical for the Stanford sieve approach is starting with all noun phrases and pronominal phrases in the whole document and then deciding how

---

[4] Source code available at https://github.com/dkt-projekt/e-NLP/tree/master/src/main/java/de/dkt/eservices/ecorenlp/modules.

to cluster them together, so that all the noun phrases referring to the same extratextual entity are in the same coreference chain. The sieves can be described as a succession of independent coreference models. Each of them selects candidate mentions and puts them together. The number of these sieves can be different depending on the task. Seven sieves are proposed for an English coreference system (Raghunathan et al. 2010), while eleven sieves are implemented for the task of finding coreference in German historic novels (Krug et al. 2015). We have currently implemented six of the seven sieves from the English system and will include additional ones in future versions of the system.

Sieve 1: Exact Match. With an exact match, noun phrases are extracted from the parse tree. Then, in a sliding window of five sentences, all noun phrases in this window are compared to each other. If they match exactly this leads to the creation of a new coreference chain. We use stemming so that minimally different word endings and differences in the article are taken into account (Table 1).

Table 1. Example for exact match

Text (de):	**Barack Obama** besuchte Berlin
	Am Abend traf **Barack Obama** die Kanzlerin
Coref:	[Barack Obama, Barack Obama]

We also account for variations in endings such as "des Landesverbandes" and "des Landesverbands der AWO" or "der Hund" and "des Hundes", and between definite and indefinite articles such as "einen Labrador" and "der Labrador".

Sieve 2: Precise Constructs. This is an implementation of precise constructs like appositive constructs, predicate nominative, role appositive and relative pronouns. Due to the different tree tags a direct application of Stanford NLP algorithms was not possible. Also missing acronym and demonym lists for German posed a challenge in completing this sieve. We therefore translated the corresponding English lists [5] into German and used them in our approach.

Sieves 3, 4, 5: Noun Phrase Head Matching. The noun phrase head matching we use is different from the one proposed in Raghunathan et al. (2010). They claim that naive matching of heads of noun phrases creates too many spurious links, because it ignores incompatible modifiers like "Yale University" and "Harvard University". Those two noun phrases would be marked as coreferent, because the head of both is "University", although the modifiers make it clear that they refer to different entities. This is why a number of other constraints are proposed. In order to utilise them we implement a coreference chain building mechanism. For example, there is a notion of succession of the words when

---

[5] https://en.wikipedia.org/wiki/List_of_adjectival_and_demonymic_forms_of_place_na mes.

chaining them together, so we cannot match head nouns or noun phrases in the antecendent cluster.

We also employ stemming in noun phrases, so we match entities such as "AWO Landesverbands" with "Landesverband", "Geschäftsführer" and "Geschäftsführers."

We also implement the sieves called "Variants of head matching" and "relaxed head matching" which require sophisticated coreference chaining.

Sieve 6: Integration of Named Entity Recognition. We use an in-house Named Entity Recognition engine based on DBpedia-Spotlight[6], that is also applied in the current version of the coreference resolution system (for example, to deal with the above mentioned "Yale University" vs. "Harvard University" issue).

German Specific Processing. Our implemented sieves include naive stemming, which means that words that vary in a few letters at the end are still considered as matching due to different case markers in German. The same holds for definite and indefinite articles, which are specific for German. Noun phrases are considered as matching although they have different articles.

An important component that is not implemented but plays a big role in coreference resolution is morphological processing for acquiring gender and number information. This component would make it possible to do pronoun matching other than our current method of merely matching the pronouns that are the same.

3.2 Statistical Approach

While we developed the rule-based approach (CoRefGer-rule), we also adapted the Stanford CoreNLP statistical system based on the Mention Ranking model (Clark and Manning 2015). We trained our coreference system on the TüBa-D/Z (Telljohann et al. 2004), and evaluated on the same dataset as SemEval 2010 Task 1 (Recasens et al. 2010). We named the system **CoRefGer-stat**. The system uses a number of features described below:

- Distance features: the distance between the two mentions in a sentence, number of mentions
- Syntactic features: number of embedded NPs under a mention, Part-Of-Speech tags of the first, last, and head word (based on the German parsing models included in the Stanford CoreNLP, Rafferty and Manning 2008)
- Semantic features: named entity type, speaker identification
- Lexical Features: the first, last, and head word of the current mention.

While the machine learning approach enables robustness and saves time in constructing sieves, the application is limited to the news domain, i.e., the domain of the training data.

---

[6] https://github.com/dbpedia-spotlight/dbpedia-spotlight.

3.3   Projection-Based Approach

In this section we outline the projection-based approach to coreference resolution. This approach is usually implemented in a low-resource language scenario, i.e., if sufficient language resources and training data are not available. Developing a coreference resolution system for any new language is cumbersome due to the variability of coreference phenomena in different languages as well as availability of high-quality language technologies (mention extraction, syntactic parsing, named entity recognition).

Crosslingual projection is a mechanism that allows transferring of existing methods and resources from one language (e.g., English) to another language (e.g., German). While crosslingual projection has demonstrated considerable success in various NLP applications like POS tagging and syntactic parsing, it has been less successful in coreference resolution, performing with 30% less precision than monolingual variants (Grishina and Stede 2017).

The projection-based approach can be implemented in one of the following two ways:

- Transferring models: Computing coreference on text in English, and projecting these annotations on parallel German text via word alignments in order to obtain German coreference model
- Transferring data: Translating German text to English, computing coreference on translated English text using English coreference model and then projecting the annotations back on to the original text via word alignment.

The "Transferring data" approach involved less overhead because new language coreference models are not generated, and proved to be more effective. We have therefore used this approach in our experiments and name it **CoRefGer-proj** system.

4   Evaluation and Case Studies

We are interested in applying reliable and robust coreference resolution for both English and German on a variety of domains from digital curation scenarios such as digital archives, newspaper reports, museum exhibits (Bourgonje et al. 2016, Rehm and Sasaki 2016):

- Mendelsohn Letters Dataset (German and English): The collection (Bienert and de Wit 2014) contains 2,796 letters, written between 1910 and 1953, with a total of 1,002,742 words on more than 11,000 sheets of paper; 1,410 of the letters were written by Erich and 1,328 by Luise Mendelsohn. Most are in German (2,481), the rest is written in English (312) and French (3).
- Research excerpts for a museum exhibition (English): This is a document collection retrieved from online archives: Wikipedia, archive.org, and Project Gutenburg. It contains documents related to Vikings; the content of this collection has been used to plan and to conceptualise a museum in Denmark.

Table 2. Summary of curations datasets.

Corpora	Language	Documents	Words	Domain
Mendelsohn	DE	2,501	699,213	Personal letters
Mendelsohn	EN	295	21,226	Personal letters
Vikings	EN	12	298,577	Wikipedia and E-books
News	DE	1,037	716,885	News articles and summaries

– Regional news stories (German): This consists of a general domain regional news collection in German. It contains 1,037 news articles, written between 2013 and 2015.

The statistics for these corpora and the standard benchmark sets are summarised in Table 2. Robustness can only be achieved if we limit the scope and coverage of the approach, i.e., if we keep the coreference resolution systems simple and actually implementable. In our use cases, a few correctly identified mentions are better than hundreds of false positives.

Evaluation is done on several datasets (standard datasets for benchmarking): CONLL 2012 for English and SemEval 2010 for German. Our goal is to determine the optimal coreference system for coreference resolution on English and German texts from digital curation scenarios (out-of-domain).

Table 3 shows the results of evaluation on CoNLL 2012 (Pradhan et al. 2012) English dataset. Two evaluation measures are employed: MUC (Vilain et al. 1995) and B-cubed (Bagga and Baldwin 1998). The MUC metric compares links between mentions in the key chains to links in the response chains. The B-cubed metric evaluates each mention in the key by mapping it to one of the mentions in the response and then measuring the amount of overlapping mentions in the corresponding key and response chains.

Table 3. Summary of evaluation of 2 coreference resolution systems on English CoNLL 2012 task across two F-1 evaluation measures.

System	MUC	B-cube
BART	45.3	64.5
Sieve	49.2	45.3
Statistical	56.3	50.4
Neural	60.0	56.8

Note that these accuracies are lower than those reported in CoNLL shared task, because the systems employed are dependent on taggers, parsers and named entity recognizers of Stanford CoreNLP and not gold standard as employed in the shared task. While we do not have any gold standard for our digital curation use-cases, a manual evaluation of a small subset of documents shows sieve-based approach to perform slightly better than the state-of-the-art statistical and neural models, most likely owing to out-of-domain applications.

For German, we experimented with different settings of the in-house CoRefGer-rule system. In Table 4, we demonstrate the performance of our Multi-Sieve (rule-based) approach on a 5000-word subset of the TüBa-D/Z corpus (Telljohann et al. 2004) using different settings of modules as follows:

- Setting 1: Whole System with all 6 sieves in place
- Setting 2: Contains all mentions but no coreference links
- Setting 3: Setting 1 minus the module that is deleting any cluster that does not contain a single mention that has been recognized as an entity
- Setting 4: Setting 1 with the module that is deleting any cluster that does not contain a single mention that has been recognized as an entity executed after the sieves have been applied.

Setting 2 assumes that our system obtained all the correct mentions and therefore tests the effectiveness of the coreference linking module only. However this setting will not work in real-life scenarios like the digital curation use cases unless we have hand-annotated corpora.

Table 4. Module-based evaluation of German sieve-based coreference (CoRefGer-rule) on different configurations across two F-1 evaluation measures.

System	MUC	B-cube
Setting 1	54.4	11.2
Setting 2	70.5	23.1
Setting 3	58.9	15.0
Setting 4	56.1	12.0

In Table 5, we present German coreference resolution results on the test set of SemEval 2010 Task 1. We also compare the performance of our three systems (CoRefGer-rule, CoRefGer-stat, CoRefGer-proj) to one other system: CorZU (Tuggener 2016). CoRefGer-stat, CoRefGer-rule, and CoRefGer-proj are the three systems we developed in this paper. Since the sieve-based approach lacks a morphological component currently, it underperforms. An error analysis of the statistical and projection-based system reveals that several features were not sufficiently discriminating for German models. We believe completing the remaining sieve will help us in training better syntactic and semantic features for the statistical system as well.

While there is no gold standard for any of our datasets from digital curation use cases, we nevertheless applied our English and German coreference resolution systems, as shown in Table 6. The sieve-based systems tend to give the best results (shown in the Table) while the statistical, neural and projection-based yield nearly 10% less entity mentions. We leave for future work a deeper investigation into this though we believe that interfacing with lexical resources such as those from WordNet may help ameliorate the out-of-domain issues.

Table 5. Summary of evaluation of coreference resolution systems on German SemEval 2010 task across 2 F-1 evaluation measures.

System	MUC	B-cube
CorZu	60.1	58.9
CoRefGer-rule	50.2	63.3
CoRefGer-stat	40.1	45.3
CoRefGer-proj	35.9	40.3

Table 6. Summary of the percentage of mentions (based on total number of words) on curation datasets for which we do not have a gold standard

Dataset	Sents.	Words	Mentions
Mendelsohn EN	21K	109K	48%
Mendelsohn DE	34K	681K	26%
Vikings EN	39K	310K	49%
News Stories DE	53K	369K	25%

4.1 Add-On Value of Coreference Resolution to Digital Curation Scenarios

Consider the following sentence:

"Then came Ray Brock for dinner. On **him** I will elaborate after my return or as soon as a solution pops up on my "Klappenschrank". Naturally, **he** sends **his** love to Esther and **his** respects to you."

A model or dictionary can only spot "Ray Brock", but, "him", "he" and "his" also refer to this entity. With the aid of coreference resolution, we can increase the recall for named entity recognition as well as potentially expand the range for event detection.

The algorithm for coreference-enabled NLP technologies is as follows:

- Input a text document, and run coreference resolution on it
- With the aid of the above, replace all occurrences of pronouns with the actual noun in full form, such that "he" and "his" are replaced with "Ray Brock" and "Ray Brock's" respectively
- Run a NLP process such as Named Entity Recognition on the new document and compare with a run without the coreference annotations.

A preliminary computation of the above algorithm shows a marked improvement on the number of entities identified (27% more coverage) by an off-the-shelf Named Entity Recogniser.

5   Conclusions and Future Work

We have performed coreference resolution in both English and German on a variety of text types and described several competing approaches (rule-based, statistical, knowledge-based).

Number and gender information is one of the core features that any coreference system uses. A major deficiency for our German rule-based system described in Sect. 3.1 is the lack of interfacing with a morphological analyser, which we leave for future work.

An interesting sieve that can also be adapted from the paper about German coreference resolution in historic novels is the semantic pass where synonyms taken from GermaNet are also taken into account for matching. They handle speaker resolution and pronoun resolution in direct speech, which makes up their tenth and eleventh sieve. They do so by using handcrafted lexico-syntactic patterns. If these patterns are only specific for their domain or if they can also be successfully applied to other domains is a point for further research.

Overall, we were able to annotate our multi-domain datasets with coreference resolution. We will be investigating how much these annotations help knowledge workers in their curation use cases.

In conclusion, we have determined that the deterministic rule-based systems, although not state-of-the-art are better choices for our out-of-domain use cases.

Acknowledgments. We would like to thank the anonymous reviewers for their insightful and helpful comments. The project Digitale Kuratierungstechnologien (DKT) is supported by the German Federal Ministry of Education and Research (BMBF), Unternehmen Region, instrument Wachstumskern-Potenzial (No. 03WKP45). More details: http://www.digitale-kuratierung.de.

References

Bagga, A., Baldwin, B.: Algorithms for scoring coreference chains. In: The First International Conference on Language Resources and Evaluation Workshop on Linguistics Coreference, pp. 563–566 (1998)

Bergler, S., Witte, R., Khalife, M., Li, Z., Rudzicz, F.: Using knowledge-poor coreference resolution for text summarization. In: Proceedings of the Document Understanding Conference (DUC 2003) (2003). http://duc.nist.gov/pubs/2003final. papers/concordia.final.pdf

Bienert, A., de Wit, W. (eds.): EMA - Erich Mendelsohn Archiv. Der Briefwechsel von Erich und Luise Mendelsohn 1910–1953. Kunstbibliothek - Staatliche Museen zu Berlin and The Getty Research Institute, Los Angeles, March 2014. http://ema. smb.museum. With contributions from Regina Stephan and Moritz Wullen, Version March 2014

Björkelund, A., Eckart, K., Riester, A., Schauffler, N., Schweitzer, K.: The extended dirndl corpus as a resource for coreference and bridging resolution. In: Proceedings of the Ninth International Conference on Language Resources and Evaluation (LREC-2014). European Language Resources Association (ELRA) (2014). http://www.lrec-conf.org/proceedings/lrec2014/pdf/891_Paper.pdf

Bourgonje, P., Moreno-Schneider, J., Nehring, J., Rehm, G., Sasaki, F., Srivastava, A.: Towards a platform for curation technologies: enriching text collections with a semantic-web layer. In: Sack, H., Rizzo, G., Steinmetz, N., Mladeni, D., Auer, S., Lange, C. (eds.) The Semantic Web: ESWC 2016 Satellite Events. LNCS, pp. 65–68. Springer International Publishing, Heidelberg (2016). https://doi.org/10.1007/978-3-319-47602-5_14. ISBN 978-3-319-47602-5

Broscheit, S., Ponzetto, S.P., Versley, Y., Poesio, M.: Extending bart to provide a coreference resolution system for German. In: Proceedings of the Seventh Conference on International Language Resources and Evaluation (LREC 2010). European Languages Resources Association (ELRA) (2010). http://aclweb.org/anthology/L10-1347

Clark, K., Manning, C.D.: Entity-centric coreference resolution with model stacking. In: Association for Computational Linguistics (ACL) (2015)

Clark, K., Manning, C.D.: Deep reinforcement learning for mention-ranking coreference models. In: Proceedings of the 2016 Conference on Empirical Methods in Natural Language Processing, pp. 2256–2262. Association for Computational Linguistics (2016). http://aclweb.org/anthology/D16-1245

Doddington, G., Mitchell, A., Przybocki, M., Ramshaw, L., Strassel, S., Weischedel, R.: The automatic content extraction (ACE) program tasks, data, and evaluation. In: Proceedings of the Fourth International Conference on Language Resources and Evaluation (LREC 2004). European Language Resources Association (ELRA) (2004). http://aclweb.org/anthology/L04-1011

Grishina, Y., Stede, M.: Multi-source annotation projection of coreference chains: assessing strategies and testing opportunities. In: Proceedings of the 2nd Workshop on Coreference Resolution Beyond OntoNotes (CORBON 2017), Valencia, Spain, pp. 41–50. Association for Computational Linguistics, April 2017. http://www.aclweb.org/anthology/W17-1506

Grishman, R., Sundheim, B.: Message understanding conference- 6: a brief history. In: COLING 1996 Volume 1: The 16th International Conference on Computational Linguistics (1996). http://aclweb.org/anthology/C96-1079

Hartrumpf, S., Glöckner, I., Leveling, J.: Coreference resolution for questions and answer merging by validation. In: Peters, C., Jijkoun, V., Mandl, T., Müller, H., Oard, D.W., Peñas, A., Petras, V., Santos, D. (eds.) CLEF 2007. LNCS, vol. 5152, pp. 269–272. Springer, Heidelberg (2008). https://doi.org/10.1007/978-3-540-85760-0_32

Klenner, M., Tuggener, D., Fahrni, A., Sennrich, R.: Anaphora resolution with real preprocessing. In: Loftsson, H., Rögnvaldsson, E., Helgadóttir, S. (eds.) NLP 2010. LNCS (LNAI), vol. 6233, pp. 215–225. Springer, Heidelberg (2010). https://doi.org/10.1007/978-3-642-14770-8_25

Krug, M., Puppe, F., Jannidis, F., Macharowsky, L., Reger, I., Weimar, L.: Rule-based coreference resolution in German historic novels. In: Proceedings of the Fourth Workshop on Computational Linguistics for Literature, Denver, Colorado, USA, pp. 98–104. Association for Computational Linguistics, June 2015. http://www.aclweb.org/anthology/W15-0711

Kummerfeld, J.K., Klein, D.: Error-driven analysis of challenges in coreference resolution. In: Proceedings of the 2013 Conference on Empirical Methods in Natural Language Processing, Seattle, Washington, USA, pp. 265–277, October 2013. http://www.aclweb.org/anthology/D13-1027

Lee, H., Chang, A., Peirsman, Y., Chambers, N., Surdeanu, M., Jurafsky, D.: Deterministic coreference resolution based on entity-centric, precision-ranked rules. Computational Linguistics **39**(4) (2013). https://doi.org/10.1162/COLI_a_00152, http://aclweb.org/anthology/J13-4004

Manning, C.D., Surdeanu, M., Bauer, J., Finkel, J., Bethard, S.J., McClosky, D.: The Stanford CoreNLP natural language processing toolkit. In: Association for Computational Linguistics (ACL) System Demonstrations, pp. 55–60 (2014). http://www.aclweb.org/anthology/P/P14/P14-5010

Werlen, L.M., Popescu-Belis, A.: Proceedings of the 2nd Workshop on Coreference Resolution Beyond OntoNotes (CORBON 2017), Chapter Using Coreference Links to Improve Spanish-to-English Machine Translation, pp. 30–40. Association for Computational Linguistics (2017). http://aclweb.org/anthology/W17-1505

Pradhan, S., Moschitti, A., Xue, N., Uryupina, O., Zhang, Y.: Conll-2012 shared task: modeling multilingual unrestricted coreference in ontonotes. In: Joint Conference on EMNLP and CoNLL - Shared Task, CoNLL 2012, Stroudsburg, PA, USA, pp. 1–40. Association for Computational Linguistics (2012). http://dl.acm.org/citation.cfm?id=2391181.2391183

Rafferty, A.N., Manning, C.D.: Parsing three German treebanks: lexicalized and unlexicalized baselines. In: Proceedings of the Workshop on Parsing German, PaGe 2008, Stroudsburg, PA, USA, pp. 40–46. Association for Computational Linguistics (2008). ISBN 978-1-932432-15-2. http://dl.acm.org/citation.cfm?id=1621401.1621407

Raghunathan, K., Lee, H., Rangarajan, S., Chambers, N., Surdeanu, M., Jurafsky, D., Manning, C.: A multi-pass sieve for coreference resolution. In: Proceedings of the 2010 Conference on Empirical Methods in Natural Language Processing, pp. 492–501. Association for Computational Linguistics (2010). http://aclweb.org/anthology/D10-1048

Recasens, M., Màrquez, L., Sapena, E., Martí, A.M., Taulé, M., Hoste, V., Poesio, M., Versley, Y.: Semeval-2010 task 1: coreference resolution in multiple languages. In: Proceedings of the 5th International Workshop on Semantic Evaluation, pp. 1–8. Association for Computational Linguistics (2010). http://aclweb.org/anthology/S10-1001

Rehm, G., Sasaki, F.: Digital curation technologies. In: Proceedings of the 19th Annual Conference of the European Association for Machine Translation (EAMT 2016), Riga, Latvia, May 2016, In print

Roesiger, I., Riester, A.: Using prosodic annotations to improve coreference resolution of spoken text. In: Proceedings of the 53rd Annual Meeting of the Association for Computational Linguistics and the 7th International Joint Conference on Natural Language Processing (Volume 2: Short Papers), pp. 83–88. Association for Computational Linguistics (2015). https://doi.org/10.3115/v1/P15-2014, http://aclweb.org/anthology/P15-2014

Stede, M.: Discourse Processing. Synthesis Lectures in Human Language Technology, vol. 15. Morgan and Claypool, San Rafael (2011)

Telljohann, H., Hinrichs, E., Kübler, S.: The tüba-d/z treebank: annotating German with a context-free backbone. In: Proceedings of the Fourth International Conference on Language Resources and Evaluation (LREC 2004). European Language Resources Association (ELRA) (2004). http://aclweb.org/anthology/L04-1096

Tuggener, D.: Incremental coreference resolution for German. Ph.D. thesis, University of Zurich (2016)

Versley, Y.: Resolving coreferent bridging in German newspaper text. Ph.D. thesis, Universität Tübingen (2010)

Versley, Y., Ponzetto, P.S., Poesio, M., Eidelman, V., Jern, A., Smith, J., Yang, X., Moschitti, A.: Bart: a modular toolkit for coreference resolution. In: Proceedings of the ACL-08: HLT Demo Session, pp. 9–12. Association for Computational Linguistics (2008). http://aclweb.org/anthology/P08-4003

Vilain, M., Burger, J., Aberdeen, J., Connolly, D., Hirschman, L.: A model-theoretic coreference scoring scheme. In: Proceedings of the 6th Conference on Message Understanding, MUC6 1995, Stroudsburg, PA, USA, pp. 45–52 (1995). Association for Computational Linguistics. ISBN 1-55860-402-2. https://doi.org/10.3115/1072399.1072405

Zelenko, D., Aone, C., Tibbetts, J.: Proceedings of the Conference on Reference Resolution and Its Applications, Chapter Coreference Resolution for Information Extraction (2004). http://aclweb.org/anthology/W04-0704

Zheng, J., Chapman, W., Crowley, R., Guergana, S.: Coreference resolution: a review of general methodologies and applications in the clinical domain. Journal of Biomedical Informatics **44**(6) (2011). https://doi.org/10.1016/j.jbi.2011.08.006, https://www.ncbi.nlm.nih.gov/pmc/articles/PMC3226856/

Word and Sentence Segmentation in German: Overcoming Idiosyncrasies in the Use of Punctuation in Private Communication

Kyoko Sugisaki[✉]

German department, University of Zurich,
Schönberggasse 9, 8001 Zurich, Switzerland
sugisaki@ds.uzh.ch

Abstract. In this paper, we present a segmentation system for German texts. We apply conditional random fields (CRF), a statistical sequential model, to a type of text used in private communication. We show that by segmenting individual punctuation, and by taking into account freestanding lines and that using unsupervised word representation (i. e., Brown clustering, Word2Vec and Fasttext) achieved a label accuracy of 96% in a corpus of postcards used in private communication.

1 Introduction

In tokenisation and sentence segmentation, a text is segmented into tokens and sentences. Word and sentence segmentation are the core components of NLP pipelines. Based on text segmentation, part of speech (POS) tagging and parsing, among other tasks, are performed.

In German texts, segmentation is de facto to classify sentence punctuation, such as periods, question marks and exclamation marks, into two categories: (A) the ends of sentences, and (B) others, such as components of abbreviations (e. g., *evtl.*, *eventuell* 'possibly'), proper names (e. g., Sat.1), numbers (e. g., 13.000) and so on. In the case of (A), the punctuation is separated from space-delimited tokens and analysed as individual tokens. In the case of (B), the punctuation constitutes a token with the preceding characters. Therefore, space-delimited tokens are not segmented further. In rare cases, punctuation that is used to mark the end of a sentence (i. e., category [A]) is a part of the token (i. e., category [B]) at the end of a sentence.[1]

Traditionally, German text segmentation systems are based on rules that contain a list of abbreviations.[2] A rule-based approach to the segmentation of German texts (Remus et al. 2016; Proisl and Uhrig 2016) is reasonable considering

---

[1] For instance, 176 (0.18%) in 95.595 sentences belong to the third category in TüBa10. An example is *usw.* at the end of a sentence.

[2] Helmut Schmid's tokenizer in TreeTagger: http://www.cis.uni-muenchen.de/~schm id/tools/TreeTagger/; Stefanie Dipper's system: https://www.linguistics.ruhr-uni-bochum.de/~dipper/resources/tokenizer.html.

© The Author(s) 2018

G. Rehm and T. Declerck (Eds.): GSCL 2017, LNAI 10713, pp. 62–71, 2018.
https://doi.org/10.1007/978-3-319-73706-5_6

Table 1. Use of punctuations (TüBa10)

Last tokens of sentences	ranking:tokens (frequency, %)
1:Period (73904, 77.31%), 2:double quotation (3849, 4.03%), 3:question mark (2921, 3.06%), 4:colon (2369, 2.48%), 5:exclamation mark (682, 0.71%), 6:semicolon (634, 0.66%), 7:parentheses(393, 0.41%), 8:ellipsis (329, 0.34%), 11:guillemet (59, 0.06%), 21:comma (26, 0.03%), 22:square bracket(26, 0.03%), 25:hypen (24, 0.03%), 37:single quote (18, 0.02%), 212:slash (6, 0.01%), else (10355, 10.83%)	
Ambiguity of punctuations tokens (A: frequency of case (A), the total number of the character, AMB(iguity):A/total(%)	
Period (A:73904 + 329*3, total:88938, AMB:84.20%), double quotation (A:3849, total:42468, AMB:9.06%), question mark (A:2921, total:3536, AMB:82.60%), colon (A:2369, total:11522, AMB:20.56%), exclamation mark (A:682, total:1424, AMB:47.89%), semicolon (A:634, total:, AMB:%), parentheses (A:393, total:5999, AMB:6.55%), guillemets (A:59, total:369, AMB:15.98%), comma (A:26, total:102425, AMB:0.02%), square bracket (A:26, total:75, AMB:34.66%), hypen (A:24, total:29863, AMB:0.08%), single quote (A:18, total:730, AMB:2.46%), slash (A:6, total:2065, AMB:0.29%)	

the complexity of the task. In a newspaper corpus (Tübinger Baumbank des Deutschen/Zeitungskorpus (Tüba-D/Z) v. 10, henceforth TüBa10, there are 1.787.801 tokens and 95.595 sentences, described in Telljohann et al. (2012)), about 91% of sentence boundaries are punctuation such as periods, colons, semi-colons and commas (Table 1). The remaining sentences end with a word. As expected, periods are the most frequently used at the ends of the sentences in TüBa10 (about 77%, cf. Table 1). Most of the periods (about 84% of all peri-ods) are used to mark the end of a sentence (Table 1). The remaining periods are parts of tokens (i. e., category [B]), of which 68 types are identified in the corpus. If we exclude token types that we can simply handle with regular expressions – that is, those with an alphabet, number, email address, web link and ellipsis – there are 27 types of abbreviations and proper names. These exceptions can be handled reasonably by listing the abbreviations and proper names.[3]

However, the task of text segmentation is not trivial if we address the fol-lowing dependencies (Palmer 2000): (1) language dependence, (2) corpus depen-dence and (3) application dependence. Thus, the segmentation of multi-lingual texts (Jurish and Würzner 2013; Kiss and Strunk 2006) is not rule-based but sta-tistical. Corpus dependence involves a wide range of text types that have various linguistic features. Lastly, the definitions of words and sentences depend on the NLP application: for example, in a machine translation, a German compound is better split into individual morphemes (El-Kahlout and Yvon 2010).

In this work, we focus on the development of a German text segmentation system that deals with the issue of corpus dependence in Palmer's term. More specifically, it has been observed – e. g., by Giesbrecht and Evert (2009) for

---

[3] However, lists of abbreviations are never complete, and need to be extended, when we use out-of-domain data.

part-of-speech-tagging and by Gildea (2001) for parsing – that a statistical model usually works for the domain and text types it has been trained for, but leaves to desire when applied to other domains and text types. In this work, we undertake domain adaptation in text segmentation, in particular, with a target domain – texts written in private communication. Typically, these texts contain many deviations from standard orthography, including idiosyncrasies in capitalisation and punctuation.

In this paper, we train text segmentation models (conditional random fields) on TüBa10 (Sect. 4) and test them on an example of a text used in private communication: a postcard corpus[4] (*Ansichtskartenkorpus*, 'picture postcard corpus', henceforth ANKO) (Sect. 5). Sections 2 and 3 provide the analysis of the use of punctuation in private communication, and describe our text segmentation system.

2   Use of Punctuation

In German, punctuation segments a text into sentences, and a sentence is segmented into words by spaces. However, these rules of thumb are not applicable in the following cases of sentence segmentation: (1) punctuation that is a part of a token with preceding characters (e. g., abbreviations); and (2) punctuation is absent. Case (1) was discussed in Sect. 1. Case (2) occurs because of freestanding lines. Freestanding lines typically end with a line break with a wide blank space or extra line spacing, and often do not end with a punctuation. Examples are titles, subtitles, addresses, dates, greetings, salutations and signatures (Official German Orthography 2006). In private communication, the rules for freestanding lines are also applied to the end of paragraphs. In addition, the following usage is common to the punctuation in a private communication: (a) repeated punctuation (e. g., *!!!, ???,......*) in order to emphasise words, phrases and sentences; and (b) the use of emotional pictograms that are typically composed of punctuation (e. g., :), ;-)) (cf. Bartz et al. (2013)).

3   Conditional Random Fields (CRF)-Based Text Segmentation

We develop a CRF-based German text segmentation system that can be applied to the types of texts used in private communication (cf. Section 2). We focus on tokenisation with punctuation and sentence segmentation. In this section, we briefly introduce CRF and define the notion of a sequence and a set of features used in the task.

---

[4] The corpus will be released in https://linguistik.zih.tu-dresden.de/ansichtskarten.

3.1   Conditional Random Fields

CRF (Lafferty et al. 2001; Sutton and McCallum 2011) is a random field (also known as undirected graph or Markov network) for conditional probability $P(y_{1:n}|x_{1:n})$, where $x_{1:n}$ is an input sequence $x_1 \ldots x_n$ and $y_{1:n}$ is an output sequence $y_1 \ldots y_n$. To calculate the conditional probability, CRF makes use of the maximum entropy model and normalizes the probability globally in a sequence:

$$P(y_{1:n}|x_{1:n}) = \frac{1}{Z(x_{1:n})} \exp \left(\sum_{n=1}^{N} \sum_{d=1}^{D} w_d f_d(x_{1:n}, y_n, y_{n-1}, n) \right)$$

$$Z(x_{1:n}) = \sum_{y_{1:n}} \exp \left(\sum_{n=1}^{N} \sum_{d=1}^{D} w_d f_d(x_{1:n}, y_n, y_{n-1}, n) \right)$$

3.2   Sequence

The CRF model learns the parameters and decodes the output based on a given sequence of input units. In our classification task, a text is a sequence of input units. We use the term *unit* to denote each atomic element in the CRF in order to differentiate it from the term *token* or *word*. We create units by splitting texts using white spaces and by separating punctuation from the attached characters. We then classify the input units into three categories: the beginning (B), intermediate (I) and end (E) of sentences. Using this notation, the chunk of a token is also marked.

We investigate how flexibly punctuation should be handled in order to be robust for domain difference, and the importance of punctuation in text segmentation. To this end, we create three types of sequences by deliberately handling the punctuation listed in Table 1 in the following three ways:

(a): Punctuation before a white space is regarded as a unit. For example, z.B. (abbreviation of *zum Beispiel* 'for example') consists of two units: z.B and .
(b): Punctuation is regarded as a unit regardless of a white space. Accordingly, z.B. consists of four units: z, ., B and .
(c): All punctuation is removed if it is followed by a white space. Accordingly, z.B. consists of one unit: z.B

Variant (a) is a setting in which the white space is well placed, and it follows standard German orthographic rules. In Variant (b), every punctuation mark is individually handled, which is expected to provide flexibility in orthographical deviations. In Variant (c), punctuation is missing in the input text.

3.3   Features

Features are key linguistic indicators that may be useful in the segmentation of sentences. In this work, we use three types of features to handle orthographic variations and unknown words in variations of types of text: forms, POS and unsupervised word representations. The following subsections describe each feature in detail (Table 2 in Appendix).

Form. Forms of units are integrated as three features: (1) unit, (2) character types of unit and (3) normalised unit. For the first feature, units are extracted as they are. In the second feature, units are categorised into alphabetic, numeric, types of special characters, and their combinations. For the third feature, units are changed to lower case.

Part of speech. POS is used as a feature in two ways: fine-grained original Stuttgart-Tübingen-TagSet (STTS, Schiller et al. (1999)) or coarse POS (CPOS), which are shown in Table 2. These two features are extracted automatically using the TreeTagger (Schmid 1999).

Brown clustering. For the first feature of unsupervised word clustering, the hierarchical classes of Brown clustering (Brown et al. 1992) are exploited. Brown clustering is a bigram-based word clustering that has been successfully integrated to improve parsing (Koo et al. 2008), domain adaptation (Candito et al. 2011) and named entity recognition (Miller et al. 2004). We ran Brown clustering on the normalised tokens (i. e., all lower case) of TüBa10 to build 100 clusters.[5] For the features, we used the first four digits and all digits in the clustering hierarchy. In the data, the grouping of named entity such as person and organization and the part of speech such as noun and verb were captured the most clearly in the first four digits of the clustering.

Word2Vec. For the second feature of unsupervised methods, we used k-means clustering in Word2Vec. Word2vec (Mikolov et al. 2013) is another kind of word representation. We ran the Word2Vec on the normalised tokens of TüBa10 to build the models.[6] To operationalise the word-embedding vectors, we further grouped them into 50 K-means clusters.[7] The resulting clusters contained a great deal of named entities.

Fasttext. For the third feature of unsupervised methods, we used k-means clustering in Fasttext. Fasttext (Bojanowski et al. 2016) is yet another kind of word representation that takes into account character n-grams (morpheme). We ran Fasttext on the normalised tokens of TüBa10 to build the models.[8] We further grouped them into 200 K-means clusters that contained a large number of German compounds.

---

[5] We used the Brown clustering implemented by P. Liang.
[6] For word2vec, we used gensim with parameters CBOW, 200 dimensions, context window 5.
[7] For K-means clustering, we used the scikit-learn.
[8] We used the fasttext with parameters, CBOW, 200 dimensions, 5 context window, 5 word ngrams.

4   Experiments

In this experiment, our goal was to develop a text segmentation model that could robustly be applied to domain difference. For the experiment, we used the TüBa10 in form of (a), (b) and (c) (cf. Sect. 3.2) with various feature configurations, and we trained and tested the CFG models using five-fold cross-validation. In the next section, we evaluate the models by applying them to a postcard corpus to test their robustness for texts generated in private communications.

Single features. In the experiment, we used data in the forms of (a), (b) and (c) with single features. First, we trained CRF models in context window 0. The results are shown in columns #1 to #12 of Table 3 in Appendix. Among the features, the character type of unit (#3) – information about capitalisation, and type of punctuation and characters – showed the best performances in sequence types (a) and (b), whereas gold STTS POS tag (#4) showed the best performance in (c). In the unsupervised methods, Brown clustering (#8/9) outperformed Word2Vec (#10) and Fasttext (#11). As expected, the sequence types (a) and (b) achieved higher accuracy than sequence (c) did. For the sequence type (c), that is, the input sequence without punctuation, all individual features did not predict the classes (B) and (E). Thus, punctuation was proven relevant in text segmentation. We extended the set of features in context window 3. However, the accuracy remained the same as in window 0.

Feature combinations. To obtain linguistic information effectively in wide contexts, we combined the features in the following two ways: (1) the same features in context 0 and 1; (2) the combination of two features (1a/1b/1c/2bT/3a4/3b/3c) in context 0. In the first setting, which combined all with all features of previous, current and next tokens, the CRF models improved with regard to class (B) and (E) (#13 in Table 3). In the second feature combination (#14), the overall accuracy was similar to that in the set of all single features (#12).

For the evaluation, we trained all single features in context 0 (#12), the combinations of the same features in context window 1 (#13) and those of various features in context window 0 (#14) without using gold POS and CPOS tags. The CRF models achieved accuracies of 0.99, 0.99 and 0.97 on the TüBa in the sequence types (a), (b) and (c), respectively. In the evaluation, we used the feature set for each input sequence type.

5   Evaluation and Conclusion

For the evaluation, we used a test set derived from a corpus of postcards (ANKO). The corpus comprised over 11,000 holiday postcards sent by post to Swiss households from 1950 to the present day. In this work, we used a sub-corpus (545 cards, 3534 sentences and 25096 tokens) that contained cards mainly written in standard German. We manually created three types of input sequences:

(I) one with text boundaries; (II) one with text and paragraph boundaries; and (III) one with text, paragraph and discourse boundaries (date, salutation, greeting and signature). We tested the final models as described in the previous section. The results are shown below:

	(I)		(II)		(III)	
	acc	F1(B, I, E)	acc	F1(B, I, E)	acc	F1(B, I, E)
(a)	.89	.73, .94, .72	.91	.80, .95, .77	.94	.88, .97, **.84**
(b)	.91	.76, .95, .44	.93	.82, .97, .61	**.96**	**.90, .98**, .82
(c)	.81	.30, .90, .42	.83	.40, .90, .55	.86	.50, .92, .73

Overall, the sequence type (b) achieved better accuracy than sequence type (a) did, which showed that orthographic deviations could be handled more effectively by segmenting punctuation individually. Clearly, the patterns of punctuation were more generally captured in (b). Furthermore, the input text type (III) achieved high accuracy. These results indicate that the annotation of a corpus with paragraphs and freestanding lines is relevant in improving the quality of the segmentation of texts used in private communication. Still, it was difficult to predict text segments without having punctuations (c) on a type of text different from the training data.[9]

As comparison, we tested a sentence segmentation system PUNKT (Kiss and Strunk 2006).[10] PUNKT is based on unsupervised methods and designed for multi-lingual text segmentation. We tested PUNKT on our ANKO test set type (III). PUNKT achieved a F1 score of 0.79 with precision 0.71 and recall 0.9. In contrast, our sentence segmentation system achieved a F1 score of 0.95 with precision 0.94 and recall 0.96, using the input format (b) and (III), that is, the best input format for tokenization.

In conclusion, we presented our German text segmentation system for texts in private communication. In future work, we will extend our text segmentation system on historical German texts.

Acknowledgments. This work is funded under SNSF grant 160238. We thank all the project members, Heiko Hausendorf, Joachim Scharloth, Noah Bubenhofer, Nicolas Wiedmer, Selena Calleri, Maaike Kellenberger, David Koch, Marcel Naef, Josephine Obert, Jan Langenhorst, Michaela Schnick to support our work. We thank two anonymous reviewers for their valuable suggestions.

---

[9] Our text segmentation system (GETS) is available: https://sugisaki.ch/tools.
[10] We used the NLTK module PUNKT.

6 Appendix

Table 2. Features

1a	**Unit**
1b	**Character type of unit**: unit form is categorised into the following classes: All characters are alphabetic, and (A) consist of just one alphabet or (B) are absent of vocal (e.g., *lg,hrzl*) or (C) all letters are uppercase or (D) only first letter is uppercase or (E) else. Or all characters are (F) numbers or (G) alphanumeric. For punctuations, (H) period, (I) comma, (J) question and exclamation mark, (K) colon and semicolon, (L) opening and (M) closing bracket, (N) opening and (O) closing quotation. For mix classes: (P) alphabets and punctuations/other special characters (e.g., *u.s.w, v.a*), (Q) numbers and punctuations/other special characters (e.g., *8.000*), or (R) else
1c	**Normalized unit**: all lower case
2a	**Fine-grained POS**: STTS tag set; In experiment, 2aG is gold standard, 2aT is TreeTagger output
2b	**Coarse POS**: Nouns, verbs, modifiers of nouns, modifiers of verbs, relative pronouns, other pronouns, articles, prepositions, postpositions, cardinal number, wh words, subordinating/infinitive conjunctions, coordinating conjunctions, spoken language markers, comma and semicolon, colon, period and question and exclamation mark, quotations, brackets, else; In experiment, 2bG is gold standard, 2bT is based on TreeTagger output
3abc	**Unsupervised methods**: Brown clustering, word2vec and fasttext, respectively. In the experiment, brown clustering is used in 4 digits (3a4) and all digits (3aA)

Table 3. Feature experiments (5-fold cross validation on TüBa10): abbreviations of features listed in Table 2; sequence type (a)(b)(c) described in Sect. 3.2; acc(uracy) = correctly predicted tokens/the total number of tokens; F1 = 2 * precision * recall/(precision + recall)

#	1	2	3	4	5	6	7	8	9	10	11	12	13	14
	1a	1b	1c	2aG	2aT	2bG	2bT	3a4	3aA	3b	3c	all	all	all
	single features in context window 0												combinations		
(a)	acc:.97*	acc:.97*	acc:.97*	acc:.96	acc:.97*	acc:.96	acc:.97*	acc:.97*	acc:.97*	acc:.91	acc:.89	acc:.98	**acc:.99**	acc:.98
	F1(B):.86	F1(B):.82	F1(B):.88*	F1(B):.81	F1(B):.83	F1(B):.81	F1(B):.83	F1(B):.84	F1(B):.81	F1(B):.47	F1(B):.08	F1(B):.91	**F1(B):.96**	F1(B):.91
	F1(I):.99*	F1(I):.98	F1(I):.99*	F1(I):.98	F1(I):.98	F1(I):.98	F1(I):.98	F1(I):.98	F1(I):.98	F1(I):.95	F1(I):.94	F1(I):.99	**F1(I):1.00**	F1(I):.99
	F1(E):.86	F1(E):.84	F1(E):.88*	F1(E):.83	F1(E):.85	F1(E):.82	F1(E):.85	F1(E):.85	F1(E):.83	F1(E):.49	F1(E):.11	F1(E):.92	**F1(E):.96**	F1(E):.91
(b)	acc:.97*	acc:.96	acc:.97*	acc:.96	acc:.97*	acc:.96	acc:.97*	acc:.97*	acc:.96	acc:.91	acc:.89	acc:.98	**acc:.99**	acc:.98
	F1(B):.82	F1(B):.79	F1(B):.85*	F1(B):.79	F1(B):.81	F1(B):.79	F1(B):.81	F1(B):.82	F1(B):.79	F1(B):.45	F1(B):.08	F1(B):.89	**F1(B):.95**	F1(B):.88
	F1(I):.98*	F1(I):.98*	F1(I):.98*	F1(I):.98*	F1(I):.98*	F1(I):.98*	F1(I):.98*	F1(I):.98*	F1(I):.98*	F1(I):.95	F1(I):.94	**F1(I):.99**	F1(I):.99	F1(I):.99
	F1(E):.83	F1(E):.81	F1(E):.85*	F1(E):.81	F1(E):.83	F1(E):.81	F1(E):.84	F1(E):.84	F1(E):.81	F1(E):.48	F1(E):.11	F1(E):.90	**F1(E):.96**	F1(E):.90
(c)	acc:.88	acc:.88	acc:.88	acc:.89*	acc:.88	acc:.88	acc:.88	acc:.88	acc:.88	acc:.88	acc:.88	acc:.96	**acc:.97**	acc:.96
	F1(B):.00	F1(B):.00	F1(B):.00	F1(B):.21*	F1(B):.00	F1(B):.02	F1(B):.00	F1(B):.00	F1(B):.00	F1(B):.00	F1(B):.00	F1(B):.82	**F1(B):.86**	F1(B):.81
	F1(I):.93	F1(I):.93	F1(I):.94*	F1(I):.94*	F1(I):.93	F1(I):.93	F1(I):.93	F1(I):.93	F1(I):.93	F1(I):.93	F1(I):.93	**F1(I):.98**	F1(I):.98	F1(I):.98
	F1(E):.00	F1(E):.00	F1(E):.00	F1(E):.22*	F1(E):.00	F1(E):.03	F1(E):.00	F1(E):.00	F1(E):.02	F1(E):.00	F1(E):.00	F1(E):.85	**F1(E):.89**	F1(E):.83

References

Bartz, T., Beißwenger, M., Storrer, A.: Optimierung des Stuttgart-Tübingen-Tagset für die linguistische Annotation von Korpora zur internetbasierten Kommunikation: Phänomene, Herausforderungen, Erweiterungsvorschläge. JLCL **28**, 157–198 (2013)

Bojanowski, P., Grave, E., Joulin, A., Mikolov, T.: Enriching word vectors with subword information. arXiv preprint arXiv:1607.04606 (2016)

Brown, P.F., Desouza, P.V., Mercer, R.L., Pietra, V.J.D., Lai, J.C.: Class-based n-gram models of natural language. Comput. Linguist. **18**(4), 467–479 (1992)

Candito, M., Anguiano, E.H., Seddah, D.: A word clustering approach to domain adaptation: effective parsing of biomedical texts. In: Proceedings of the 12th International Conference on Parsing Technologies, pp. 37–42 (2011)

El-Kahlout, I.D., Yvon, F.: The pay-offs of preprocessing for German-English statistical machine translation. In: Proceedings of the 7th International Workshop on Spoken Language Translation (IWSLT), pp. 251–258 (2010)

Giesbrecht, E., Evert, S.: Is part-of-speech tagging a solved task? An evaluation of POS taggers for the German web as corpus. In: Proceedings of the Fifth Web as Corpus Workshop (WAC5), pp. 27–35 (2009)

Gildea, D.: Corpus variation and parser performance. In: Proceedings of the EMNLP, pp. 167–202 (2001)

Jurish, B., Würzner, K.M.: Word and sentence tokenization with Hidden Markov Models. JLCL **28**, 61–83 (2013)

Kiss, T., Strunk, J.: Unsupervised multilingual sentence boundary detection. Comput. Linguis. **32**(4), 485–525 (2006)

Koo, T., Carreras, X., Collins, M.: Simple semi-supervised dependency parsing. In: Proceedings of the ACL/HLT, pp. 595–603 (2008)

Lafferty, J., McCallum, A., Pereira, F.C.N.: Conditional random fields: probabilistic models for segmenting and labeling sequence data. In: Proceedings of the 18th International Conference on Machine Learning 2001, pp. 282–289 (2001)

Mikolov, T., Chen, K., Corrado, G., Dean, J.: Efficient estimation of word representations in vector space. arXiv preprint arXiv:1301.3781 (2013)

Miller, S., Guinness, J., Zamanian, A.: Name tagging with word clusters and discriminative training. In: Proceedings of HLT-NAACL vol. 4, 337–342 (2004)

Palmer, D.D.: Tokenisation and Sentence Segmentation. CRC Press, Boca Raton (2000)

Proisl, T., Uhrig, P.: Somajo: state-of-the-art tokenization for German web and social media texts. In: Proceedings of the 10th Web as Corpus Workshop (WAC-X) and the EmpiriST Shared Task, pp. 57–62 (2016)

Remus, S., Hintz, G., Benikova, D., Arnold, T., Eckle-Kohler, J., Meyer, C.M., Mieskes, M., Biemann, C.: EmpiriST: AIPHES robust tokenization and POS-tagging for different genres. In: Proceedings of the 10th Web as Corpus Workshop, pp. 106–114 (2016)

Schiller, A., Teufel, S., Stöckert, C., Thielen, C.: Guidelines für das tagging deutscher textcorpora mit STTS (Kleines und großes Tagset). Technical report. Universität Stuttgart, Universität Tübingen, Stuttgart, Germany (1999)

Schmid, H.: Improvements in part-of-speech tagging with an application to German. In: Armstrong, S., Church, K., Isabelle, P., Manzi, S., Tzoukermann, E., Yarowsky, D. (eds.) Natural Language Processing Using Very Large Corpora. Text, Speech and Language Technology, pp. 13–26. Springer, Dordrecht (1999). https://doi.org/10.1007/978-94-017-2390-9_2

Sutton, C., McCallum, A.: An introduction to conditional random fields. Found. Trends Mach. Learn. **4**(4), 267–373 (2011)

Telljohann, H., Hinrichs, E.W., Sandra, K., Heike, Z., Kathrin, B.: Stylebook for the Tübingen treebank of written German (TüBa-D/Z). Technical report. Universität Tübingen (2012)

Fine-Grained POS Tagging of German Social Media and Web Texts

Stefan Thater[✉]

Department of Language Science and Technology, Universität des Saarlandes,
Saarbrücken, Germany
stth@coli.uni-saarland.de

Abstract. This paper presents work on part-of-speech tagging of
German social media and web texts. We take a simple Hidden Markov
Model based tagger as a starting point, and extend it with a distributional approach to estimating lexical (emission) probabilities of out-of-
vocabulary words, which occur frequently in social media and web texts
and are a major reason for the low performance of off-the-shelf taggers
on these types of text. We evaluate our approach on the recent *EmpiriST
2015 shared task* dataset and show that our approach improves accuracy
on out-of-vocabulary tokens by up to 5.8%; overall, we improve state-of-
the-art by 0.4% to 90.9% accuracy.

1 Introduction

Part-of-speech (POS) tagging is a standard component in many linguistic processing pipelines, so its performance is likely to impact the performance of all subsequent steps in the pipeline, such as morphological analysis or syntactic parsing.
In the newswire domain, modern POS taggers can reach accuracy scores beyond
97%, close to human performance (Manning 2011). For "non-standard" texts like
social media or web texts, however, tagger performance is usually much lower.
For the *EmpiriST 2015 shared task* dataset considered in this paper, Beißwenger
et al. (2016) report accuracy scores of 80–82% for off-the-shelf taggers.

One important reason for this decline in accuracy is that datasets which
are large enough to train a tagger are typically from the newswire domain. For
social media and web texts, no large training sets are available. At the same
time, these texts differ substantially from newswire text. They contain a lot of
"bad" language (Eisenstein 2013) such as misspellings, phrasal abbreviations or
intentional orthographical variations as well as phenomena like contractions or
interaction words which are not covered by standard tagsets.

On a technical level, the problem can be traced back, at least to some extent,
to out-of-vocabulary ("unknown") words which do not occur in the training set.
Giesbrecht and Evert (2009) observe that typical web texts contain, compared to
newswire texts, more unknown words, and that tagger performance on unknown
words is much lower. We make similar observations for the dataset considered
in this paper.

G. Rehm and T. Declerck (Eds.): GSCL 2017, LNAI 10713, pp. 72–80, 2018.
https://doi.org/10.1007/978-3-319-73706-5_7

One way to address this problem is to add small amounts of manually annotated in-domain data to existing (out-of-domain) training sets when training the tagger. For German, this approach has been explored by Horbach et al. (2014) and Neunerdt et al. (2014). The approach is appealing, as it is conceptually very simple, easy to implement and quite effective. Yet, it can only address part of the problem, as many words remain out-of-vocabulary. Another approach is to exploit distributional similarity information about unknown words. The underlying observation is that distributionally similar words tend to belong to the same lexical class, so POS information of out-of-vocabulary words can be derived from distributionally similar in-vocabulary words (Schütze 1995). Several approaches to POS tagging of various kinds of non-standard texts that exploit this idea have been proposed in the past few years. Gimpel et al. (2011) train a CRF-based tagger using features derived from a reduced co-occurrence matrix; Owoputi et al. (2013), Ritter et al. (2011) and Rehbein (2013) use clustering to derive features to train a discriminative tagger model. Prange et al. (2015) use distributional similarity information to learn a POS lexicon for out-of-vocabulary tokens, and combine it with a Hidden Markov Model (HMM) based tagger.

In this paper, we present an approach that is conceptually similar to the one of Prange et al. (2015) but which uses distributional similarity information to estimate emission probabilities of the HMM, rather than deriving an external POS lexicon. Results on the *EmpiriST 2015 shared task* dataset (Beißwenger et al. 2016) show that our approach improves accuracy on out-of-vocabulary words by up to 5.8%; overall, we improve state-of-the-art by 0.4% to 90.9% accuracy.

2   Model

We briefly present the underlying tagger model in Sect. 2.1 before presenting our distributional approach to estimating lexical probabilities for out-of-vocabulary tokens in Sect. 2.2. Section 2.3 describes the lookup procedure implemented by the tagger.

2.1   Baseline Model

We use a second order Hidden Markov Model to implement our baseline tagger. To tag a given input sequence $w_1 \ldots w_n$ of words, we calculate

$$\arg\max_{t_1,\ldots,t_n} \left[\prod_{i=1}^{n} P(t_i \mid t_{i-1}, t_{i-2}) P(w_i \mid t_i) \right] P(t_{n+1} \mid t_n)$$

where $t_1 \ldots t_n$ are elements of the tagset and t_{-1}, t_0 and t_{n+1} are additional tags marking the beginning and the end of the sequence, respectively.

Our implementation closely follows Brants (2000). Transition probabilities $P(t_i \mid t_{i-1}, t_{i-2})$ are computed using a linear combination of unigrams, bigrams and trigrams, which are estimated from a tagged training corpus using maximum likelihood. For the tokens in the training corpus, we estimate emission probabilities $P(w_i \mid t_i)$ using maximum likelihood and for out-of-vocabulary tokens emission probabilities are estimated based on the word's suffix. Our implementation differs slightly from (Brants 2000) in that we use, for purely practical reasons, a maximal suffix length of 5 instead of 10 in the computation of suffix distributions, and that we do not maintain different suffix distributions for uppercase and lowercase words.

2.2 Distributional Smoothing

We use a large, automatically POS-tagged corpus and estimate $P(w \mid t)$ by considering all contexts in which w occurs in the corpus, and estimating the emission probability of w based on the emission probability of all in-vocabulary words w' that occur in the same contexts as w. We set:

$$P(t \mid w) = \sum_{w'} \sum_{C} P(t \mid w') \, P(w' \mid C) \, P(C \mid w) \qquad (1)$$

where w' ranges over all in-vocabulary words in the manually annotated training corpus used to train the baseline model and C ranges over all n-grams consisting of the POS tags of the two words on either side of an unknown word w in the automatically tagged corpus. $P(t \mid w')$ is the probability of a tag t of an in-vocabulary word w', $P(w' \mid C)$ is the probability that w' occurs in a given context C and $P(C \mid w)$ is the probability of context C given an out-of-vocabulary word w. The probabilities are estimated on the automatically tagged corpus using maximum likelihood. Following recommendations by Prange et al. (2015), we consider only contexts in which the two surrounding words are in-vocabulary; the idea is that in-vocabulary tokens are tagged with much higher precision and thus give us more reliable context information.

While using (1) to estimate emission probabilities of out-of-vocabulary tokens improves tagger performance beyond the baseline model, (1) is still somewhat noisy. We further improve tagger performance by combining (1) with a second distribution $P(t \mid w)$ which estimates the probability of a tag t of an unknown word w based on the suffix of w. In principle, we could simply use the corresponding distribution of the baseline tagger, but it turns out that the following approach works much better:

$$P(t \mid w) = \sum_{w'} \sum_{s} P(t \mid w') \, P(w' \mid s) \, P(s \mid w) \qquad (2)$$

where s ranges over all possible suffixes. The distributions $P(s \mid w)$ and $P(w' \mid s)$ are estimated on the type level, $i.e.$, $P(s \mid w) = 1$ if s is a suffix of w, 0 otherwise, and $P(w' \mid s) = \frac{1}{n}$, where n is the number of types with suffix s.

We combine (1) and (2) using multiplication, re-normalize the result and apply Bayes' theorem to obtain the final emission probabilities $P(w \mid t)$.

2.3 Lookup

Our tagger implements the following lookup strategy: When reading in a token w, we first try to look up w in the lexicon; if that fails, we redo the lookup with w mapped to lower case; if that fails, we consult the distributional lexicon; as a fallback, we use the suffix lexicon of the baseline tagger.

We follow common practice and normalize all numerical expressions (sequences of digits) into a single token type. To improve tagger performance on social media texts, we additionally normalize all tokens beginning with an "@" or "#".

3 Evaluation

We evaluate our approach on the dataset of the *EmpiriST 2015 shared task on automatic linguistic annotation of computer-mediated communication and social media* (Beißwenger et al. 2016) and compare it to the two systems that performed best on the share task as baselines.

3.1 Datasets

EmpiriST. This dataset has been provided by the *EmpiriST 2015 shared task*. It has been compiled from data samples considered representative for two types of corpus data. The *CMC* subset consists of selections of microposts from Twitter, a subset of the *Dortmund Chat Corpus* (Beißwenger 2013), threads from Wikipedia talk pages, WhatsApp interactions and blog comments. The *Web* subset consists of selections of websites and blogs covering various genres and topics like hobbies and travel, Wikipedia articles on topics like biology and botany and Wikinews on topics like IT security and ecology. The dataset is split into two parts, one for training and one for testing. The *CMC* subset consists of 5109 tokens for training and 5234 tokens for testing; the *Web* subset consists of 4944 tokens for training 7568 tokens for testing.

The dataset has been annotated using the "STTS IBK" tagset (Beißwenger et al. 2015), which is based on the STTS tagset (Schiller et al. 1999). STTS is the standard tagset for German. It distinguishes 11 parts of speech which are subdivided into 54 subcategories. STTS IBK adds 16 new tags for phenomena that occur frequently in social media texts, such as interaction words, addressing terms or contractions.

Schreibgebrauch. This dataset has been provided by (Horbach et al. 2015) and has been used as additional in-domain training data by the best-performing system of the *EmpitiST shared task* (Prange et al. 2016). It consists of manual annotations of forum posts of the German online cooking community http://www.chefkoch.de, a subset of the *Dortmund Chat-Korpus* and microposts from Twitter. In total, the annotated dataset consists of 34 173 tokens. Since the dataset has been annotated with a tagset that differs in some details from STTS IBK, Prange et al. (2016) re-annotated the dataset so that it matches the annotation scheme and guidelines of the shared task. We use the re-annotated version in our experiments.

We also use the complete *Chefkoch* corpus from which the annotated subset was selected to train lexical probabilities of out-of-vocabulary tokens. The corpus contains 470M tokens and covers a relatively large range of everyday topics.

TIGER. The TIGER corpus (Brants et al. 2004) is one of the standard corpora used for German POS tagging. It consists of 888 238 tokens which have been semi-automatically annotated with POS information, using the standard STTS tagset.

3.2 Experimental Setup

We train two different models: The TE model is trained on a combination of the TIGER corpus and the EmpiriST training set. The TES model additionally uses the *Schreibgebrauch* dataset. Since the two in-domain datasets are very small compared to TIGER, we follow Prange et al. (2016) and oversample them by a factor of 5. We automatically annotate the *Chefkoch* corpus using each of the two tagger models to estimate emission probabilities for out-of-vocabulary words as described in Sect. 2.2.

3.3 Results

Figure 1 shows the results of our approach on the EmpiriST evaluation dataset. We consider two different configurations for each of our two models: TE/BL and TES/BL use suffix-based emission probabilities of the baseline tagger for out-of-vocabulary tokens, while TE/DS and TES/DS use distributional smoothing. To set the results into perspective, we compare our models to two state-of-the-art approaches: *UdS* refers to the system of Prange et al. (2016), which performed

Model	CMC	Web	Overall
TE/BL	86.78	92.47	89.63
TES/BL	87.89	92.72	90.31
TE/DS	87.08	93.22	90.15
TES/DS	**88.38**	93.34	**90.86**
UDE	86.07	92.10	89.09
UdS	87.33	**93.55**	90.44

Fig. 1. Accuracy comparison for different configurations of our tagger and the two best performing shared task models on the EmpiriST test set.

best in the EmpiriST shared task. The tagger is based on a Hidden Markov Model trained on EmpiriST, *Schreibgebrauch* and TIGER and uses distributional information obtained from the *Chefkoch* corpus to automatically learn a POS dictionary. *UDE* refers to the system of Horsmann and Zesch (2016). The tagger is based on Conditional Random Fields (CRFs) trained on EmpiriST and TIGER and was the best system in the shared task that does not use any in-domain data in addition to the training data provided by the shared task. In addition to standard features of a CRF-based tagger, the system uses word cluster information from Twitter messages, a POS lexicon and a morphological lexicon.

We compare our TE model to the UDE system and the TES model to the UdS system. Figure 1 shows that already our baseline configurations outperform state of the art (except UdS on *Web*). This is particularly surprising when comparing TES to UdS on *CMC*, since both models are based on trigram HMMs trained on the same datasets. To some extent, the difference can be explained by our use of simple patterns for @- and #-expressions, but we note that even without these patterns our basic tagger still outperforms UdS on *CMC* by 0.2%.

We also see that distributional smoothing is effective across all four configurations. On the *CMC* subset, the performance gain increases quite substantially for the TES model compared to the TE model (+0.49 *vs.* +0.30). This is to be expected, since the emission probabilities are derived from an automatically annotated corpus, which is tagged with higher accuracy when the TES model is used. For the *Web* subset, the performance gain is even larger. The relative performance gain is a bit lower for the TES model (+0.62) compared to the TE model (+0.75), which can be explained by the fact that the TES model generally performs better than the TE model on out-of-vocabulary items; see Sect. 3.4 below for details.

Overall, our tagger improves state-of-the-art substantially. Our best configuration (TES/DS) outperforms the previous best system by 0.42% accuracy.

3.4 Performance on Unknown Words

In a second experiment, we investigate the performance of our distributional smoothing approach in more detail. We split the test set into three parts— in-vocabulary tokens (IV), out-of-vocabulary tokens covered by our distributional smoothing approach (OOV/DS) and out-of-vocabulary tokens which do not occur in the *Chefkoch* corpus and are thus dealt with using suffix probabilities only (OOV/BL)—and measure accuracy of our models on these three subsets separately. Figure 2 shows, for each of the three subsets, the number of tokens in the subset, the performance of the DS models and the performance gain of the DS models over the corresponding BL models, for both TE and TES. We see that distributional smoothing is very effective and improves accuracy over the baseline by 7–8%, except for the TE model on the *CMC* subset where

	TE/DS *vs.* TE/BL			TES/DS *vs.* TES/BL		
	CMC		Web	CMC		Web
IV	4589	90.1 (+0.03)	6624 94.9 (+0.05)	4732 90.3 (+0.02)	6682 94.9 (+0.03)	
OOV/DS	472	65.0 (+2.97)	629 83.0 (+8.11)	343 71.7 (+7.29)	581 84.0 (+7.40)	
OOV/BL	173	67.1 (+0.58)	315 77.8 (+0.95)	159 66.7 (+0.00)	305 77.7 (+0.66)	

Fig. 2. Accuracy comparison of the DS and BL models for in- (IV) and out-of-vocabulary (OOV) tokens on the *CMC* and the *Web* subset. The rows give, for each group, the number of tokens, the accuracy of the DS model and the accuracy gain of the DS model over the BL model.

we obtain only a moderate improvement of approx. 3%. Overall, the improvement over the baseline is 5.1% (TE) and 5.8% (TES) on all out-of-vocabulary tokens.

4   Conclusions

In this paper, we presented work on part-of-speech tagging of German social media and web texts, using a fine grained tagset. Our tagger is based on a simple trigram Hidden Markov Model, which we extend with a distributional approach to estimating emission probabilities of out-of-vocabulary tokens. While technically very simple, our tagger is very effective and outperforms, or comes very close to, state-of-the-art systems even in the baseline configuration without distributional smoothing. Using distributional smoothing improves accuracy of out-of-vocabulary tokens by up to 5.8%. Overall, we improve state-of-the-art by 0.4% to 90.9% accuracy.

References

Beißwenger, M.: Das Dortmunder Chat Korpus. Zeitschrift für Germanistische Linguistik **41**(1), 161–164 (2013)

Beißwenger, M., Bartz, T., Storrer, A., Westpfahl, S.: Tagset und Richtlinie für das Part-of-Speech-Tagging von Sprachdaten aus Genres internetbasierter Kommunikation. EmpiriST 2015 guideline document (2015)

Beißwenger, M., Bartsch, S., Evert, S., Würzner, K.-M.: EmpiriST 2015: a shared task on the automatic linguistic annotation of computer-mediated communication and web corpora. In: Proceedings of the 10th Web as Corpus Workshop, pp. 44–56 (2016). http://aclweb.org/anthology/W16-2606

Brants, S., Dipper, S., Eisenberg, P., Hansen, S., Koenig, E., Lezius, W., Rohrer, C., Smith, G., Uszkoreit, H.: TIGER: linguistic interpretation of a German corpus. J. Lang. Comput. **2**(4), 597–620 (2004). Special Issue

Brants, T.: TnT - a statistical part-of-speech tagger. In: Proceedings of the Sixth Conference on Applied Natural Language Processing, pp. 224–231 (2000)

Eisenstein, J.: What to do about bad language on the internet. In: Proceedings of the 2013 Conference of the North American Chapter of the Association for Computational Linguistics: Human Language Technologies, pp. 359–369 (2013). http:// aclweb.org/anthology/N13-1037

Giesbrecht, E., Evert, S.: Is part-of-speech tagging a solved task? An evaluation of POS taggers for the German web as corpus. In: Proceedings of the Fifth Web as Corpus Workshop, San Sebastian, Spain, pp. 27–35 (2009)

Gimpel, K., Schneider, N., O'Connor, B., Das, D., Mills, D., Eisenstein, J., Heilman, M., Yogatama, D., Flanigan, J., Smith, N.A.: Part-of-speech tagging for Twitter: annotation, features, and experiments. In: Proceedings of the 49th Annual Meeting of the Association for Computational Linguistics: Human Language Technologies, pp. 42–47 (2011). http://aclweb.org/anthology/P11-2008

Horbach, A., Steffen, D., Thater, S., Pinkal, M.: Improving the performance of standard part-of-speech taggers for computer-mediated communication. In: Proceedings of the 12th Edition of the KONVENS Conference, vol. 1, pp. 171–177 (2014)

Horbach, A., Thater, S., Steffen, D., Fischer, P.M., Witt, A., Pinkal, M.: Internet corpora: a challenge for linguistic processing. Datenbank Spektrum 15(1), 41–47 (2015)

Horsmann, T., Zesch, T.: LTL-UDE@EmpiriST 2015: tokenization and PoS tagging of social media text. In: Proceedings of the 10th Web as Corpus Workshop, pp. 120–126 (2016). http://aclweb.org/anthology/W16-2615

Manning, C.D.: Part-of-speech tagging from 97% to 100%: is it time for some linguistics? In: Gelbukh, A.F. (ed.) CICLing 2011. LNCS, vol. 6608, pp. 171–189. Springer, Heidelberg (2011). https://doi.org/10.1007/978-3-642-19400-9_14

Neunerdt, M., Reyer, M., Mathar, R.: Efficient training data enrichment and unknown token handling for POS tagging of non-standardized texts. In: Proceedings of the 12th edition of the KONVENS conference, vol. 1, pp. 186–192 (2014)

Owoputi, O., O'Connor, B., Dyer, C., Gimpel, K., Schneider, N., Smith, N.A.: Improved part-of-speech tagging for online conversational text with word clusters. In: Proceedings of the 2013 Conference of the North American Chapter of the Association for Computational Linguistics: Human Language Technologies, pp. 380–390 (2013). http://aclweb.org/anthology/N13-1039

Prange, J., Thater, S., Horbach, A.: Unsupervised induction of part-of-speech information for OOV words in German internet forum posts. In: Proceedings of the 2nd Workshop on Natural Language Processing for Computer-Mediated Communication/Social Media (2015)

Prange, J., Horbach, A., Thater, S.: UdS-(retrain|distributional|surface): improving POS tagging for OOV words in German CMC and web data. In: Proceedings of the 10th Web as Corpus Workshop, pp. 63–71 (2016). http://aclweb.org/anthology/ W16-2608

Rehbein, I.: Fine-grained POS tagging of German tweets. In: Gurevych, I., Biemann, C., Zesch, T. (eds.) GSCL 2013. LNCS (LNAI), vol. 8105, pp. 162–175. Springer, Heidelberg (2013). https://doi.org/10.1007/978-3-642-40722-2_17

Ritter, A., Clark, S., Mausam, Etzioni, O.: Named entity recognition in tweets: an experimental study. In: Proceedings of the 2011 Conference on Empirical Methods in Natural Language Processing, pp. 1524–1534 (2011). http://aclweb.org/anthology/ D11-1141

Schiller, A., Teufel, S., Stöckert, C., Thielen, C.: Guidelines für das Tagging deutscher Textcorpora mit STTS. Technical report, Institut für maschinelle Sprachverarbeitung, Universität Stuttgart and Seminar für Sprachwissenschaft, Universität Tübingen (1999). http://www.ims.uni-stuttgart.de/forschung/ressourcen/lexika/TagSets/stts-1999.pdf

Schütze, H.: Distributional part-of-speech tagging. In: Seventh Conference of the European Chapter of the Association for Computational Linguistics (1995). http://aclweb.org/anthology/E95-1020

Developing a Stemmer for German Based on a Comparative Analysis of Publicly Available Stemmers

Leonie Weissweiler[✉] and Alexander Fraser

Center for Information and Language Processing,
LMU Munich, Munich, Germany
{weissweiler,fraser}@cis.lmu.de

Abstract. Stemmers, which reduce words to their stems, are important components of many natural language processing systems. In this paper, we conduct a systematic evaluation of several stemmers for German using two gold standards we have created and will release to the community. We then present our own stemmer, which achieves state-of-the-art results, is easy to understand and extend, and will be made publicly available both for use by programmers and as a benchmark for further stemmer development.

1 Introduction

1.1 The Stemming Task

In Information Retrieval (IR), an important task is to not only return documents that contain the exact query string, but also documents containing semantically related words or different morphological forms of the original query word (Manning et al. 2008, p. 57).

This is achieved by a stemming algorithm.

A stemming algorithm is a computational procedure which reduces all words with the same root (or, if prefixes are left untouched, the same stem) to a common form, usually by stripping each word of its derivational and inflectional suffixes (Lovins 1968).

Thus, the purpose of a stemmer is not to find the morphologically correct root for a word, but merely to reduce it to a form it shares with all words that are sufficiently semantically related to be considered relevant to a search engine query for one of them. The exact nature of that form is irrelevant.

1.2 Motivation

Stemming for German is naturally a task that attracts less attention then stemming for English. There are, however, a number of different available stemmers

© The Author(s) 2018
G. Rehm and T. Declerck (Eds.): GSCL 2017, LNAI 10713, pp. 81–94, 2018.
https://doi.org/10.1007/978-3-319-73706-5_8

for German, the most popular of which are the Snowball German stemmer, developed by the team of Martin Porter, and the stemmer developed by Caumanns (1999). Available stemmers are fairly different in terms of their algorithms and their approaches to stemming, with solutions ranging from recursive stripping of just a few characters to identifying prefixes and suffixes from a pre-compiled list. Of all the stemmers presented here the Snowball stemmer is the only one for which an official implementation is available. For the others, the implementations that are used in NLP toolkits are by third parties, and, as we will show, sometimes contain flaws not intended by the original authors.

At the same time, we are not aware of any comprehensive evaluation of stemmer performance for German. The main goal of this paper is therefore to supply such a study in order to enable NLP programmers to make an informed choice of stemmer. We also want to improve existing stemmers and therefore present a new state-of-the-art stemmer, which we will make freely available in Perl, Python and Java. So a secondary goal is to make a clean and simple implementation of our stemmer available for programmers. Finally, we will release the two gold standards we have developed, which can act as a benchmark for future stemmer development work.

1.3 Summary of Work

We looked at five available stemmers for German and compared their algorithms.

We then automatically compiled two different gold standards from the morphological information in the CELEX2 data (Baayen et al. 1995) for German. They aim to represent two slightly different opinions on what stemming should be. One was compiled by stripping away morphemes that had not been assigned their own wordclass and the other using the wordform to lemma matching in the CELEX2 data.

We then evaluate the stemmers on the two gold standards, computing precision, recall and f-measure in a cluster-based evaluation that evaluated performance based on which words were stemmed to the same stem (and not how the stems actually looked, which is not relevant in most applications of stemming, as we discussed above).

Based on the results of our evaluation, we developed a new stemmer called CISTEM which is simpler and more aggressive than the previously existing stemmers. We show that CISTEM performs better than the previously existing stemmers.

2 Existing Stemmers for German and Related Work

2.1 German Stemmers

In this section we provide an overview of the German stemmers that we studied, briefly outlining their availability and the algorithms used. We show the differences between them with the example shown in Fig. 1, where we stemmed the

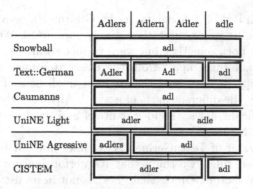

	Adlers	Adlern	Adler	adle
Snowball	adl			
Text::German	Adler	Adl		adl
Caumanns	adl			
UniNE Light	adler		adle	
UniNE Agressive	adlers	adl		
CISTEM	adler			adl

Fig. 1. Comparison of all stemmers using the words "Adler" (eagle), "Adlers" (eagle, genitive case), "Adlern" (eagles, dative case), "adle" (inflected form of "to ennoble")

word "Adler" (eagle). We show the stem produced and the other words reduced to the same stem for each stemmer. All stemmers except Text::German have the same preprocessings steps which are lowercasing the word and replacing umlauts with their normalized vowel versions (e.g., ü is replaced with ue). These steps will therefore not be mentioned below.

Snowball. In 1996, Martin Porter developed the Snowball stemmer for English (Porter 1980). It became by far the most widely used stemmer for English. The Snowball team has developed stemmers for many European languages, which are included as a set in important natural language processing toolkits such as NLTK (Bird et al. 2009) for Python or Lingua::Stem for Perl.

The Snowball German stemmer is an adaptation of the original English version and thus restrains itself to suffix-stripping. It defines two regions R1 and R2, where R1 "is the region after the first non-vowel following a vowel, or is the null region at the end of the word if there is no such non-vowel" and R2 is defined in the same way, with the difference that the definition is applied inside of R1. After defining R1 and R2 Snowball deletes a number of suffixes if they appear in R1 or R2. It does not do this recursively but instead in three steps, in each of which at most one suffix can be stripped. The first two steps strip fairly common suffixes like "ern" or "est", while the third step strips derivational suffixes, e.g., "isch" or "keit", which are fairly uncommon.

In our example, the Snowball stemmer correctly places "Adlers" (eagle, genitive case), "Adlern" (eagles, dative case) and "Adler" (eagle) together in the stem "adl". However, it also incorrectly stems "adle", which is the first person singular of "adeln" (to ennoble) to "adl". This is because the length restriction on how short stems can become is defined in terms of R1 and R2, as explained above, and in this example, R1 for all four words is the part after "adl".

Text::German. The stemmer in the Perl CPAN Module Text::German was, as far as we could find out, developed in 1996 at the Technical University of

Darmstadt by Ulrich Pfeifer, following work by Gudrun Putze-Meier for which no reference is available. It is not currently actively supported. We made a number of efforts to contact both scientists but were unsuccessful.

What sets Text::German apart from the other stemmers examined here is the fact that it strips prefixes, and that it uses small lists of prefixes, suffixes and roots to identify the different parts of a word. Although the implementation in CPAN has significant flaws, the idea is novel and produced good results, as can be seen in Sect. 3.3.

While the behaviour of Text::German is at times difficult to understand due to its binary-encoded rules, we think that its performance on our example is primarily due to two factors. One is that "ers" is not in its list of suffixes, which is why "Adlers" is stemmed to itself. The other is that it does not lowercase stems, which results in "adle" (correctly) being stemmed seperately.

Caumanns. The stemmer proposed by Caumanns (1999) is unique in two ways. One is that it uses recursive suffix stripping of the character sequences "e", "s", "n", "t", "em", "er" and "nd", which are the letters out of which every declensional suffix for German is built. The other is that it strips "ge" before and after the word, which makes it one of the two stemmers that stem prefixes. It also substitutes "sch", "ch", "ei" and "ie" with special characters so they are not separated and replaces them back at the end of the stemming process.

In our example, the Caumanns stemmer conflates all four words to the same stem "adl". This is because of the recursive suffix stripping and because its length constraint is not producing words shorter than three characters, which is why "adle" was stemmed to "adl" which is exactly three characters long.

UniNe. The UniNE stemmer, developed by Savoy (2006) from the University of Neuchatel in 2006, has an aggressive and a light stemming option.

Light Option. The light option merely attempts to strip plural morphemes. After the standard Umlaut substitutions, it strips one of "nen", "se", "e" before one of "n", "r" and "s" or one of "n", "r" and "s" at the end of the word. As only one of these options can take effect, it is a very conservative stemmer.

In the "Adler" example, the stemmer stems "Adlers" and "Adlern" to "adler" and "Adler" and "adle" to "adle". It does not go further because it removes at most two letters and doesn't strip suffixes recursively.

Aggressive Option. The aggressive option goes through a number of suffix stripping steps, which always depend on the length of the word. The difference with the other stemmers is that UniNE has two groups of stripping operations and at most one out of each group is executed. Also, its conditions for stripping "s" and "st" are very similar to those of the Snowball stemmer, which defines a list of consonants that are valid s- and st-endings respectively and have to occur before the "s" or "st" so that the consonant in question is stripped.

This stemmer's main problem in our example is that it stems "Adlers" to itself because "r" is not included in its list of valid s-endings which have to occur before "s" for it to be stripped.

2.2 Evaluation Studies

The literature on the comparative evaluation of stemmers for German is relatively sparse. Braschler and Ripplinger (2003) compared the NIST stemmer and the commercial Spider stemmer with two baselines of simply not stemming and morphological segmentation based on unsupervised machine learning and morpho-syntactic analysis. They found precision improvements of up to 23% points and recall improvements of up to 12% points for the NIST stemmer over no stemming compared to 20% points improvement in precision and 30% points in recall for the commercial Spider Stemmer. Savoy (2006) tested their UniNE stemmer and the Snowball German stemmer in an information retrieval system and found that the UniNE stemmer improved the Mean Average Precision (MAP) by 8.4% points while the Snowball stemmer improved it by 12.4% points against a baseline without any stemming.

Our evaluation is based on two gold standards which we will make publicly available, allowing them to act as a benchmark for future work on German stemming.

3 Evaluation

3.1 Runtime Analysis

The runtimes that can be seen in Table 2 are averaged over 10 runs of each stemmer. The Snowball implementation used was our own implementation in Perl which we did in order to better compare the Snowball stemmer to the others (it should be noted that the official implementation of the stemmer is in Martin Porter's own programming language Snowball, compiled to C code, which will therefore, in practice, be much faster than implementations in Perl). For the UniNE stemmer, we used the implementation in the CPAN module Lingua::Stem::UniNE::DE, with slight modifications of our own with regards to the use of a module, and for the Caumanns Stemmer we used our own Perl implementation, which was fairly difficult to implement because the paper of Caumanns (1999) doesn't clearly state a definitive algorithm, instead describing main ideas and then making suggestions for improvements (Fig. 2).

To assess average runtime, we then stemmed a corpus of 624029 words on each stemmer using a single threaded Perl 5.8.18 program on a Xeon ×7560 2,26 GHz running openSUSE ten times and computed the mean runtime. As can be seen in the table, the runtimes of the Caumanns, UniNE and CISTEM stemmers are fairly similar, while Snowball takes about twice and Text::German nearly three times as long

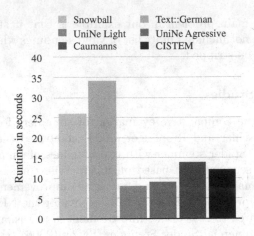

Fig. 2. A comparison of stemmer runtimes. 624029 words were stemmed by each stemmer using a single threaded Perl 5.8.18 program on a Xeon ×7560 2,26 GHz Processor running openSUSE

3.2   Gold Standard Development

We compiled two different gold standards. The reason for this is that exactly which words belong to the same stem is something that is difficult for people to agree on. The question of whether, for example, "billig" (cheap) belongs together with "billigen" (to approve) seen from an IR perspective, is a difficult one because the adjective "billig" also exists in the sense of "something worthy of approval". Therefore, our hope is that the two gold standards will capture the different ends of this spectrum where one end, when in doubt, puts words in a cluster together and the other doesn't. Having two gold standards capturing this distinction enables us to be more objective in our evaluation.

For the first gold standard, we used the morphological information in CELEX 2. It gives the flat segmentation into morphemes and annotates each morpheme with its word class, and X if no word class applies. This should be equivalent to the distinction between lexical and grammatical morphemes. We then stripped the morphemes annotated with X to form the stem. For the second gold standard, we simply used the fact that every wordform in CELEX2 is assigned a lemma, and used that lemma as the wordform's stem. In each case, we then grouped the wordforms by stem according to the principle that the exact stem is irrelevant as long as the cluster makes sense. The resulting gold standards are 30951 stems large in the case of gold standard 1 and 47852 stems for gold standard 2. From each, we took a random sample of 1000 stems and used those as gold standards for our evaluation. To avoid overfitting, we changed the samples several times while developing CISTEM, including after the end of development for the final evaluation.

As you can see in Fig. 1, there are differences between the gold standards. For the "absurd" example, gold standard 2 classified "absurd" as a different lemma than "absurditäten" (absurdities) and thus put them in two seperate stems while gold standard 1 sees them as having the same stem. The difference is even more pronounced in the second example, where the first gold standard has one stem for "relativier" (relative), one for "Relativismus" (a theory in philosophy), one for "Relativität" (the general noun for relative) and one for "relativistisch" (relative, but only in the context of Einstein's theory of relativity).

From an information retrieval point of view, one would consider "Relativismus" and "relativistisch" as belonging in one stem that relates to the theory of Relativity, and the other two stems as belonging in another stem. Overall, gold standard 2 is much more likely to seperate words into several different stems while gold standard 1 is more likely to group them into a single stem. This makes sense considering gold standard 2 thinks in lemmata, e.g., in a dictionary one would like to have seperate entries for "Relativismus" and "Relativität" while gold standard 1 groups them together because neither "ismus" nor "tät" are lexical stems that can be assigned a word class.

This confirms our hopes that the two gold standards would capture two ways of looking at stemming. Gold standard 1 represents a more aggressive-stemming-friendly view and gold standard 2 a more conservative one. Personally, we consider gold standard 1 on the whole to be more suitable for stemmer evaluation, but arguments could also be made for the opposite point of view. For this reason, both gold standards are included in the following evaluation (Table 1).

Table 1. Two examples for the differences between the two gold standards

Gold standard 1	Gold standard 2
– absurderen absurdestem [...] absurditäten absurdität	– absurderen absurdestem absurder absurden [...]
	– absurditäten absurdität
– relativem relatives [...] relativistischerer [...] relativität relativitäten	– relativieret relativiertest [...]
	– relativität relativitäten
	– Relativismus
	– relativistischsten relativistischen [...]

3.3 Evaluation

We stemmed the Celex2 corpus. We then went through each of the stems from the gold standard (1000 stems large) and matched them with a stem from the stemmed corpus depending on how many of the words belonging to these two stems matched. For each stem of the gold standard, we computed precision, recall and f1-measure and then computed the average of each of those metrics to form the overall evaluation results for that gold standard. The results can be seen in Table 2 and are illustrated in Figs. 3a, b and 4.

Table 2. Evaluation results of different stemmers using our two gold standards, each of which is for the same 1000 stems (note that CISTEM is our new stemmer which will be introduced later in the paper)

Gold standard 1						
Stemmer	Snowball	Text::German	Caumanns	UniNE Light	UniNE Aggressive	CISTEM
Precision	96.17%	97.56%	96.76%	**98.39%**	97.37%	96.83%
Recall	83.78%	79.29%	9.43%	67.69%	80.29%	**89.73%**
F1	89.55%	87.48%	92.95%	80.20%	88.01%	**93.15%**
Gold standard 2						
Stemmer	Snowball	Text::German	Caumanns	UniNE Light	UniNE Aggressive	CISTEM
Precision	85.89%	96.00%	92.26%	**96.43%**	94.50%	92.43%
Recall	86.61%	86.97%	96.17%	70.91%	83.81%	**96.45%**
F1	86.25%	91.27%	94.17%	81.72%	88.83%	**94.40%**

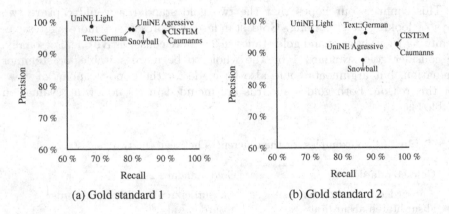

(a) Gold standard 1   (b) Gold standard 2

Fig. 3. Precision - recall values on the two gold standards

3.4   Results

The most surprising result of our evaluation was that the difference between the two gold standards was not as pronounced as we expected, considering that they represent the two ends of the spectrum of what one wants a stemmer to do. The two gold standards agree on the best stemmer and the worst stemmer in terms of precision, recall and f-measure. As can be seen in Fig. 4, these measures differ for the three middle ranked stemmers Snowball, Text::German and UniNE Aggressive.

The difference between the two gold standards is shown most clearly in their assessment of the Snowball stemmer. This is to be expected as the stripping of clearly derivational suffixes like "lich" or "ung" matches the stemming concept of gold standard 1 quite closely, where suffixes like these suffixes are removed. This explains why, while Snowball achieved the lowest precision on both gold standards, the gap to the next best precision is much lower in gold standard 1

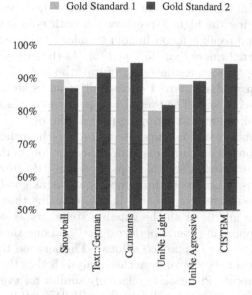

Fig. 4. F1-measures on both gold standards

(just 0.59% points) than in gold standard 2 (6.37% points), where Snowball's aggressive stemming is much more likely to affect precision negatively.

Being one of the more conservative stemmers, Text::German scores significantly higher on gold standard 2. On gold standard 1, it achieves fourth place in both recall and f-measure and third place in precision of the existing stemmers. We attribute this mainly to the fact that Text::German stems at most one suffix from a small list which doesn't include derivational suffixes, which is guaranteed to hurt recall on gold standard 1 because gold standard 1 requires more than just very conservative suffix stripping. This same fact results in a relatively good score on gold standard 2, achieving the second place (when compared with previously existing stemmers) in all metrics. We think that the main problem that hurts its performance on both gold standards should be that Text::German identifies prefixes from a list, nearly all of which are clearly lexical, and strips them according to a complicated set of rules. From an IR standpoint, the stripping of lexical prefixes which clearly change the word's meaning is suboptimal.

The Caumanns stemmer achieves first place (with respect to already existing stemmers) in both gold standards in recall and f-measure. The gap to the other stemmers' values is about 3% points on both gold standards. An interesting point here is that while precision is significantly higher than recall on gold standard 1, the opposite is true for gold standard 2. This points to the Caumanns stemmer having achieved a middle line between both gold standards' concepts of stemming. The stemmer is more conservative than gold standard 1 and more radical than gold standard 2. This, together with the large gap in performance to the competitors, makes the Caumanns stemmer the best stemmer for German we have seen so far.

The light option of the UniNE stemmer, as expected, scores last in recall and f-measure while having the highest precision on both gold standards. The gap between precision and recall is larger in gold standard 1 (more than 30% points) than in gold standard 2 (more than 25% points). As the express goal of the light option is to merely strip affixes denoting plural, the lack of recall is naturally more pronounced in gold standard 1 because it requires stemmers to be more radical in order to score well. The overall bad performance is not surprising, as stemming entails more than just stripping plural suffixes.

The agressive option of the same stemmer, on the other hand, achieves mediocre results, coming in third place in f-measure on both gold standards. While it does strip suffixes in several steps, it doesn't do so recursively, which is why it makes sense that the performance is about as good as the Snowball stemmer, which has a similar approach. It also explains that the performance is somewhere in the middle of all the existing stemmers as we have seen that recursive suffix stripping in general performs best and one-time stripping worst, because UniNE Aggressive's approach is located in between these two ideas.

The clearest lesson to be drawn from this analysis is that the problem of existing stemmers is in recall. Precision is relatively similar for every stemmer, only varying by 2.22% points while recall varies by 21.04% points in gold standard 1. The discrepancy is similarly pronounced in gold standard 2, where precision varies by 10.54% points and recall by 25.54% points. Because of the nature of f1-measure, recall therefore decides the stemmer's f-measure ranking: in gold standard 1, the recall order from best to worst exactly mirrors that of f-measure and in gold standard 2, only the positions of Snowball and UniNE Aggressive, the two middle stemmers, are reversed. The mean precision in gold standard 1 is 97.13% and the mean recall is 82.04%. The mean precision in gold standard 2 is 92.62% and the mean recall is 87.45%. Not only does recall vary much more, it is also consistently much lower than precision.

4   Development

4.1   CISTEM Development

Following the insight that recall is the most promising area for stemmer development, we focused on improving recall over existing stemmers with CISTEM. As starting point, we used the Caumanns stemmer, as it was the best performing stemmer of our evaluation, and tried to improve on it. We tried several changes and evaluated each of them seperately to improve f-measure. One feature of the Caumanns stemmer that we deleted was the substitution of "z" for "x", which improved precision slightly in gold standard 1 and changed no other metrics. We also found that stripping "ge" before and after the word after the suffix stripping, as proposed by Caumanns, didn't work well. The version of this that delivered the best performance was stripping "ge" as a prefix, before the suffix stripping and only if the remaining word is at least four characters long. This is consistent with the requirement for suffix stripping that the resulting word needs to be at least four characters long, while the Caumanns stemmer undercuts that requirement by removing "ge" after the suffix stripping without checking the length of the result. Interestingly, introducing a new variable that measures true length (necessary because substitutions of multiple characters by one character, e.g., "ei" by "%" make the word shorter than it actually is) hurt performance quite clearly. We also deleted the substitution of "ch" by "§" because we found it hurt recall on gold standard 2 and changed nothing on gold standard 1. The length constraint on the stripping of "nd", which was at least five remaining characters in the Caumanns stemmer, was changed to at least six, which doesn't only improve performance but also makes the algorithm simpler as "nd" is now stripped in the same step as "em" and "er". Our other contribution was to give the steps a definitive order, which had not been clear in the Caumanns paper and led to subtle flaws in third-party implementations we tried.

The resulting algorithm, which can be seen in Fig. 5, is simpler than the Caumanns stemmer and easy to understand and implement. We will also offer a context-insensitive version which ignores case for "t"-stripping because the original Caumanns stemmer's performance is drastically worse when using a corpus of only lowercase words, which might be necessary in some contexts, but would lead to the stemmer never stripping "t".

4.2   Final Evaluation

CISTEM shows slight improvements over the Caumanns stemmer in both precision and recall. The difference in recall is more pronounced, which is consistent with our goal of removing some constraints of the Caumanns stemmer to improve recall.

If we look back to the example in Fig. 1, we can see that CISTEM stems the four words correctly. It stems "adle" to "adl", which is the same stem that Caumanns assigned it, but stems the other three words to "adler" because the length requirement for stripping "er" is that the resulting stem will be longer than five characters (not four characters).

The main advantage of CISTEM over other stemmers available is that we have a definitive algorithm shown in Fig. 5. The algorithm is bug free, the order is fixed and we will make it available in a range of programming languages to prevent flawed third-party implementations.

We hope that our new stemmer CISTEM will be useful in a wide range of applications. In addition, stemming-based segmentation of German has recently been shown to be effective in reducing vocabulary in neural machine translation (Huck et al. 2017). So we will additionally provide a version of the algorithm which segments words, rather than stemming them.

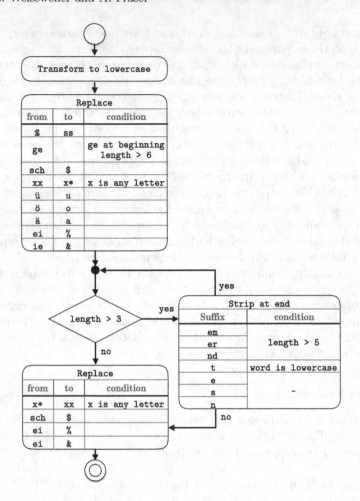

Fig. 5. The CISTEM algorithm

5 Conclusion

We presented two gold standards for stemming which represent two different views on stemming. We then evaluated five existing stemmers for German on those gold standards and discussed the results. Finally, we presented our own stemmer, which improves on the stemmer of Caumanns and achieves state-of-the-art results on both gold standards.

One of the main problems in stemmer development is the divide between the stemmers that are published and those that are actually used in NLP applications. The Snowball stemmer continues to be most widely used because it is the default stemmer for most NLP libraries and offers stemmers for a wide range of European languages.

For this reason, we will publish official implementations in a range of programming languages, starting with Perl, Python and Java. We are also planning to release our gold standards in the hope that they will be used in further work on stemming for German. The code and gold standards will be made available at https://www.github.com/LeonieWeissweiler/CISTEM, and we hope to also be included in some standard NLP packages in the future. Other future work would be to find other ways of building a gold standard for stemming in order to have one definitive gold standard where words are clustered exactly as they should be for information retrieval. We were often obstructed in our development by having to show improvements in both gold standards for every change, which could be avoided by having just one gold standard. Another more unconventional idea would be implementing a small rule-learning system that suggests new rules for the stemmer based on their effectiveness in matching a gold standard or when used actively in a working IR system.

Acknowledgments. This project has received funding from the European Union's Horizon 2020 research and innovation programme under grant agreement No. 644402 (HimL). This project has received funding from the European Research Council (ERC) under the European Union's Horizon 2020 research and innovation programme (grant agreement No. 640550).

References

Baayen, R.H., Piepenbrock, R., Gulikers, L.: The CELEX Lexical Database (Release 2) on [CD-ROM]. Linguistic Data Consortium, University of Pennsylvania, Philadelphia, PA (1995)

Bird, S., Klein, E., Loper, E.: Natural Language Processing with Python. O'Reilly Media, Sebastopol (2009). ISBN 0596516495, 9780596516499

Braschler, M., Ripplinger, B.: Stemming and decompounding for German text retrieval. In: Sebastiani, F. (ed.) ECIR 2003. LNCS, vol. 2633, pp. 177–192. Springer, Heidelberg (2003). https://doi.org/10.1007/3-540-36618-0_13. http://dl.acm.org/citation.cfm?id=1757788.1757806. ISBN 3-540-01274-5

Caumanns, J.: A fast and simple stemming algorithm for German words. Technical report Nr. tr-b-99-16. Fachbereich Mathematik und Informatik, Freie Universität Berlin, Oktober 1999

Huck, M., Riess, S., Fraser, A.: Target-side word segmentation strategies for neural machine translation. In: Proceedings of the Second Conference on Machine Translation (WMT), Copenhagen, Denmark, September 2017

Lovins, J.B.: Development of a stemming algorithm. Mech. Transl. Comput. Linguist. 11(1), 22–31 (1968)

Manning, C.D., Raghavan, P., Schütze, H.: Introduction to Information Retrieval. Cambridge University Press, Cambridge (2008). http://nlp.stanford.edu/IR-book/pdf/irbookprint.pdf

Porter, M.F.: An algorithm for suffix stripping. Program **14**(3), 130–137 (1980)

Savoy, J.: Light stemming approaches for the French, Portuguese, Germanand hungarian languages. In: Proceedings of the 2006 ACM Symposium on Applied Computing, pp. 1031–1035. ACM, New York (2006). http://doi.acm.org/10.1145/1141277.1141 523. ISBN 1-59593-108-2

Negation Modeling for German Polarity Classification

Michael Wiegand[1(✉)], Maximilian Wolf[1], and Josef Ruppenhofer[2]

[1] Spoken Language Systems, Saarland University, 66123 Saarbrücken, Germany
michael.wiegand@lsv.uni-saarland.de, mwolf@coli.uni-saarland.de
[2] Institute for German Language, 68161 Mannheim, Germany
ruppenhofer@ids-mannheim.de

Abstract. We present an approach for modeling German negation in open-domain fine-grained sentiment analysis. Unlike most previous work in sentiment analysis, we assume that negation can be conveyed by many lexical units (and not only common negation words) and that different negation words have different scopes. Our approach is examined on a new dataset comprising sentences with mentions of polar expressions and various negation words. We identify different types of negation words that have the same scopes. We show that already negation modeling based on these types largely outperforms traditional negation models which assume the same scope for all negation words and which employ a window-based scope detection rather than a scope detection based on syntactic information.

1 Introduction

Negation is one of the most central linguistic phenomena. Therefore, negation modeling is essential to various common tasks in natural language processing, such as relation extraction (Sanchez-Graillet and Poesio 2007), recognition of textual entailment (Harabagiu et al. 2006) and particularly sentiment analysis (Wiegand et al. 2010). In the latter task, negation typically inverts the polarity of polar expressions. For example, in (1), the negated positive polar expression *like* conveys negative polarity.

(1) I do [**not** [like]⁺]⁻ this new Nokia model.

While most research on negation has been carried out on English language data, little research has looked into the behaviour of negation in German. This is surprising since German negation is even harder to handle than English negation. For example, since German displays a more flexible word order than English, the German negation word *nicht (not)* may appear both left (2) or right (3) of a polar expression it modifies. In English, however, there is a strong tendency of a negation word to precede the polar expression it negates (1).

(2) Der Kuchen ist [**nicht** [köstlich]⁺]⁻.
(*The cake is not delicious.*)

G. Rehm and T. Declerck (Eds.): GSCL 2017, LNAI 10713, pp. 95–111, 2018.
https://doi.org/10.1007/978-3-319-73706-5_9

(3) Ich [[mag]$^+$ den Kuchen **nicht**]$^-$.
      (*I do not like the cake.*)

To make the task even more difficult, there are not only function words, such as the particle *nicht (not)*, to express negation but also content words, such as verbs (4), nouns (5) or adjectives (6). (2)–(6) also show that these different negation word types have different scopes.

(4) [[Dieses Bemühen]$^+$ **scheiterte**^{verb}]$^-$.
      (*These efforts failed.*)
(5) [Das **Scheitern**^{noun} [dieser Bemühungen]$^+$]$^-$ war vorhersehbar.
      (*The failure of these efforts was foreseeable.*)
(6) Angesichts [dieser **gescheiterten**^{adj} [Bemühungen]$^+$]$^-$ ist nun ein Umdenken erforderlich.
      (*These failed efforts now require a change of thinking.*)

In this paper, we follow a **rule-based approach** to negation modeling for fine-grained sentiment analysis that largely draws information from lexicons. We focus on the **task** of identifying the scope of negation words with regard to polarity classification. In other words, given a mention of a negation word and a polar expression, we want to automatically determine whether the negation word negates the polar expression.

We do *not* claim to have full knowledge of all German negation words. (Given that content words can perform implicit negation, we assume the overall vocabulary of negation words to be fairly large.) Instead, we propose a typology of negation words and assign a characteristic scope to each type. Therefore, we provide a *formalism* that is able to compute the respective scope of every possible negation word, once the negation word has been assigned to its respective type.

Our approach heavily relies on syntactic knowledge, particularly information contained in a dependency parse. We demonstrate that the analyses that state-of-the-art parsers produce for German are insufficient for our task and require further normalization.

The **contributions** of this paper are:

- We present the first comprehensive study on German negation modeling for fine-grained sentiment analysis.
- Instead of having one generic scope for all types of negation words, we formulate different types of scopes for different types of negation words.
- We substantially go beyond negation (function) words, that is, we also consider negation verbs, nouns and adjectives.
- We introduce a new dataset[1] comprising German sentences in which negation words are manually annotated with respect to the polar expressions they negate.
- We publicly release a tool[1] for fine-grained German sentiment analysis that implements our proposed approach.

---

[1] Available under https://github.com/artificial-max/polcla.

2   Data and Annotation

In order to evaluate negation in context, we built a small focused dataset comprising sentences with negation. In order to keep the annotation effort manageable, we extracted those sentences in which a negated polar expression is likely. We therefore extracted from a corpus only those sentences in which both some negation word co-occurs with at least one polar expression according to the **sentiment lexicon** of the PolArt-system (Klenner et al. 2009). In order not to bias the scope of the negation in those sentences we did not impose any restriction regarding the relation between negation words and polar expressions. In order to recognize negation words, we also created a **negation lexicon**. For that we used several resources. On the one hand, we used all negation expressions from the PolArt system. In addition, we translated a large list of English negation verbs to German and also manually added morphologically related nouns and adjectives if existent, e.g., for the verb *stagnieren* (*stagnate*) we would also add the noun *Stagnation* (*stagnation*).

In total, we sampled 500 sentences from the *DeWaC*-corpus (Baroni et al. 2009). We manually annotated *every* polar expression in those sentences. (Note that we did not only annotate those polar expressions we could automatically identify with the help of the PolArt sentiment lexicon.) We also marked every negation word in case it negates a polar expression. The dataset comes in *TIGER/SALSA format* (Erk and Padó 2004). Figure 1 illustrates the annotation of our dataset. Polar expressions evoke a frame *SubjectiveExpression*. If a polar expression is negated, then its negation word is labeled as a frame element *Shifter* of its frame.[2]

Of the 500 sentences, we removed 67 sentences which contained obvious errors (i.e., misspellings, grammatical mistakes or incorrect sentence boundaries). We excluded those sentences since the methods we are going to examine rely on a correct syntactic parse. Erroneous sentences are likely to produce spurious syntactic analyses.

On a sample of 200 sentences, we measured an interannotation agreement of $\kappa = 0.87$ which can be considered substantial (Landis and Koch 1977).

Table 1 provides some statistics of our dataset. Even though every sentence contains at least one polar expression (in most cases there is more than one) and a negation word, there are only 282 cases in which a polar expression is within the scope of the negation word, i.e., it is actually negated. This shows that it is not trivial to determine whether a polar expression has been negated. It is also worth pointing out that a negation is as likely to precede the polar expression it negates as to follow it.

---

[2] In our annotation, we refer to negation words as *shifters* since, in computational linguistics, this is the preferred term. A shifter need not fully invert the polarity of a polar opinion but just *shifts* it into the opposite direction (Polanyi and Zaenen 2006). For example, *faded optimism* does not mean that *optimism* is completely absent but it means that the current amount has substantially decreased – which can be interpreted as a negative opinion.

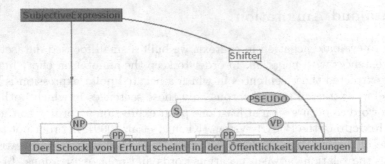

Fig. 1. Example sentence annotation from dataset (translation: *The shock of Erfurt seems to have faded away in the public*). Polar expressions (e.g., *Schock*) evoke a frame *SubjectiveExpression*; the word that negates a polar expression (e.g., *verklungen*) is assigned the frame element *Shifter* of that frame.

Table 1. Statistics of negation detection dataset.

Property	Freq
Number of sentences	433
Number of polar expressions	979
Number of sentences with negated polar expression	282
Number of negation words *left* of polar expression	142
Number of negation words *right* of polar expression	140

3   Baselines

3.1   Baseline I: Window-Based Scope

Our first baseline applies a simple window-based approach for the scope detection of negation. It is inspired by various works from polarity classification on English language data (Wilson et al. 2005; Wiegand et al. 2009). One considers as scope a span of n tokens around the negation word. While on English data it typically suffices to scan only the tokens succeeding the negation word, on German data we check three different window types: one that assumes the polar expression to succeed the negation word, one that assumes the polar expression to precede the negation word and one in which both directions are examined.

Figure 2 shows the performance of those different window-based scopes on our dataset. It shows that for German negation one needs to look into both directions. (This is in line with our statistic from Table 1.) All of the three window types have their maximum at $n = 4$. In our forthcoming experiments, we use the best window-based scope (i.e., using both directions at $n = 4$) as a baseline.

3.2   Baseline II: Clause-Based Scope

Our second baseline models the scope on the basis of syntactic information. Instead of using a window of fixed size, we scan all words in the same clause in

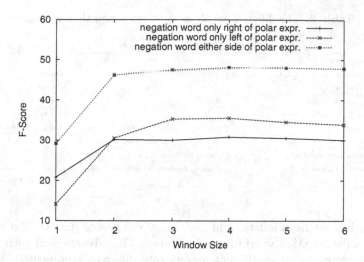

Fig. 2. Illustration of window-based scope using different window sizes.

which the negation word occurs for a polar expression. Typically the scope of a negation never exceeds clause boundaries. For example, in (7) the negation word *niemand (nobody)* does not negate the polar expression *entsetzlich (appalling)* in the subordinate clause. From a linguistic perspective, this scope is more adequate than the window-based approach.

(7)   [**Niemand** wird etwas zu dem Ereignis sagen wollen]$_{main\_clause}$, [weil es sich dabei um eine [entsetzliche]$^-$ Angelegenheit handelt]$_{subordinate\_clause}$.
      (*Nobody will want comment on this incident, since it is an appalling affair.*)

4   Our Approach

Our approach is fundamentally different to the previous baselines in the sense that we define individual scopes for different types of words. Our framework allows arbitrary scopes to be defined for every possible negation word. A scope is defined in terms of a grammatical relation. For instance, we could formulate that the *subject* of the negation word *aufhören (subside)* is the expression that is negated as in (8).

(8)   [[Die Schmerzen]$^-$ **hören auf**^{verb}]$^+$.
      (*The pain subsides.*)

We do not have the knowledge to explicitly enumerate the scope for every possible negation word from our negation lexicon (Sect. 2). Instead, we grouped words with similar scope characteristics (Sects. 4.1 and 4.2) and assigned one scope which satisfies the entire group of words.

Our framework allows the specification of a **priority scope list** for a negation word, i.e., a list with more than one argument position (see also Table 2). We

Table 2. The different negation word types and their scopes.

Type	Example negation words	Priority scope list
Negation function words		
Negation adverbs and indefinite pronouns	*nie, niemals, kein, kaum*	clause
Negation particle	*nicht*	governor
Negation prepositions	*ohne, gegen*	dependent
Negation content words		
Negation adjectives	*weniger, gescheitert, korrigierbar*	subj, attr-rev
Negation nouns	*Abschaffung, Linderung, Zerstörung*	gmod, objp-*
Negation verbs intransitive	*enden, nachlassen, verschwinden*	subj
Negation verbs transitive	*ablegen, lindern, senken, vermindern*	objg, obja, objd, objc, obji, s, objp-*, subj

process such a list from left to right and apply the first argument that matches for the specific negation word in some sentence. The advantage of such a list is that there are negation words that may negate different arguments. The flexibility we gain with priority lists is essential for identifying the correct scope of negation content words as we will explain in Sect. 4.2.

We do not claim that our proposed approach perfectly models the scope of every German negation word. But we show that with relatively little lexical knowledge we can largely outperform a traditional approach that treats all negation words in the same way. Therefore, our proposed method should be regarded as a strong baseline for future research.

Table 2 summarizes the different negation word types that we discuss in detail below.

4.1 Scope for Negation Function Words

The type of lexical units most commonly associated with negation are negation particles such as *nicht (not)*, negation adverbs, such as *niemals (never)*, indefinite pronouns, such as *kein (no)*, and a few prepositions, such as *ohne (without)*. Even though these negation words only constitute a handful of lexical units, they are known to have a large impact. This is due to the fact that these words are function words which entails that they occur frequently. We call these words **negation function words**. Regarding the scope of those words, we distinguish between three types.

Negation Adverbs and Indefinite Pronouns. These negation words exhibit similar behaviour (in terms of scope) as sentential adverbs. As a consequence, these negation words have a wide scope. It is the entire clause in which they are embedded (9) and (10). We use the same definition as we applied for our baseline in Sect. 3.2.

(9) [Noch **nie** wollte Kiew [Frieden]$^+$]$^-$.
    (*Kiev never wanted peace.*)
(10) [**Kein** Mensch möchte sie dabei [unterstützen]$^+$]$^-$.
    (*No one wants to support them with that.*)

Negation Particle. The particle *nicht (not)* has a narrow scope. We only include the word which it *governs* in the dependency graph (11).

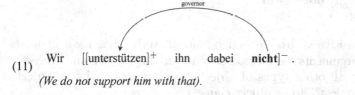

(11)   Wir   [[unterstützen]$^+$ ihn   dabei   **nicht**]$^-$.

   (We do not support him with that).

Negation Prepositions. Negation prepositions also have a narrow scope. However, unlike the negation particle, their scope does not include the words which they govern but which are their dependents (hence the reverse relation), e.g., *Hass (hatred)* in (12).

(12) Wir bauen eine Welt ganz [**ohne**^{prep} [Hass]$^-_{dependent}$]$^+$.
   (*We create a world without hatred.*)

4.2   Scope for Negation Content Words

In the following, we describe the remaining words, all of which are content words. We therefore refer to these words as **negation content words**.

Negation Nouns. Negation nouns typically reverse the polarity of two types of dependents, either a genitive modifier (13) or a prepositional object (14). Note that we leave the preposition underspecified so that it can match any potential preposition.

(13) Das Gericht beschloss [die **Aufhebung**^{noun} [der Strafe]$^-_{gmod}$]$^+$.
   (*The court decided to lift the sentence.*)
(14) Qi Gong dient auch zur [**Vorbeugung**^{noun} [vor Krankheiten]$^-_{objp\text{-}vor}$]$^+$.
   (*Qi gong is also used for preventing diseases.*)

Negation Adjectives. There are two different major constructions in which adjectives may occur. Adjectives may be used *predicatively* or *attributively*. Negation adjectives may occur in both constructions. Therefore, polar expressions negated by an adjective may be in two different argument positions, namely a noun in subjective position in the predicative case (15) or a noun that is modified by an attributive adjective (16).

(15) [[Diese Bemühungen]$^+_{subj}$ sind **gescheitert**$^{pred\text{-}adj}$]$^-$.
   (*These efforts failed.*)
(16) Das sind alles [**korrigierbare**$^{attr\text{-}adj}$ [Fehler]$^-_{attr\text{-}rev}$]$^+$.
   (*These are recoverable errors.*)

Negation Verbs. For this study, we distinguish between between two major types of verb groups, *transitive verbs* and *intransitive verbs*. In the case of transitive negation verbs, it is the object that is negated (17), while for intransitive verbs, it is the subject that is negated (18).

(17) Dieses Medikament [**lindert**$^{transitive\_verb}$ [die Schmerzen]$^-_{obja}$]$^+$.
(*This drug cures the pain.*)

(18) [[Die Schmerzen]$^-_{subj}$ **hören auf**$^{intransitive\_verb}$]$^+$.
(*The pain subsides.*)

Note that by transitive verbs, we understand all verbs that have *at least* two arguments. By arguments we do not only mean the subject and (direct) accusative object but all other types of objects, for instance, a dative object (19), a prepositional object (20) or object clause (21).

(19) Die Menschheit [**entging**$^{transitive\_verb}$ [einer Katastrophe]$^-_{objd}$]$^+$.
(*Mankind averted disaster.*)

(20) Wir [**kämpfen**$^{transitive\_verb}$ [gegen dieses Problem]$^-_{objp-gegen}$ **an**]$^+$.
(*We fight against this problem.*)

(21) Ich [**bezweifle**$^{transitive\_verb}$, [dass dies eine gute Idee ist]$^+_{objc}$]$^-$.
(*I doubt that this is a good idea.*)

For sentences where negation verbs have more than one object, the ordering of the objects on our *priority scope list* decides which type of object is given priority. For example, in case of ditransitive verbs, the accusative object is more likely to be negated than the dative object (22).

(22) Das [**ersparte**$^{transitive\_verb}$ [uns]$_{objd}$ [viel Ärger]$^-_{obja}$]$^+$.
(*This saved us a lot of trouble.*)

In principle, the arguments of verbs to be negated could be most adequately described in terms of *semantic roles*. In the terminology of *FrameNet* (Baker et al. 1998), we are basically looking for THEME or PATIENT; in the terminology of *PropBank* (Palmer et al. 2005) it is *A1*. Unfortunately, automatic semantic role labeling for German is still in its infancy. As a consequence, we need to approximate semantic roles with dependency relations.

Using a priority scope list also partly allows us to model sense ambiguity. Some verbs may be used both intransitively and transitively. We simply add the subject position at the end of the priority list. This allows the German negation verb *abnehmen (take from/decrease)* to negate its accusative object in (23) while it negates its subject in (24).

(23) Sie [**nahm**$^{transitive\_verb}$ ihm [ein große Last]$^-_{obja}$ **ab**]$^+$.
(*She took a great burden from him.*)

(24) [[Seine Wut]$^-_{subj}$ **nahm**$^{intransitive\_verb}$ deutlich **ab**]$^+$.
(*His anger notably decreased.*)

4.3   Normalization of the Dependency Graph

Our previous examples have shown that in order to model scope, we largely rely on a syntactic analysis, particularly on a dependency parse. We employ *ParZu* (Sennrich et al. 2009). We chose that particular parser because of its fine-grained label inventory which is essential for our approach. Still, our rules cannot be immediately applied to the original output of that parser.

Our rules are defined for active-voice constructions. The parse for passive-voice constructions would be misleading since ParZu provides dependency structures that describe the surface structure. For example, in (25) we would not be able to correctly establish the scope of *bremsen* over *Fortschritt*, since it is marked as the *surface subject*. By normalizing the dependency relation labels to active voice (i.e., the deep structure), as indicated by (26), however, our rules work correctly since *Fortschritt* becomes an accusative object. It would be uneconomical to directly operate on the surface representation as it would mean writing redundant rules for negation scopes.

(25) [[Der Fortschritt]$^{+}_{subj\_surface}$ wurde [von der Kirche]$_{objp-\_surface}$ stets **gebremst**^{verb}]$^{-}$.
(Progress was held off by the church.)

(26) [[Der Forschritt]$^{+}_{obja\_deep}$ wurde [von der Kirche]$_{subj\_deep}$ stets **gebremst**^{verb}]$^{-}$.
(Progress was held off by the church.)

Another major problem is that for several tensed predicates, such as *wird versiegt sein (will be faded away)* in (27), ParZu adds several auxiliary edges accommodating the auxiliary verbs of the predicate. As a consequence, a full verb and its arguments may no longer be directly related. For instance, in (27) the negation verb *versiegen (dry up)* is not directly related to its polar subject *Zuversicht (confidence)*. Neither is the adjective *korrigierbar (recoverable)* in (29) directly related to its polar subject *Fehler (error)*. In a further normalization step we, therefore, remove the edges involving the auxiliary verbs so that the full verb and its argument (28) or the predicate adjective and its argument (30) are directly connected.

(27) [[Diese Zuversicht]$^{+}$ wird **versiegt**$^{-}$ sein .
plain parse output by ParZu

(28) [[Diese Zuversicht]$^{+}$ wird **versiegt**$^{-}$ sein .
normalized dependency relations

(29) [[Der Fehler]$^{-}$ ist **korrigierbar**$^{+}$.
plain parse output by ParZu

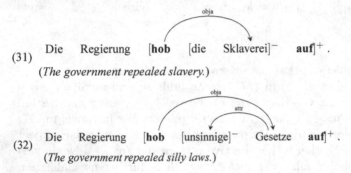

(30)   [[Der   Fehler]⁻   ist   **korrigierbar**]⁺ .

normalized dependency relations

4.4   Scope Expansion

Most of our negation rules assume the negation word and the polar expression
it negates to be in a direct syntactic relation (31). However, there are also cases
of negation in which there is no such direct relationship. For example, in (32)
the polar expression that is negated is not the accusative object of the negation verb but its attributive adjective. To account for this, we implemented a
scope expansion where also indirect relationships are allowed (i.e., we include the
dependents of the words that match the direct syntactic relation).

(31)   Die   Regierung   [**hob**   [die   Sklaverei]⁻   **auf**]⁺ .

(*The government repealed slavery.*)

(32)   Die   Regierung   [**hob**   [unsinnige]⁻   Gesetze   **auf**]⁺ .

(*The government repealed silly laws.*)

5   Experiments

5.1   Intrinsic Evaluation on Negation Dataset

In this section, we evaluate on the dataset we specially created for the task
of German negation detection for fine-grained sentiment analysis (Sect. 2). The
task is to identify for each polar expression the negation word in whose scope it
falls.

Since the focus of our work is neither to automatically detect polar expressions nor to detect negation words, in our first set of experiments, we consider
them as given. That is, we read them off from the gold standard. The specific
task therefore becomes to decide whether a *given* polar expression is negated by
a *given* negation word.

Table 3 compares the different negation detection approaches. It clearly shows
that our proposed method outperforms the two baseline methods, that is, the
window-based approach (Sect. 3.1) and the clause-based approach (Sect. 3.2).
Table 3 also displays the performance of our proposed method with some components, i. e., scope expansion (Sect. 4.3) or normalization (Sect. 4.4) switched
off. The table shows that, clearly, both functionalities have a beneficial effect.

With regard to the normalization, however, the active-voice conversion only contributes a minor share to the overall performance. So it is the conflation of relation edges in the dependency graph (27)–(30) that has the biggest impact.

Table 3. Comparison of different approaches.

Approach	Prec	Rec	F1
Window-based (baseline I)	42.13	55.97	48.08
Clause-based (baseline II)	38.89	60.07	47.21
Proposed method	67.22	**60.45**	**63.65**
Proposed meth. w/o scope expansion	69.65	52.24	59.70
Proposed meth. w/o any normalization	**71.54**	34.70	46.73
Proposed meth. w/o active-voice norm	66.80	60.07	63.26

Table 4 compares different verb rules. First, we evaluate a set of single verb rules, that is, we ignore the distinction between transitive and intransitive verbs. The performance is fairly competitive if we use the largest possible priority list *objg, obja, objd, objc, obji, s, objp-*, subj*. If we distinguish between transitive and intransitive verbs but only have two atomic rules and have no priority list (i. e., *obja* vs. *subj*), then this is worse than having only one rule but a priority list (i. e., *objg, obja, objd, objc, obji, s, objp-*, subj*). From that we conclude that many negated polar expressions are realized as a type of object but not necessarily as an accusative object (i. e., *obja*). Accounting for intransitive verbs has a relatively marginal impact, since the scores of *1 rule: obja* and *2 atomic rules: obja for trans.; subj for intrans.* are not that far apart. We assume that the reason for this is that (deep) subjects are relatively rarely negated.

Our previous experiments all assumed knowledge of polar expressions in a sentence as given. We now want to examine how performance changes if we detect all polar expressions automatically. For this experiment, we employ the sentiment lexicon of the PolArt system (Klenner et al. 2009). The detection of polar expressions based on a lexicon has two disadvantages. Firstly, all existing sentiment lexicons only have a limited coverage. Secondly, lexicon look-up does

Table 4. Impact of different **verbs** rules.

Approach	Prec	Rec	F1
1 rule: *obja*	**78.00**	43.66	55.98
1 rule: *objg, obja, objd, objc, obji, s, objp-**	67.29	53.73	59.75
1 rule: *objg, obja, objd, objc, obji, s, objp-*, subj*	64.82	**61.19**	62.96
2 atomic rules: *obja* for trans.; *subj* for intrans.	75.78	45.52	56.88
2 verb rules as proposed in Table 2	67.22	60.45	**63.65**

not account for word-sense ambiguity, that is, some words may only convey subjectivity in certain contexts (Akkaya et al. 2009).

Table 5 compares the performance of our two baselines and our proposed method based on the manual detection of polar expressions and the automatic detection of those expressions. It comes as no surprise that the performance of classifiers based on the automatic detection is lower than that using manual detection. However, by and large, the difference between our three approaches to determine the scope of negation is similar on both detection types. In other words, no matter how the polar expressions are detected, our proposed method to determine negation always largely outperforms the two baseline classifiers.

We refrain from carrying out a similar experiment by detecting negation words automatically, since our dataset is biased towards the negation words we know. Moreover, inspection of our data revealed that polar expressions tend to be much more ambiguous than negation words.

Table 5. Comparison of manual and automatic detection of polar expressions.

Approach	Manual detection			Automatic detection		
	Prec	Rec	F1	Prec	Rec	F1
Window-based (baseline I)	42.1	56.0	48.1	26.0	35.5	30.0
Clause-based (baseline II)	38.9	60.1	47.2	23.2	37.7	28.7
Proposed method	67.2	60.5	63.7	46.8	35.8	40.6

5.2   Extrinsic Evaluation on Sentence-Level Polarity Classification

In this section, we evaluate our negation modeling approach on the task of sentence-level polarity classification. The task is to correctly classify the overall polarity of a given sentence.

We consider two datasets: the *Multi-layered Reference Corpus for German Sentiment Analysis (MLSA)* (Clematide et al. 2012) and the *Heidelberg Sentiment Treebank (HeiST)* (Haas and Versley 2015). MLSA contains 270 sentences from the DeWaC Corpus (Baroni et al. 2009) which is a collection of German-language documents of various genres obtained from the web. HeiST contains 1184 sentences from German movie reviews.

We run two types of evaluations: a **three-class setting** in which the sentences are to be labeled as either *positive, negative* or *neutral*, and a **two-class setting** where we remove the neutral instances and the classifier just has to distinguish between *positive* and *negative* polarity. For HeiST, we remove 253 (neutral) sentences in the two-class setting while for MLSA, we remove 91 sentences.

The polarity classification algorithm we follow is kept simple. For each sentence we sum the scores associated with the polar expressions occurring in that sentence according to the sentiment lexicon of the PolArt-system. In case a polar expression is within the scope of a negation, we move the polarity score in the opposite direction by the absolute value of 1.3. This is an adhoc-value, however, it complies with the recent elicitation study from Kiritchenko and Mohammad (2016) in that the score of a negated polar expression should not be represented as its inverse. This is since a negated polar expression (e. g., *not excellent*) is less polar intense than a (plain) polar expression of the opposite polarity with the same polar intensity (e. g., *abysmal*). Our scoring is illustrated in Table 6. The final sentence-level polarity is derived from the sign of the sum of scores that we computed.

Table 6. Illustration of negation scores.

Expression	Score	Score of negation
exzellent (excellent)	+1.0	−0.3
ausreichend (sufficient)	+0.5	−0.8
umstritten (controversial)	−0.5	+0.8
miserable (abysmal)	−1.0	+0.3

In our experiments, we examine two different configurations, one where no negation modeling is considered and another where our proposed negation modeling is incorporated. We evaluate in terms of macro-average precision, recall and F-score. Table 7 shows the evaluation on HeiST while Table 8 shows the evaluation on MLSA. In both cases our proposed negation modeling outperforms the polarity classifier in which no negation modeling is incorporated.

Table 7. Polarity classification on HeiST.

Approach	2 Classes			3 Classes		
	Prec	Rec	F1	Prec	Rec	F1
w/o negation	65.3	52.9	58.4	50.8	50.8	50.8
with negation	67.3	54.7	60.3	52.0	51.9	52.0

Table 8. Polarity classification on MLSA.

Approach	2 Classes			3 Classes		
	Prec	Rec	F1	Prec	Rec	F1
w/o negation	78.7	74.5	76.6	51.1	50.6	50.8
with negation	80.9	77.4	79.1	51.0	51.6	51.3

6   Related Work

The most notable work dealing with different types of negation is Wilson et al. (2005) who point out that there are other words expressing negation than the commonly associated negation (function) words *not, never, no* etc. Since that work is carried out on English data, the scope modeling is kept simple using a window-based approach. Recently, Socher et al. (2013) proposed the usage of Recursive Neural Tensor Network (RNTN) for sentiment analysis. RNTN is a compositional sentence-level polarity classifier providing polarity values for each node in a constituency parse of a sentence. The authors claim that this method allows learning negation directly from labeled training data without explicit knowledge of negation words and their scopes. However, there has been no empirical examination of how reliably RNTN actually models negation. Moreover, that approach only produced results inferior to conventional SVMs trained on bag of words on German data (Haas and Versley 2015). For a detailed summary of negation modeling in sentiment analysis, we refer the reader to Wiegand et al. (2010).

Next to sentiment analysis, negation modeling has also been studied in the biomedical domain. Most of this work focuses on supervised classification on the (English) BioScope corpus (Szarvas et al. 2008), such as Morante et al. (2008) or Zou et al. (2013). The approach which is mostly related to ours is, however, the descriptive work by Morante (2010) who analyzes the individual negation words within the BioScope corpus and their scopes. This is one of the very few prominent research efforts that explicitly enumerates the different scopes of different negation words.

As far as German NLP is concerned, we are only aware of two research efforts that address negation. PolArt (Klenner et al. 2009) is a system that carries out sentence-level polarity classification. It matches polar expressions from a sentiment lexicon and then computes the sentence-level polarity compositionally on the basis of rules operating on syntactic constituents. This algorithm also incorporates negation modeling. However, the underlying lexicon only includes 22 polar shifters of which the majority are negation function words. The scope detection is further restricted by the fact that syntactic information is drawn from a chunk parser which only produces very flat output structures. Our work substantially differs from Klenner et al. (2009) in that we devised a framework for negation scope detection that is able to handle many more types of negation words and allows the specification of individual scopes. Moreover, unlike Klenner et al. (2009), we employ a dependency parser and further normalize its output. So, we exploit much more accurate syntactic information.

Cotik et al. (2016) propose a method for negation detection in clinical reports. This approach is an adaptation of *NegEx* (Chapman et al. 2001) which is simple negation detection algorithm that operates on a set of negation cues embedded in lexical patterns (i. e., word token sequences). This method operates on the string level, that is, unlike our approach no form of syntactic parsing is considered. Cotik et al. (2016) consider a set of 167 German negation phrases. These are highly domain-specific phrases most of which include a common negation function word, e. g., *trifft fuer den Patienten* **nicht** *zu (does* **not** *apply for the patient)* or **keine** *Beschwerden ueber (* **no** *complaints of)*. Due to the domain specificity of that approach, the negation cues and the scope detection mechanism cannot be applied to our dataset. Our approach also differs from Cotik et al. (2016) in that it is aimed at processing *unrestricted* text.

7 Conclusion

We presented an approach for modeling German negation in open-domain fine-grained sentiment analysis. Unlike most previous work in sentiment analysis, we assume that negation can be conveyed by many lexical units and that different negation words have different scopes.

We examined our approach on a new dataset comprising sentences with mentions of polar expressions and various negation words. We identify different types of negation words that have similar scopes. We showed that negation modeling based on these types largely outperforms traditional negation models assuming the same scope for all negation words no matter whether a window-based or clause-based scope is employed.

Our proposed method is only a first approximation of a more advanced negation handling for German. By making our implementation publicly available, we hope to stimulate further research in that direction using our new tool as a basis.

Acknowledgements. The authors would like to thank Stephanie Köser for annotating the dataset presented in this paper. The authors were partially supported by the German Research Foundation (DFG) under grants RU 1873/2-1 and WI 4204/2-1.

References

Akkaya, C., Wiebe, J., Mihalcea, R.: Subjectivity word sense disambiguation. In: Proceedings of the Conference on Empirical Methods in Natural Language Processing (EMNLP), Singapore, pp. 190–199 (2009)

Baker, C., Fillmore, C., Lowe, J.: The Berkeley FrameNet project. In: Proceedings of the International Conference on Computational Linguistics and Annual Meeting of the Association for Computational Linguistics (COLING/ACL), Montréal, Quebec, Canada, pp. 86–90 (1998)

Baroni, M., Bernardini, S., Ferraresi, A., Zanchetti, E.: The WaCky wide web: a collection of very large linguistically processed web-crawled corpora. Lang. Resour. Eval. **43**(3), 209–226 (2009)

Chapman, W.W., Bridewell, W., Hanbury, P., Cooper, G., Buchanan, B.: A simple algorithm for identifying negated findings and diseases in discharge summaries. J. Biomed. Inform. **34**, 301–310 (2001)

Clematide, S., Gindl, S., Klenner, M., Petrakis, S., Remus, R., Ruppenhofer, J., Waltinger, U., Wiegand, M.: MLSA – a multi-layered reference corpus for German sentiment analysis. In: Proceedings of the Conference on Language Resources and Evaluation (LREC), Istanbul, Turkey, pp. 3551–3556 (2012)

Cotik, V., Roller, R., Xu, F., Uszkoreit, H., Budde, K., Schmidt, D.: Negation detection in clinical reports written in German. In: Proceedings of the COLING-Workshop on Building and Evaluating Resources for Biomedical Text Mining (COLING-BioTxtM), Osaka, Japan, pp. 115–124 (2016)

Erk, K., Padó, S.: A powerful and versatile xml format for representing role-semantic annotation. In: Proceedings of the Conference on Language Resources and Evaluation (LREC), Lisbon, Portugal, pp. 799–802 (2004)

Haas, M., Versley, Y.: Subsentential sentiment on a shoestring: a crosslingual analysis of compositional classification. In: Proceedings of the Human Language Technology Conference of the North American Chapter of the ACL (HLT/NAACL), Denver, CO, USA, pp. 694–704 (2015)

Harabagiu, S., Hickl, A., Lacatusu, F.: Negation, contrast and contradiction in text processing. In: Proceedings of the National Conference on Artificial Intelligence (AAAI), Boston, MA, USA, pp. 755–762 (2006)

Kiritchenko, S., Mohammad, S.: The effect of negators, modals, and degree adverbs on sentiment composition. In: Proceedings of the Workshop on Computational Approaches to Subjectivity and Sentiment Analysis (WASSA), San Diego, CA, USA (2016)

Klenner, M., Fahrni, A., Petrakis, S.: PolArt: a robust tool for sentiment analysis. In: Proceedings of the Nordic Conference on Computational Linguistics (NoDaLiDa), Odense, Denmark, pp. 235–238 (2009)

Landis, R., Koch, G.: The measurement of observer agreement for categorical data. Biometrics **33**(1), 159–174 (1977)

Morante, R.: Descriptive analysis of negation cues in biomedical texts. In: Proceedings of the Conference on Language Resources and Evaluation (LREC), Valletta, Malta, pp. 1429–1436 (2010)

Morante, R., Liekens, A., Daelemans, W.: Learning the scope of negation in biomedical texts. In: Proceedings of the Conference on Empirical Methods in Natural Language Processing (EMNLP), Honolulu, pp. 715–724 (2008)

Palmer, M., Gildea, D., Kingsbury, P.: The proposition bank: an annotated corpus of semantic roles. Comput. Linguist. **31**(1), 71–106 (2005)

Polanyi, L., Zaenen, A.: Contextual valence shifters. In: Shanahan, J.G., Qu, Y., Wiebe, J. (eds.) Computing Attitude and Affect in Text: Theory and Applications, pp. 1–10. Springer, Heidelberg (2006). https://doi.org/10.1007/1-4020-4102-0_1

Sanchez-Graillet, O., Poesio, M.: Negation of protein-protein interactions: analysis and extraction. Bioinformatics **23**(13), i424–i432 (2007)

Sennrich, R., Schneider, G., Volk, M., Warin, M.: A new hybrid dependency parser for German. In: Proceedings of the German Society for Computational Linguistics and Language Technology (GSCL), Potsdam, Germany, pp. 115–124 (2009)

Socher, R., Perelygin, A., Wu, J., Chuang, J., Manning, C., Ng, A., Potts, C.: Recursive deep models for semantic compositionality over a sentiment treebank. In: Proceedings of the Conference on Empirical Methods in Natural Language Processing (EMNLP), Seattle, WA, USA, pp. 1631–1642 (2013)

Szarvas, G., Vincze, V., Farkas, R. Csirik, J.: The BioScope corpus: annotation for negation, uncertainty and their scope in biomedical texts. In: Proceedings of Current Trends in Biomedical Natural Language Processing (BioNLP), Columbus, OH, USA, pp. 38–45 (2008)

Wiegand, M., Balahur, A., Roth, B., Klakow, D., Montoyo, A.: A survey on the role of negation in sentiment analysis. In: Proceedings of the Workshop on Negation and Speculation in Natural Language Processing, Uppsala, Sweden, pp. 60–68 (2010)

Wiegand, M., Klenner, M., Klakow, D.: Bootstrapping polarity classifiers with rule-based classification. Lang. Resour. Eval. **47**(4), 1049–1088 (2013)

Wilson, T., Wiebe, J., Hoffmann, P.: Recognizing contextual polarity in phrase-level sentiment analysis. In: Proceedings of the Conference on Human Language Technology and Empirical Methods in Natural Language Processing (HLT/EMNLP), Vancouver, BC, Canada, pp. 347–354 (2005)

Zou, B., Zhou, G., Zhu, Q.: Tree kernel-based negation and speculation scope detection with structured syntactic parse features. In: Proceedings of the Conference on Empirical Methods in Natural Language Processing (EMNLP), Seattle, WA, USA, pp. 968–976 (2013)

Processing German: Named Entities

NECKAr: A Named Entity Classifier for Wikidata

Johanna Geiß, Andreas Spitz[✉], and Michael Gertz

Institute of Computer Science, Heidelberg University, Im Neuenheimer Feld 205,
69120 Heidelberg, Germany
{geiss,spitz,gertz}@informatik.uni-heidelberg.de

Abstract. Many Information Extraction tasks such as Named Entity
Recognition or Event Detection require background repositories that
provide a classification of entities into the basic, predominantly used
classes LOCATION, PERSON, and ORGANIZATION. Several available knowledge bases offer a very detailed and specific ontology of entities that
can be used as a repository. However, due to the mechanisms behind
their construction, they are relatively static and of limited use to IE
approaches that require up-to-date information. In contrast, Wikidata is
a community-edited knowledge base that is kept current by its userbase,
but has a constantly evolving and less rigid ontology structure that does
not correspond to these basic classes. In this paper we present the tool
NECKAr, which assigns Wikidata entities to the three main classes of
named entities, as well as the resulting Wikidata NE dataset that consists of over 8 million classified entities. Both are available at http://
event.ifi.uni-heidelberg.de/?page_id=532.

1 Introduction

The classification of entities is an important task in information extraction (IE)
from textual sources that requires the support of a comprehensive knowledge
base. In a standard workflow, a Named Entity Recognition (NER) tool is used
to discover the surface forms of entity mentions in some input text. These surface forms then have to be disambiguated and linked to a specific entity in a
knowledge base (entity linking) to be useful in subsequent IE tasks. For the
latter step of entity linking, suitable entity candidates have to be selected from
the underlying knowledge base. In the general case of linking arbitrary entities, information about the classes of entity candidates is advantageous for the
disambiguation process and for pruning candidates. In more specialized cases,
only a subset of entity mentions may be of interest, such as toponyms or person
mentions, which requires the classification of entity mentions. As a result, the
classification of entities in the underlying knowledge base serves to support the
linking procedure and directly translates into a classification of the entities that
are mentioned in the text, which is a necessary precondition for many subsequent tasks such as event detection (Kumaran and Allan 2004) or document
geolocation (Ding et al. 2000).

G. Rehm and T. Declerck (Eds.): GSCL 2017, LNAI 10713, pp. 115–129, 2018.
https://doi.org/10.1007/978-3-319-73706-5_10

There is a number of knowledge bases that provide such a background repository for entity classification, predominantly DBpedia, YAGO, and Wikidata (Färber et al. 2017). While these knowledge bases provide semantically rich and fine-granular classes and relationship types, the task of entity classification often requires associating coarse-grained classes with discovered surface forms of entities. This problem is best illustrated by an IE tasks that has recently gained significant interest in particular in the context of processing streams of news articles and postings in social media, namely event detection and tracking, e.g., (Aggarwal and Subbian 2012; Sakaki et al. 2010; Brants and Chen 2003). Considering an event as *something that happens at a given place and time between a group of actors* (Allan 2012), the entity classes PERSON, ORGANIZATION, LOCATION, and TIME, are of particular interest. While surface forms of temporal expressions are typically normalized by using a temporal tagger (Strötgen and Gertz 2016), dealing with the classification of the other types of entities often is much more subtle. This is especially true if one recalls that almost all available NER tools tag named entities only at a very coarse-grained level, e.g., Stanford NER (Finkel et al. 2005), which predominately uses the classes LOCATION, PERSON, and ORGANIZATION.

The objective of this paper is to provide the community with a dataset and API for entity classification in Wikidata, which is tailored towards entities of the classes LOCATION, PERSON, and ORGANIZATION. Like knowledge bases with inherently more coarse or hybrid class hierarchies such as YAGO and DBpedia, this version of Wikidata then supports entity linking tasks at state-of-the-art level (Geiß and Gertz 2016; Spitz et al. 2016b), but links entities to the continuously evolving Wikidata instead of traditional KBs. As we outline in the following, extracting respective sets of entities from a KB for each such class is by no means trivial (Spitz et al. 2016a), especially given the complexity of simultaneously dealing with multi-level class and instance structures inherent to existing KBs, an aspect that is also pointed out by Brasileiro et al. (2016). However, there are several reasons to chose Wikidata over other KBs. First, especially when dealing with news articles and social media data streams, it is crucial to have an up-to-date repository of persons and organizations. To the best of our knowledge, at the time of writing this paper, the most recent version of DBPedia was published in April 2016, and the latest evaluated version of YAGO in September 2015, whereas Wikidata provides a weekly data dump. Even though all three KBs (Wikidata, DBpedia, and YAGO3) are based on Wikipedia, Wikidata also contains information about entities and relationships that have not been simply extracted from Wikipedia (YAGO and DBpedia extract data predominantly from infoboxes) but collaboratively added by users (Müller-Birn et al. 2015). Although the latter feature might raise concerns regarding the quality of the data in Wikidata, for example due to vandalism (Heindorf et al. 2015), we find that the currentness of information far outweights these concerns when using Wikidata as basis for a named entity classifying framework and as a knowledge base in particular. While Wikidata provides a SPARQL interface for direct query access in addition to the weekly dumps, this method of accessing the data has several downsides. First, the interface is not designed for speed and is thus ill suited for entity extraction or linking tasks in large corpora, where many lookups are necessary. Second, and more importantly, the continually evolving content of Wikidata prevents reproducability of scientific results if the online SPARQL access is used, as the versioning is unclear and it is impossible to recreate experimental conditions. Third, we find that the hierarchy and structure in Wikidata is (necessarily) complicated and does not lend itself easily to creating coarse class hierarchies on the fly without substantial prior investigation into the existing hierarchies. Here, NECKAr provides a stable, easy to use view of classified Wikidata entities that is based on a selected Wikidata dump and allows reproducible results of subsequent IE tasks.

In summary, we make the following contributions: We provide an easy to use tool for assigning Wikidata items to commonly used NE classes by exclusively utilizing Wikidata. Furthermore, we make one such resulting Wikidata NE dataset available as a resource, including basic statistics and a thorough comparison to YAGO3.

The remainder of the paper is structured as follows. After a brief discussion of related work in the following section, we describe our named entity classifier in detail in Sect. 3 and present the resulting Wikidata NE dataset in Sect. 4. Section 5 gives a comparison of our results to YAGO3.

2   Related Work

The DBpedia project[1] extracts structured information from Wikipedia (Auer et al. 2007). The 2016-04 version includes 28.6M entities, of which 28M are classified in the DBpedia Ontology. This DBpedia 2016-04 ontology is a directed-acyclic graph that consists of 754 classes. It was manually created and is based on the most frequently used infoboxes in Wikipedia. For each Wikipedia language version, there are mappings available between the infoboxes and the DBpedia ontology. In the current version, there are 3.2M persons, 3.1M places and 515,480 organizations. To be available in DBpedia, an entity needs to have a Wikipedia page (in at least one language version that is included in the extraction) that contains an infobox for which a valid mapping is available.

YAGO3 (Mahdisoltani et al. 2015), the multilingual extension of YAGO, combines information from 10 different Wikipedia language versions and fuses it with the English WordNet. YAGO[2] concentrates on extracting facts about entities, using Wikipedia categories, infoboxes, and Wikidata. The YAGO taxonomy is constructed by making use of the Wikipedia categories. However, instead of using the category hierarchy that is "barely useful for ontological purposes" (Suchanek et al. 2007), the Wikipedia categories are extracted, filtered and parsed for noun phrases in order to map them to WordNet classes. To include the multilingual categories, Wikidata is used to find corresponding English category names. As

---

[1] http://wiki.dbpedia.org.

[2] http://www.yago-knowledge.org.

a result, the entities are assigned to more than 350K classes. YAGO3, which is extracted from Wikipedia dumps of 2013–2014, includes about 4.6M entities.

Both KBs solely depend on Wikipedia. Since it takes some time to update or create the KBs, they do not include up-to-date information. In contrast, the current version of Wikidata can be directly queried and a fresh Wikidata dump is available every week. Another advantage is that Wikidata does not rely on the existence of an infobox or Wikipedia page. Entities and information about the entities can be extracted from Wikipedia or manually entered by any user, meaning that less significant entities that do not warrant their own Wikipedia page are also represented. Since Wikipedia infoboxes are partially populated through templates from Wikidata entries, extracting data from infoboxes instead of Wikidata itself adds an additional source of errors. Furthermore, unless all language versions of Wikipedia are used as a source, such an approach would even limit the amount of retrieved information due to Wikidata's inherent multi-lingual design as the knowledge base behind all Wikipedias (Vrandečić and Krötzsch 2014).

For completeness, Freebase[3] should be mentioned as a fourth available knowledge base that has historically been used as a popular alternative to YAGO and DBpedia. However, efforts have recently been taken to merge it entirely into Wikidata (Pellissier Tanon et al. 2016). Given the need for current, up-to-date entity information in many event-related applications, the fact that Freebase is no longer actively maintained and updated means that it is increasingly ill-suited for such tasks.

3   The NECKAr Tool

The Named Entity Classifier for Wikidata (NECKAr) assigns Wikidata entites to the NE classes PERSON, LOCATION, and ORGANIZATION. The tool, which is available as open source code (see the URL in the Abstract), is easy to use and only requires a minimum setup of Python3 packages as well as an instance of a MongoDB.

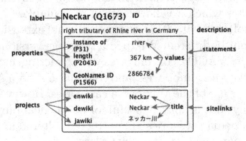

Fig. 1. Wikidata data model

---

[3] https://developers.google.com/freebase/.

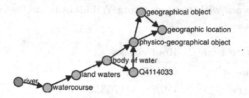

Fig. 2. Class hierarchy for *river*, generated with the Wikidata Graph Builder

3.1 Wikidata Data Model

Wikidata[4] is a free and open knowledge base that is intended to serve as central storage for all structured data of Wikimedia projects. The data model of Wikidata consists primarily of two major components: *items* and *properties*. Items represent all *things* in human knowledge. Each item corresponds to a clearly identifiable concept or object, or to an instance of a concept or object. For example, there is one item for the concept *river* and one item *Neckar*, which is an instance of a river. In this context, a concept is the same as a class. All items are denoted by numerical identifiers prefixed with a Q, while properties have numerical identifiers prefixed with P. Properties (P) connect items (Q) to values (V). A pair (P, V) is called a statement. A property classifies the value of a statement. Figure 1 shows a simplified entry for the item *Neckar*. Here P2043 describes that the value 367 km has to be interpreted as the length of the river. Both items and properties have a label, a description, and (multilingual) aliases. Property entries in Wikidata do not have statements, but data types that restrict what can be given as a properties value. These data types include items, external identifiers (e.g., ISBN codes), URLs, geographic coordinates, strings, or quantities, to name a few.

When we are interested in the classification of items, we require the knowledge which item is an instance of which class. Class membership of an item is predominately modelled by the property *instance of* (P31). For example, consider the statement Q1673:P31:Q4022 (Neckar is an instance of river), in which *Q4022* can be seen as a class. Classes can be subclasses of other classes, e.g., *river* is a subclass of *watercourse*, which is a subclass of *land water*. Figure 2 shows the subclass graph for *river*.

The property *subclass of* (P279) is transitive, meaning that since *Neckar* is an instance of *river*, which is a subclass of *watercourse*, *Neckar* is implicitly also an instance of *watercourse*. Due to this transitivity rule, in Wikidata there is no need to specify more than the most specific statement[5]. In other words, there is no statement that directly specifies *Neckar* to be a *geographic location*. Thus, we cannot simply extract items that are instances of the general classes. There are, for example, only 1,733 items that are direct instances of *geographic location*. Instead, we need to extract the transitive hull, that is, all items that are an

---

[4] http://www.wikidata.org.
[5] http://www.wikidata.org/wiki/Help:Basic_membership_properties.

Table 1. Location types with corresponding Wikidata root classes for location types and number of subclasses

Location type	Root classes	# sub
Continent	Q5107	1
Country	Q6256, Q1763527, Q3624078	50
State	Q7275	173
Settlement	Q486972	1224
City	Q515	126
Sea	Q165	12
River	Q4022	34
Mountain	Q8502	81
Mountain range	Q1437459	18
Territorial entity	Q15642541	3,657

instance of any subclass of the general class (henceforward *root classes*). There are several tools available to show and query the class structure of Wikidata. For Fig. 2 we used the *Wikidata Graph Builder* (WGB)[6] to visualize the class tree. For NECKAr, we make use of the SPARQL based *Wikidata Query Service*[7] to extract all subclasses of a root class, e.g., *geographic location*. Once the subclasses of a root class are identified, we can extract all items that are instances of these subclasses.

The task is then to find root classes that, together with their subclasses, best represent the predominately used NE classes LOCATION, ORGANIZATION, and PERSON. In the following we describe how items of these classes are extracted and what kind of information we store for each item. For all items, we store the Wikidata ID, the label (the most common name for an item), the links to the English and German Wikipedia, and the description.

3.2 Location Extraction

To extract all locations from Wikidata, we use the root class *geographic location* (Q222-1906). This class is very large and includes 23,383 subclasses[8]. For each location item, we extract the following statements: coordinate location (P625), population (P1082), country (P17), and continent (P30). Additionally, we assign a location type if an item is an instance of a subclass of the root class for that location type (see Table 1).

In this large set of subclasses of *geographic location*, we encounter several problems. For example, *Food* is a subclass of *geographic location*. *Food* is connected to *geographic location* by a path of length 3 (*food → energy*

---

[6] http://angryloki.github.io/wikidata-graph-builder/.

[7] https://query.wikidata.org.

[8] In the following, all class and subclass sizes are as of February 22, 2017.

Table 2. Example of classified entities

neClass	LOCATION		ORGANIZATION		PERSON
id	Q1796771		Q81230		Q76658
norm_name	Köthen		Siemens		Frank-Walter Steinmeier
description	capital of the district of Anhalt-Bitterfeld Saxony-Anhalt		Engineering and electronics conglomerate		politician
en Wikipedia	Köthen (Anhalt)		Siemens		Frank-Walter Steinmeier
location type	city, settlement	instance of	concern, bus. enterprise	occupation	politician, jurist, lawyer
population	26,384				
continent	Europe	CEO	Joe Kaeser	gender	male
country	Germany		Klaus Kleinfeld	dob	1956-01-05
coordinate	51.75	founder	Ernst Werner von Siemens	dod	none
	11.916666666667			alias	Steinmeier
GeoNames	2885237	inception	1847-10-01		
		HQ	Munich		
		country	Germany		
		website	www.siemens.com		

storage → storage → geographic location). We cannot simply limit the allowed path length since there are other subclasses with a greater path length that we consider a valid location. For example the shortest path for *village of Japan* has a length of 4 (*village of Japan → municipality of Japan → municipality → human settlement → geographic location*). In this case we decided to exclude the subtree for *Food*, which reduces the number of subclasses considerably to 13,445. However, there might be other subclasses that are not considered a proper location (e.g., *Arcade Video Game* with the path: *arcade video game → arcade game machine → computing platform → computing infrastructure → infrastructure → construction → geographical object → geographic location*). For the time being we only exclude the *Food* subclasses. The identified location items can be filtered for a certain application by using the location type or by only using items for which a coordinate location is given.

3.3 Organization Extraction

The root class *organization* (Q43229) includes 4811 subclasses, such as *nonprofit organization, political organization, team, musical ensemble, newspaper*, or *state*. For each item in this category, we extract additional information such as country (sovereign state of this item, P17), founder (P112), CEO (P169), inception (P571), headquarter location (P159), instance of (P31), official website (P856), and official language (P37).

3.4 Person Extraction

To extract all real world persons from Wikidata, we only use the class *human* (Q5) instead of a list of subclasses. In Wikidata, a more specific classification of a person is usually given by the occupation property or by having several *instance of* statements. All items with the statement *is instance of human* are classified as person. Fictional characters, such as *Homer Simpson* or *Harry Potter* and deities that are not also classified as human, are not extracted. For each person, we gather some basic information: date of birth (dob) (P569), date of death (dod) (P570), gender (P21), occupation (P106), and alternative names.

3.5 Extracting Links to Other Knowledge Bases

In addition to the above information, we also record identifiers for the items in other publicly available databases (Wikipedia, DBPedia, Integrated Authority File of the German National Library, Internet Movie Database, MusicBrainz, GeoNames, OpenStreet Map). This information is represented in Wikidata as statements and can be extracted analogously to the examples above.

3.6 Extraction Algorithm

The named entity classes to be extract can be specified in a configuration file. For each chosen class of named entities, the process then works as described in the following. First, the subclasses of the root class are extracted using the Wikidata SPARQL API. The output of this step is a list of subclasses, from which the invalid subclasses are excluded. For locations, we also generate lists of subclasses for the specific location types. The tool then searches the Wikidata dump (stored in a local MongoDB) for all items that are instances of one of the subclasses in the list and extracts the common features (id, label, description, Wikipedia links). Depending on the named entity class, additional information (see above) is extracted, and for locations, the list of location type subclasses is used to assign a location type. This data is then stored in a new, intermediary MongoDB collection. In a subsequent step, we extract for each item the identifiers that link them to the other databases as described in Sect. 3.5 and store them in a separate collection. In the last step, the data is exported to CSV and JSON files for ease of use.

4 Wikidata NE Dataset

The Wikidata NE dataset[9] was extracted using the NECKAr tool. For the version that we discuss in this paper, we extracted entities from the Wikidata dump from December 5, 2016, which includes 24,580,112 items and 2,910 distinct properties.

In total, we extracted and classified 8,842,103 items, of which 51.8% are locations, 37.6% persons, and 10.6% organizations. Table 2 shows examples for each named entity class, including the class specific additional information.

---

[9] http://event.ifi.uni-heidelberg.de/?page_id=532.

4.1   Location Entities

Of the 4,582,947 identified locations, 51% have geographic coordinates. Location types are extracted for 93% of the location items (see Table 3)

Table 3. Number of entities for location types

Type	Continent	Country	State	City	Territorial entity
Count	10	2,496	4,330	25,470	1,805,213
Type	Settlement	Sea	River	Mountain range	Mountain
Count	1,983,860	183	199,991	7,814	229,853

Most of the classified locations are settlements and territorial entities. We find over 2,400 countries: although there currently are only 206 countries[10], Wikidata also includes former countries like the Roman Empire, Ancient Greece, or Prussia.

4.2   Person Entities

We extracted 3,322,217 persons, of which 78% are male, 15% female, while for 7% of the persons another gender or no gender is specified. Occupations are given for 66% of the person items, where the largest group are *politicians*, followed by *football players* (see Table 4).

Table 4. The five most frequent occupations

#	1	2	3	4	5
Occupation	Politician	Assoc. football player	Actor	Writer	Painter
Count	312,571	206,142	170,291	127,837	99,060

Wikidata covers mostly persons from recent history, so 70% of the persons for whom a birth date is given (over 2,5M persons) were born in the 20th century, while around 20% were born in the 19th century.

4.3   Organization Entities

936,939 items were classified as organizations, of which 11% are business enterprises. Table 5 shows the top 5 organization types. Where possible, we also extracted the country in which the organization is based. Figure 3 shows a heatmap of the number of organizations per country. Most organizations are based in the U.S.A., followed by France and Germany. This is partially due to the fact that *commune of France* and *municipality of Germany* are subclasses of *organization*.

---

[10] https://en.wikipedia.org/wiki/List_of_sovereign_states.

Table 5. The five most frequent organization types

#	Type	Count
1	Business enterprise	102,129
2	Band	58,996
3	Commune of France	38,387
4	Primary school	36,078
5	Association football club	31,257

Fig. 3. Heatmap of organization frequency by country

4.4 Assignment to More Than One Class

400,856 Wikidata items are assigned to more than one NE class by NECKAr. The vast majority of this subset (over 99%) are members of the two classes *location* and *organization*. This is mainly caused by a subclass overlap between the root classes *geographic location* and *organization*. In total, they share 1,310 subclasses, e.g., *hospital*, *state* or *library* and their respective subclasses. We do not favour one class over the other, because both interpretations are possible, depending on the context. There are also items that have several *instance of* statements, which in six cases leads to an assignment to all three classes, e.g., *Jean Leon* is described as instance of human and instance of winery, which is a subclass of both organization and geographic location. There are 116 items that are classified as person and location or person and organization, which is again caused by multiple *instance of* statements. In contrast to the subclass overlap between root classes, these cases are caused by incorrect user input into Wikidata.

5   Comparison to YAGO3

In order to get an estimate of the quality of the NECKAr tool, we compare the resulting Wikidata NE dataset to the currently available version of YAGO (Version 3.0.2). When using the YAGO3 hierarchy to classify YAGO3 entities, we find 1,745,219 distinct persons (member of YAGO3 class wordnet_person_100007846), 1,267,402 distinct locations (member of YAGO class yagoGeoEntity) and 481,001 distinct organisations (member of YAGO class wordnet_social_group_107950920) for a total of 3,493,622 entities in comparison to the 8,8M entities in the Wikidata NE dataset (see Table 6).

YAGO3 entities can be linked to Wikidata entries via their subject id, which corresponds to Wikipedia page names. If a YAGO3 entity is derived from a non-English Wikipedia, the subject id is prefixed with the language code. For 3,430,065 YAGO3 entities we find a corresponding entry in Wikidata (1,715,305 persons, 1,250,409 locations and 464,351 organization). This subset is the basis for our comparison in the following.

To assess the quality of NECKAr, the well-known IR measures F_1-score, precision (P) and recall (R) are used. Precision is a measure for exactness, that is, how many of the classified entities are classified correctly. Recall measures completeness and gives the fraction of correctly classified entities of all given entities. F_1 is the harmonic mean of P and R. The measures are defined as:

$$F_1 = 2 * \frac{P * R}{P + R} \qquad P = \frac{TP}{TP + FP} \qquad R = \frac{TP}{TP + FN} \qquad (1)$$

Here, TP (true positives) is the number of YAGO3 entites, that NECKAr assigns to the same class, while FP (false positives) is the number of YAGO3 entities that are falsely assigned to that class. $TP + FP$ represents the number of entities assigned to that class by NECKAr. FN (false negatives) is the number of YAGO3 entities in a given class that NECKAr does not assign to that class (these entities might be assigned to a different class or to no class). Thus, $TP + FN$ is the number of YAGO3 entities in a given class. Using these standard metrics, we receive a overall F_1-score of 0.88 with $P = 0.90$ and $R = 0.86$ (see Table 7). The lower recall is due to the fact that NECKAr does not classify all entries that are a *person*, *location*, or *organization* entity in YAGO3. Only about 88% of the YAGO3 entites that correspond to Wikidata entries are classified. For example, NECKAr does not find *Pearson*, a town in Victoria, Australia, because the Wikidata entry does not include any *is instance of* relation. This is true for 290,905 of the 387,259 entities (75.12%) that are not classified by NECKAr. Some entities are missed by NECKAr entirely for a couple of reasons. In some cases, the correct *is instance of* relation is not given in Wikidata. In others, a relevant subclass or property may not have been included. Finally, since YAGO3 was automatically extracted and not every fact was checked for correctness it contains some erroneous claims or classifications. For example, some overview articles in Wikipedia are classified as entities in YAGO, such as *Listed_buildings_in* or *Index of*. The original evaluation of YAGO3 lists the fraction of incorrect facts that it contains as 2% (Mahdisoltani et al. 2015).

5.1 Location Comparison

For LOCATION, NECKAr achieves a F_1-score of 0.88 (P = 0.93 and R = 0.84). 170,860 YAGO3 locations were not classified, of which 81% have no entry in Wikidata for the *instance of* property.

Table 6. Number of entities per class in the Wikidata NE dataset created by NECKAr, YAGO3 and the intersection of YAGO3 and Wikidata

NE	NECKAr	Yago3	Yago3 ∩ WD
LOC	4,582,947	1,267,402	1,250,409
PER	3,322,217	1,745,219	1,715,305
ORG	0936,939	0481,001	0464,351

Table 7. Evaluation results (F_1 score, Precision (P) and Recall (R)) for the Wikidata NE dataset created by NECKAr in comparison to YAGO3

NE	F_1	P	R
LOC	0.88	0.93	0.84
PER	0.97	0.99	0.95
ORG	0.57	0.54	0.60
All	0.88	0.90	0.86

Of the entities that are assigned to a different class, NECKAr classified 97,6% as ORGANIZATION instead of LOCATION. Most of these entities (85%) are radio or television stations for which a classification into either class is a matter of debate. These items are described in Wikidata as *instance of radio station* or *television station* which are subclasses of ORGANIZATION. The majority of the FPs for locations are assigned by NECKAr to two classes (ORGANIZATION and LOCATIONS), whereas in YAGO3 these are only organizations.

5.2   Person Comparison

For entities of class PERSON, we receive the highest F_1-score of all classes with 0.97 (P $= 0.99$ and R $= 0.95$). Most of the entities that NECKAr assigned to a different class (90% to ORGANIZATION, 10% to LOCATION) are bands or musical ensembles which are classified as ORGANIZATION.

5.3   Organization Comparison

The class ORGANIZATION shows the lowest F_1-score $= 0.57$ (P $= 0.54$ and R $= 0.60$). The low precision is caused by the high number of false positives. As discussed in the previous section, many entities that are classified as Persons or Locations by YAGO3 are classified as organizations by NECKAr. The low recall is due to the fact that 156,926 YAGO3 organizations were not identified by NECKAr. Again, the majority of these items (82%) has no *is instance of* relation in Wikidata, so NECKAr was not able to classify them. The reason for the missing 18% warrants future investigation in more detail, as it is possible that an important subclass or property was excluded. 29,625 YAGO3 organizations were assigned to another class, 96% to LOCATION and 4% to PERSON.

Many of these items are constituencies or administrative units, which could be seen as organizations and/or locations.

In summary, we find that the application of NECKAr to Wikidata produces a set of classified entities that is comparable in quality to a well known and widely used knowledge base. However, in contrast to existing knowledge bases, which are not updated regularly, NECKAr can be used to extract substantially more entities and up-to-date lists of persons, locations and organizations. Since NECKAr can be applied to weekly dumps of Wikidata, it can be used to extract a reproducible resource for subsequent IE tasks.

6   Conclusion and Future Work

In this paper, we introduced the NECKAr tool for assigning NE classes to Wikidata items. We discussed the data model of Wikidata and its class hierarchy. The resulting NE dataset offers the simple classification of entities into locations, organizations and persons that is often used in IE tasks. The datasets includes basic, class specific information on each item and links them to other linked open data sets. The clear and lightweight structure makes the dataset a valuable and easy to use resource. Much of the original more fine grained classification is preserved and can be used to create application-specific subsets. A comparison to YAGO3 showed that NECKAr is able to create state-of-the-art lists of entities with the added advantage of providing larger and more recent data.

Based on these results, we are further investigating the Wikidata class hierarchy in order to reduce the number of incorrect or multiple assignments. We are also working on an automated process to provide the Wikidata NE dataset on a monthly basis. In future releases of NECKAr, we plan to include the option of choosing between a Wikidata dump and the SPARQL API as source for obtaining the entity data.

References

Aggarwal, C.C., Subbian, K.: Event detection in social streams. In: Proceedings of the Twelfth SIAM International Conference on Data Mining, ICDM, pp. 624–635 (2012). https://doi.org/10.1137/1.9781611972825.54

Allan, J.: Topic Detection and Tracking: Event-based Information Organization, vol. 12. Springer Science & Business Media, New York (2012). https://doi.org/10.1007/978-1-4615-0933-2

Auer, S., Bizer, C., Kobilarov, G., Lehmann, J., Cyganiak, R., Ives, Z.G.: DBpedia: a nucleus for a web of open data. In: Aberer, K., et al. (eds.) ASWC/ISWC-2007. LNCS, vol. 4825, pp. 722–735. Springer, Heidelberg (2007). https://doi.org/10.1007/978-3-540-76298-0_52

Brants, T., Chen, F.: A system for new event detection. In: Proceedings of the 26th Annual International ACM SIGIR Conference on Research and Development in Information Retrieval, pp. 330–337 (2003). http://doi.acm.org/10.1145/860435.860495

Brasileiro, F., Almeida, J.P.A., de Carvalho, V.A., Guizzardi, G.: Applying a multi-level modeling theory to assess taxonomic hierarchies in Wikidata. In: Proceedings of the 25th International Conference on World Wide Web, WWW Companion Volume, pp. 975–980 (2016). http://doi.acm.org/10.1145/2872518.2891117

Ding, J., Gravano, L., Shivakumar, N.: Computing geographical scopes of web resources. In: Proceedings of 26th International Conference on Very Large Data Bases, VLDB, pp. 545–556 (2000). http://www.vldb.org/conf/2000/P545.pdf

Färber, M., Bartscherer, F., Menne, C., Rettinger, A.: Linked data quality of DBpedia, Freebase, OpenCyc, Wikidata, and YAGO. Semant. Web 9(1), 77–129 (2018). https://doi.org/10.3233/SW-170275

Finkel, J.R., Grenager, T., Manning, C.D.: Incorporating non-local information into information extraction systems by Gibbs sampling. In: Proceedings of the 43rd Annual Meeting of the Association for Computational Linguistics, ACL, pp. 363–370 (2005). http://aclweb.org/anthology/P/P05/P05-1045.pdf

Geiß, J., Gertz, M.: With a little help from my neighbors: person name linking using the Wikipedia social network. In: Proceedings of the 25th International Conference on World Wide Web, WWW Companion Volume, pp. 985–990 (2016). http://doi.acm.org/10.1145/2872518.2891109

Heindorf, S., Potthast, M., Stein, B., Engels, G.: Towards vandalism detection in knowledge bases: corpus construction and analysis. In: Proceedings of the 38th International ACM SIGIR Conference on Research and Development in Information Retrieval, pp. 831–834 (2015). http://doi.acm.org/10.1145/2766462.2767804

Kumaran, G., Allan, J.: Text classification and named entities for new event detection. In: Proceedings of the 27th Annual International ACM SIGIR Conference on Research and Development in Information Retrieval, pp. 297–304 (2004). http://doi.acm.org/10.1145/1008992.1009044

Mahdisoltani, F., Biega, J., Suchanek, F.M.: YAGO3: a knowledge base from multilingual Wikipedias. In: Seventh Biennial Conference on Innovative Data Systems Research, CIDR (2015). http://cidrdb.org/cidr2015/Papers/CIDR15_Paper1.pdf

Müller-Birn, C., Karran, B., Lehmann, J., Luczak-Rösch, M.: Peer-production system or collaborative ontology engineering effort: what is Wikidata? In: Proceedings of the 11th International Symposium on Open Collaboration, pp. 20:1–20:10 (2015). http://doi.acm.org/10.1145/2788993.2789836

Pellissier Tanon, P., Vrandečić, D., Schaffert, S., Steiner, T., Pintscher, L.: From freebase to Wikidata: the great migration. In: Proceedings of the 25th International Conference on World Wide Web, WWW, pp. 1419–1428 (2016). http://doi.acm.org/10.1145/2872427.2874809

Sakaki, T., Okazaki, M., Matsuo, Y.: Earthquake shakes Twitter users: real-time event detection by social sensors. In: Proceedings of the 19th International Conference on World Wide Web, WWW, pp. 851–860 (2010). http://doi.acm.org/10.1145/1772690.1772777

Spitz, A., Dixit, V., Richter, L., Gertz, M., Geiß, J.: State of the union: a data consumer's perspective on Wikidata and its properties for the classification and resolution of entities. In: Wiki, Papers from the 2016 ICWSM Workshop (2016a). http://aaai.org/ocs/index.php/ICWSM/ICWSM16/paper/view/13200

Spitz, A., Geiß, J., Gertz, M.: So far away and yet so close: augmenting toponym disambiguation and similarity with text-based networks. In: Proceedings of the Third International ACM SIGMOD Workshop on Managing and Mining Enriched Geo-Spatial Data, GeoRich@SIGMOD, pp. 2:1–2:6 (2016b). http://doi.acm.org/10.1145/2948649.2948651

Strötgen, J., Gertz, M.: Domain-Sensitive Temporal Tagging. Synthesis Lectures on Human Language Technologies. Morgan & Claypool Publishers, San Rafael (2016). https://doi.org/10.2200/S00721ED1V01Y201606HLT036

Suchanek, F.M., Kasneci, G., Weikum, G.: YAGO: a core of semantic knowledge. In: Proceedings of the 16th International Conference on World Wide Web, WWW, pp. 697–706 (2007). http://doi.acm.org/10.1145/1242572.1242667

Vrandečić, D., Krötzsch, M.: Wikidata: a free collaborative knowledgebase. Commun. ACM **57**(10), 78–85 (2014). http://doi.acm.org/10.1145/2629489

Investigating the Morphological Complexity of German Named Entities: The Case of the GermEval NER Challenge

Bettina Klimek[✉], Markus Ackermann, Amit Kirschenbaum, and Sebastian Hellmann

AKSW/KILT Research Group, InfAI, University of Leipzig, Leipzig, Germany
{klimek,ackermann,amit,hellmann}@informatik.uni-leipzig.de

Abstract. This paper presents a detailed analysis of Named Entity Recognition (NER) in German, based on the performance of systems that participated in the GermEval 2014 shared task. It focuses on the role of morphology in named entities, an issue too often neglected in the NER task. We introduce a measure to characterize the morphological complexity of German named entities and apply it to the subset of named entities identified by all systems, and to the subset of named entities none of the systems recognized. We discover that morphologically complex named entities are more prevalent in the latter set than in the former, a finding which should be taken into account in future development of methods of that sort. In addition, we provide an analysis of issues found in the GermEval gold standard annotation, which affected also the performance measurements of the different systems.

1 Introduction

Despite initiatives to improve Named Entity Recognition (NER) for German such as in challenges as part of CoNLL 2003[1] and GermEval 2014[2], a noticeable gap still remains between the performance of NER systems for German and English. Pinpointing the cause of this gap seems to be an impossible task as the reasons are manifold and in addition difficult to realize due to their potentially granular (and subtle) nature as well as their inter-relatedness. However, we can name several aspects that might have an influence: (1) lack of linguistic resources suitable for German, (2) less demand (and interest) for improving the quality of NER systems for German, (3) variance of annotation guidelines and annotator consensus, (4) different NER problem definitions, (5) inherent differences between both language systems, (6) quality of provided data and source material, (7) etc. Studying the degree of impact for each of these factors

---

[1] CoNLL 2003 Challenge Language-Independent Named Entity Recognition, http://www.cnts.ua.ac.be/conll2003/ner/.

[2] GermEval 2014 Named Entity Recognition Shared Task, https://sites.google.com/site/germeval2014ner/, see also (Benikova et al. 2014a).

© The Author(s) 2018
G. Rehm and T. Declerck (Eds.): GSCL 2017, LNAI 10713, pp. 130–145, 2018.
https://doi.org/10.1007/978-3-319-73706-5_11

as a whole revokes any attempt to apply scientific methods for error analysis. However, a systematic investigation of linguistic aspects of proper nouns, i.e., named entities in technical terms[3], in German can reveal valuable insights on the difficulties and the improvement potential of German NER tools. Such an aspect is the morphological complexity of proper nouns. Due to its greater morphological productivity and variation, the German language is more difficult to analyze, offering additional challenges and opportunities for further research. The following list highlights a few examples:

- More frequent and extensive compounding requires correct token decompounding to identify the named entity (e.g., *Bibelforscherfrage* - 'bible researchers' question').
- Morphophonologically conditioned inner modifications are orthographically reflected and render mere substring matching ineffective (e.g., *außereuropäisch(Europa)* - 'non-European').
- Increased difficulty in identifying named entities which occur within different word-classes after derivation (e.g., *lutherischen*, an adjective, derived from the proper noun *Martin Luther*).

These observations support the hypothesis that morphological alternations of proper nouns constitute another difficulty layer which needs to be addressed by German NER systems in order to reach better results. Therefore, this paper presents the results of a theoretic and manual annotation and evaluation of a subset of the GermEval 2014 Corpus challenge task dataset. This investigation focuses on the complexity degree of the morphological construction of named entities and shall serve as reference point that can help to estimate whether morphological complexity of named entities is an aspect which impacts NER and if it should be considered when creating or improving German NER tools. During the linguistic annotation of the named entity data, issues in the GermEval gold standard (in the following "reference annotation") became apparent and, hence, were also documented in parallel to the morphological annotation. Even though an analysis of the reference annotations was originally not intended, it is presented as well because it effects the measures of tool performance.

The rest of the paper is structured as follows. Section 2 presents an overview of related work in German NER morphology and annotation analysis. The corpus data basis and the scope of the analysis are described in Sect. 3. The main part constitutes Sect. 4, where in Sect. 4.1 the morphological complexity of German named entities is investigated and in Sect. 4.2 the distribution of morphologically complex named entities in the dataset is presented. Section 5 then explains and examines six different annotation issues that have been identified within the GermEval reference annotation. This part also discusses the outcomes. The paper concludes with a short summary and a prospect of future work in Sect. 6.

---

[3] From a linguistic perspective *named entities* are encoded as *proper nouns*. In this paper both terms are treated synonymously.

2   Related Work

The performance of systems for NER is most often assessed through standard metrics like precision and recall, which measure the overall accuracy of matching predicted tags to gold standard tags. NER systems for German are no exception in this respect. In some cases the influence of difference linguistic features is reported, e.g., part of speech (Reimers et al. 2014) or morphological features (Capsamun et al. 2014; Schüller 2014). The closest to our work, and the only one, to the best of our knowledge, which addresses linguistic error analysis of NER in German is that of Helmers (2013). The study examined different systems for NER, namely, TreeTagger (Schmid 1995), SemiNER (Chrupała and Klakow 2010), and the Stanford NER (Finkel and Manning 2009) trained on German data (Faruqui and Padó 2010). Helmers (2013) applied these systems to the German Web corpus CatTle.de.12 (Schäfer and Bildhauer 2012) and inspected the influence of different properties on NER in a random sample of 100 true positives and 100 false negatives. It reports the odd-ratios for false classification for each of the properties. It was found that, e.g., named entities written exclusively in lower case were up to 12.7 times more likely to be misidentified, which alludes the difficulty of identifying adjectives derived from named entities. Another relevant example was named entities labelled as "ambiguous", i.e., which have a non-named entity homonym as in the case of named entities derived from a common noun phrase. In this case three out of four NER systems were likely to not distinguish named entities from their appellative homonyms with an odd-ratio of up to 13.7. Derivational suffixes harmed the identification in one classifier but inflectional suffixes seemed not to have similar influence. In addition, abbreviations, special characters and terms in foreign languages were features which contributed to false positive results. In comparison with this study, ours addresses explicitly the effect of the rich German morphology on NER tasks.

Derczynski et al. (2015) raise the challenges of identifying named entities in microblog posts. In their error analysis the authors found that the errors were due to several factors: capitalization, which is not observed in tweets; typographic errors, which increase the rate of OOV to 2–2.5 times more compared to newsire text; compressed form of language, which leads to using uncommon or fragmented grammatical structures and non-standard abbreviations; lack of context, which hinders word disambiguation. In addition, characteristics of microblogs genre such as short messages, noisy and multilingul content and heavy social context, turn NER into a difficult task.

Benikova et al. (2015) describe a NER system for German, which uses the NoSta-D NE dataset (Benikova et al. 2014a) for training as in the GermEval challenge. The system employs CRF for this task using various features with the result that word similarity, case information, and character n-gram had the highest impact on the model performance. Though the high morphological productivity of German was stressed in the dataset description as well as in the companion paper for the conference (Benikova et al. 2014a), this method did not address it. What is more, it excluded partial and nested named entities which were, however, used in the GermEval challenge.

As this overview shows, linguistic error analysis is of great importance for the development of language technologies. Error analysis performed for NER tasks has been mostly concentrated on the token level, since this is the focus of most NER methods. However, our analysis differs in that it investigates specifically the role that morphology plays in forming named entities given that German is a language with rich morphology and complex word-formation processes.

3 Data Basis and Approach

3.1 GermEval 2014 NER Challenge Corpus

In order to pursue the given research questions we decided to take the Nosta-D NE dataset (Benikova et al. 2014b) included in the GermEval 2014 NER Challenge as the underlying data source of our investigations. The GermEval challenges were initiated to encourage closing the performance gap for NER in German compared to similar NER annotations for English texts. GermEval introduced a novelty compared to previous challenges, namely, additional (sub-) categories have been introduced indicating if the named entity mentioned in a token is embedded in compounding. Altogether, the named entity tokens could be annotated for the four categories *person*, *location*, *organisation* and *other* together with the information if the token is a compound word containing the named entity (e.g., LOCpart) or a word that is derived from a named entity (e.g., PERderiv). In addition it highlights a second level of 'inner' named entities (e.g., the person "Berklee" embedded in the organisation "Berklee College of Music"). Though the latter was addressed earlier, e.g., in Finkel and Manning (2009), it has been generally almost neglected. For detailed information about the GermEval NER Challenge, its setup, and the implemented systems we refer to Benikova et al. (2014a). Out of the eleven systems submitted to the challenge, only one considered morphological analyses (Schüller 2014) systematically. The best system, however, albeit utilizing some hand-crafted rules to improve common schemes of morphological alterations, did not model morphological variation systematically.

Besides a considerable volume of manual ground truth (31300 annotated sentences), the challenge data favourably was based upon well-documented, predefined guidelines[4]. This allowed us to create our complimentary annotations and to (re-)evaluate a subset of the original challenge ground truth along the same principles as proposed by the guidelines. Table 1 shows example sentences annotated for named entities (which can also be multi-word named entities

---

[4] The guidelines describing the categorization choice and classification of named entity tokens can be consulted in the following document: https://www.linguistik. hu-berlin.de/de/institut/professuren/korpuslinguistik/forschung/nosta-d/nosta-d-ner-1.5 (revision 1.0 effective for GermEval is referenced in https://sites.google. com/site/germeval2014ner/data).

consisting of more than one token) and their expected named entity types according to the provided GermEval reference annotation.

Table 1. Example of reference data from the GermEval provided annotated corpus.

Sentence	NE type
1951 bis 1953 wurde der nördliche Teil als Jugendburg des **Kolpingwerkes** gebaut	OTH
Beschreibung Die **Kanadalilie** erreicht eine Wuchshöhe von 60 bis 180 cm und wird bis zu 25 cm breit	LOCpart
Um 1800 wurde im ehemaligen **Hartung'schen** Amtshaus eine Färberei eingerichtet	PERderiv
1911 wurde er Mitglied der **sozialistischen Partei**, aus der er aber ein Jahr später wieder austrat	ORG

3.2 GermEval 2014 System Predictions

In order to obtain insights on the distribution of morphological characteristics of ground truth named entities which were successfully recognized by the systems (true positives) compared to ground truth named entities which were not recognized or categorized correctly[5] (false negatives), we requested the system prediction outputs of GermEval participants from the challenge organizers[6].

Based on the best predictions[7] submitted for each system, we computed (1) the subset of ground truth named entities that all systems recognized (i.e., the true positive intersection, TPi; 1008 named entities) and (2) analogously the subset of ground truth named entities that none of the systems was able to recognize correctly (false negative intersection, FNi; 692 named entities). As performance of participating systems varied widely, we also analyzed (3) the false negatives of Hänig et al. (2014) (FN ExB; 1690 named entities).

3.3 Scope of the Analyses

The three mentioned data subsets were created to pursue two analysis goals: first, to investigate to what extent German named entities occur in morphologically altered forms and how complex these are and second, to report and evaluate issues we encountered in the GermEval reference annotations. The first investigation constitutes the main analysis and targets the question of whether there

---

[5] We adopted the criteria of the official Metric 1 of Benikova et al. (2014a).

[6] We kindly thank the organizers for their support by providing these and also thank the challenge participants that agreed to have them provided to us and shared with the research community as a whole.

[7] according to F_1-measure.

is a morphological gap in German NER. The second examination evolved out of annotation difficulties during the conduction of the first analysis. Even though not intended, we conducted the analysis of the reference annotation issues and present the results because the outcomes can contribute to the general research area of evaluating NER tools' performances.

The three data subsets build the foundation for both examination scopes. To obtain insights into the morphological prevalence and complexity of German named entities, the annotation was conducted according to the following steps: First, the annotator looked at those named entities in the datasets, which deviated from their lexical canonical form (in short LCF) which is the morphologically unmarked form. From gaining an overview of these named entities, linguistic features have been identified that correspond to the morphological segmentation steps which were applied to these morphologically altered named entities (see Sect. 4.1 for a detailed explanation). These linguistic features enable a measurement of the morphological complexity of a given named entity token provided by the reference annotation (i.e., the source named entity, in short SNE), e.g., "Kolpingwerkes" or "Kanadalilie" in Table 1. This measurement, however, required a direct linguistic comparison of the SNEs to their corresponding LCF form (i.e., their target named entity, in short TNE, e.g., "Kolpingwerk" and "Kanada"). Since the reference annotations provided only SNE tokens but no TNE data, a second annotation step was performed in which, all TNEs of the three subsets were manually added to the morphologically altered SNEs respectively[8]. In the third and last step the SNE has been annotated for its morphological complexity based on the numbers of different morphological alterations that were tracked back.

During the second and the third step of the morphological complexity annotation, problematic cases occurred in which a TNE could not be identified for the SNE given in the reference annotation. The reasons underlying these cases have been subsumed under six different annotation issues (details on these are explained in Sect. 5.1), which can significantly affect the performance measure of the tested GermEval NER systems. Therefore, if a SNE could not be annotated for morphological complexity, the causing issue was annotated for this SNE according to the six established annotation issues.

All three created GermEval data subsets have been annotated manually by a native German speaker and linguist and have been partially revised by a native German Computer Scientist while the code for the import and statistics was developed[9].

---

[8] The choice of a TNE included also the consideration of the four classification labels PER, LOC, ORG and OTH provided together with the SNE.

[9] The entire annotations of the morphological complexity of the named entities as well as the identified reference annotation error types can be consulted in this table including all three data subsets: https://raw.githubusercontent.com/AKSW/germeval-morph-analysis/master/data/annotation_imports/compl-issues-ann-ranks.tsv.

4 Morphological Complexity of German NE Tokens

4.1 Measuring Morphological Complexity

Morphological variation of named entity tokens has been considered as part of the GermEval annotation guidelines. I.e., next to the four named entity types, a marking for SNEs being compound words or derivates of a TNE has been introduced (e.g., LOCderived or ORGpart). While this extension of the annotation of named entity tokens implies that German morphology impacts NER tasks, it does not indicate which morphological peculiarities actually occur. The linguistic analysis investigating morphologically altered SNEs revealed that SNEs exhibit a varying degree of morphological complexity. This degree is conditioned by the morphological inflection and/or word-formation steps that have been applied to a SNE in order to retrace the estimated TNE in its LCF. The resulting formalization of these alternation steps is as follows:

$$L \in \{\mathcal{C}_k \mathcal{D}_l \mid k, l \in \mathbb{N}\} \times \mathcal{P}(\{c, m, f\}) \text{ where}$$

$\mathcal{C}_k$ denotes that k compounding transformations were applied
$\mathcal{D}_l$ denotes that l derivations were applied
c denotes that resolving the derivation applied to the SNE resulted in a word-class change between SNE and TNE
m denotes that the morphological transformation process applied encompasses an inner modification of the TNE stem compared to its LCF
f denotes that the SNE is inflected.

For convenience, we will omit the tuple notation and simplify the set representation of c and f: $\mathcal{C}_1 \mathcal{D}_2 f, \mathcal{C}_1 \mathcal{D}_1 cmf, \mathcal{C}_3 \mathcal{D}_0 \in L$. In order to obtain the differing levels[10] of morphological complexity for named entities, we went through the identified morphological transformation steps always comparing the given SNE in the test set with the estimated TNE in its LCF. It is defined that all named entities annotated with a complexity other than $\mathcal{C}_0 \mathcal{D}_0$ are morphologically relevant and all named entities with a complexity satisfying $\mathcal{C} + \mathcal{D} \geq 1$ (i.e., involving at least one compounding relation or derivation) are morphologically complex, i.e., these require more than one segmentation step in the reanalysis of the SNE to the TNE in its LCF.

Thus, the SNE token can be increasingly complex, if it contains the TNE within a compound part of a compound or if the TNE is embedded within two derivations within the SNE. An example illustrating the morphological segmentation of the SNE "Skialpinisten" is given in Fig. 1. It shows each segmentation step from the SNE back to the TNE in its LCF in detail and illustrates how deeply German named entities can be entailed in common nouns due to morphological transformations. Overall, the annotation of the three subsets revealed

---

[10] Although, we use the term level to simplify formulations, no strict ordering between the different possible configurations for the aforementioned formalization of complexity is presupposed.

27 levels of morphological complexity for German named entities. The appendix holds a comprehensive listing in Table 4 of these levels together with examples taken from the corpus[11].

SNE = Skialpinisten

(1) remove inflectional suffix
f: Skialpinist + en

(2) split up compounding parts
C: Ski + Alpinist

(3) resolve derivation #1
D_1c: Alpinist$_N$ ⟶ alpin$_{ADJ}$

(4) resolve derivation #2
D_2cm: alpin$_{ADJ}$ ⟶ Alpen$_N$

TNE = Alpen (LOCderived)
overall complexity: C_1D_2cmf

Fig. 1. Example segmentation for annotating the SNE "Skialpinist" with the estimated TNE "Alpen".

4.2 Distribution of Morphologically Complex NE Tokens

Based on our systematization of complexity, we defined more focused complexity criteria such as $C > 0$ and 'has m' (i.e., inner modification occurred) to complement the criteria morphologically relevant and morphologically complex introduced in Sect. 4.1. Figure 2 shows comparative statistics of the prevalence of named entities matching these criteria for the TPi, FNi and FN ExB[12]. In general, morphologically relevant and morphologically complex named entities are much more prevalent among the false negatives. With respect to the more focused criteria, the strongest increases occur for $C > 0$, $D > 0$ and 'is inflected'. In line with the definition of the criterion c, we observe $P(D > 0 \mid c) = 1$. I.e., the occurrence of c in a complexity assignment strictly implies that at least one derivation was applied. The observation of a strong association between inner modification and derivation processes ($P(D > 0 \mid m) = 0.86$) also is in line with intuitive expectations for German morphology.

Figure 3 presents the same comparative statistics between TPi and FNi for the named entities grouped according to their reference classification. In general morphological alteration is more common in named entities annotated with the types PER and LOC. Further, we find lower variance of increase of $C > 0$ across the classes compared to $D > 0$, which is much more common in LOC named

---

[11] Note, that more levels can be assumed but no occurrences were found in the annotated subsets.

[12] The Scala and Python source code used to prepare the annotations, gather statistics and generate the plots is available at: https://github.com/AKSW/germeval-morph-analysis.

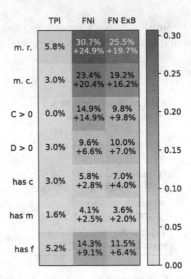

Fig. 2. Prevalence of morphological complexities satisfying specified criteria. Colors encode magnitude of increase of the FN subset compared to the TPi. (m.r. = morph. relevant, m.c. = morph. complex). (Color figure online)

entities (+20.9%) and PER named entities (+12.8%) than in named entities classified ORG and OTH (increase ≤2%). The statistics partitioned by named entity type also reveal that the only types morphologically complex named entities in the TPi subset are LOC named entities with derivations. Analogous statistics between TPi and FN ExB showed similar trends and were omitted for brevity[13].

4.3   Morphological Complexity in Context of NER System Errors

Interestingly, the LOC and PER named entities, that were found to be morphologically complex most often on the one hand are, conversely, the ones covered best by the top GermEval systems according to Benikova et al. (2014a). However, these classes were also deemed more coherent in their analysis, a qualitative impression we share with respect to variety of occurring patterns for morphological alterations. Also, since the morphological complexity of named entities is also one of many factors determining its difficulty to be spotted and typed correctly (besides, e.g., inherent ambiguity of involved lexial semantics), this might indicate that these two categories might still simply be the ones potentially benefiting most from more elaborate modelling of effects of morphological alteration, as the reported F1 of approx. 84% for LOC and PER still indicates space for improvements.

  Further, 19 morphologically complex named entities in FNi could be found, whose TNE was identical with a TNE from the TPi. For example, all systems

---

[13] The corresponding plot is available at: https://github.com/AKSW/germeval-morph-analysis/blob/master/plots/phrase-partitioned-stats-FalseNegExB.pdf.

	PER	LOC	ORG	OTH
m. r.	FN:35.8% TP:1.6% +34.2%	FN:45.2% TP:8.5% +36.7%	FN:28.1% TP:5.2% +22.9%	FN:19.9% TP:0.0% +19.9%
m. c.	FN:29.4% TP:0.0% +29.4%	FN:41.4% TP:5.2% +36.2%	FN:15.8% TP:0.0% +15.8%	FN:12.8% TP:0.0% +12.8%
C > 0	FN:18.3% TP:0.0% +18.3%	FN:18.5% TP:0.0% +18.5%	FN:14.4% TP:0.0% +14.4%	FN:11.1% TP:0.0% +11.1%
D > 0	FN:12.8% TP:0.0% +12.8%	FN:26.1% TP:5.2% +20.9%	FN:1.4% TP:0.0% +1.4%	FN:1.8% TP:0.0% +1.8%
has c	FN:7.3% TP:0.0% +7.3%	FN:15.9% TP:5.2% +10.7%	FN:0.7% TP:0.0% +0.7%	FN:1.3% TP:0.0% +1.3%
has m	FN:5.5% TP:0.0% +5.5%	FN:9.6% TP:2.7% +6.8%	FN:2.1% TP:0.0% +2.1%	FN:0.9% TP:0.0% +0.9%
has f	FN:14.7% TP:1.6% +13.1%	FN:17.2% TP:7.4% +9.8%	FN:16.4% TP:5.2% +11.3%	FN:10.6% TP:0.0% +10.6%

0.05 0.10 0.15 0.20 0.25 0.30 0.35

Fig. 3. Prevalence of morphological complexities satisfying specified criteria, grouped be named entity type. Each cell presents ratios in the FNi, the TPi and respective increase. Colors encode magnitude of increase. (m.r. = morph. relevant, m.c. = morph. complex). (Color figure online)

were able to correctly assign LOC-deriv to 'polnischen' (TNE = 'Polen'), however no system was able to recognize 'austropolnischen' (same TNE). Analogously, there is 'Schweizer' in TPi, but 'gesamtschweizerischen' in FNi (common TNE: 'Schweiz'). There were 38 additional morphologically complex named entities in FN ExB with a corresponding TPi named entity sharing the TNE, e.g., 'Japans' (TP) vs. 'Japan-Aufenthaltes' (FN). For all of these pairs, it appears plausible to assume that the difficulty for the corresponding false negative can be attributed to a large extend to the morphological complexity, as simpler variants posed no hindrances to any of the tested systems[14]. For the ExB system, these kind of false negatives constitute 3.4% of all false negatives, which could be viewed raw estimation of potential increase in recall if hypothetically morphological complexity of named entities would be mitigated entirely. It should also be noted that the reported occurrence counts of these pairs for ExB are lower bounds, since not all of its true positives had been annotated at the time of writing.

---

[14] Still we also acknowledge that several factors of lexical semantics, syntax etc. influence how challenging it is to spot a specific NE occurrence in context and more systematic analysis of these factors would be needed to attribute the error to morphological causes with certainty.

5 Reference Annotation Related Issues

5.1 Reference Annotation Issue Types

During the annotation for morphological complexity issues arose with regard to the GermEval reference annotations which led to various difficulties.

Table 2. Encountered issues pertaining to GermEval reference annotations.

Issue	Example	Prevalence
NOT DERIVED	SNE = *Kirgisische* (LOC-deriv) with TNE = *Kirgistan*	94 (31.5%)
WRONG NE TYPE	SNE = *barocker* (ORG-deriv) with TNE = *Barock*, "Baroque" is an epoch, it should have been annotated as OTH-deriv	62 (20.8%)
WRONG SPELLING	SNE = *Freiburg/31:52* with TNE = *Freiburg*	51 (17.1%)
No NE	SNE = *Junta* - "Junta" is a common noun, there is no TNE	18 (6.0%)
INVALID REFERENCE	SNE = *Was ist theoretische Biologie?* - this is a HTML link label, which is not related to any NE	7 (2.4%)
TNE UNCLEAR	SNE = *Köln/Weimar/Wien* - TNE is unclear, unknown to which of the three named entities is referred to	66 (22.2%)

Overall, six reference annotation issues have been identified and all three subsets have been annotated for these issues (also cf. Table 2):

Issue #1 NOT DERIVED: A significant number of SNEs with the type LOCderived is morphologically not derived from the location TNE but from the inhabitant noun, e.g., "Kirgisisch" is not derived from "Kirgistan" but from "Kirgise".

Issue #2 WRONG NE TYPE: This issue refers to SNEs which are correctly identified, but are assigned to the wrong named entity category.

Issue #3 WRONG SPELLING: SNEs annotated with this issue are either incorrectly spelled or tokenized.

Issue #4 No NE: This issue holds for SNEs, which turn out to be only common nouns in the sentences they occur.

Issue #5 INVALID REFERENCE: SNEs referring to book/film titles, online references or citations which are incomplete, wrong or the online reference is a title for a website given by some person but not the real title or URL.

Issue #6 TNE UNCLEAR: This issue summarizes reasons for preventing a TNE of being identifiable form a given SNE, i.e., it is not possible to morphologically decompose the SNE to retrieve the TNE or there are more than one TNEs included in the SNE.

If NOT DERIVED, NO NE, INVALID REFERENCEor TNE UNCLEARoccur for a named entity, assignment of a morphological complexity level becomes impossible. Consequently, the corresponding named entities (189) were excluded from the complexity statistics presented in Sects. 4.2 and 4.3. WRONG NE TYPEand WRONG SPELLING, on the other hand, albeit also implying difficulties for NER systems, do not interfere with identifying the TNE (and thus the complexity level). Hence, such named entities were not excluded.

5.2   Distribution and Effects of Annotation Issues

Table 2 provides, in addition to examples for the aforementioned categories of annotation issues, their total prevalence across TPi and FN ExB (subsuming FNi). Table 3 additionally indicates the distribution of issue occurrences in comparison between the subsets. Overall, occurrence of annotation issues are about three times more likely in the false negative sets compared to TPi, a trend in a similar direction as for the occurrence of morpholoically complex named entities.

Table 3. Frequencies of occurrence of annotation issues by category and subset. Percentages in parentheses are relative frequencies for the corresponding subset.

Issue	TPi	FNi	FN ExB
#1	41 (4.07%)	18 (2.60%)	53 (3.14%)
#2	0 (0.00%)	30 (4.34%)	62 (3.67%)
#3	1 (0.10%)	24 (3.47%)	50 (2.96%)
#4	1 (0.10%)	10 (1.45%)	17 (1.01%)
#5	0 (0.00%)	4 (0.58%)	7 (0.41%)
#6	0 (0.00%)	19 (2.75%)	66 (3.91%)
All	43 (4.27%)	105 (15.17%)	255 (15.09%)

It appears questionable to count named entities with WRONG NE TYPE, NO NE and INVALID REFERENCE that have not been recognized by any NER system as a false negative, as these named entities do not actually constitute named entities as defined by the guidelines (analogously for true positives). Thus, we projected the M1 performance measures on the test split for the ExB system disregarding these named entities[15]. The adjustment results in discounting five

---

[15] Due to lack of complete screening of all true positives of ExB for annotation issues we linearly interpolated the exemption of one true positive according to TPi to the exemption of five true positives for all true positives of that system.

false positives and 44 false negatives, result in an increase in recall by 0.48% and F1 by 0.34%. Although, this change is not big in absolute magnitude, it can still be viewed relevant considering that the margin between the to best systems at GermanEval was merely 1.28% for F1 as well Benikova et al. (2014a).

6   Conclusion

This study presented an analysis of German NER as reflected by the performance of systems that participated in the GermEval 2014 shared task. We focused on the role of morphological complexity of named entities and introduced a method to measure it. We compared the morphological characteristics of named entities which were identified by none of the systems (FNi) to those identified by all of the systems (TPi) and found out that FNi named entities were considerably more likely to be complex than the TPi ones (23.4% and 3.0% respectively). The same pattern was detected also for the system which achieved the best evaluation in this shared task. These findings emphasize that morphological complexity of German named entities correlates with the identification of named entities in German text. This indicated that the task of German NER could benefit from integrating morphological processing.

We further discovered annotation issues of named entities in the GermEval reference annotation for which we provided additional annotation. We believe that the presented outcomes of this annotation can help to improve the creation of NER tasks in general.

As a future work, we would like to extend our annotation to analyze how these issues affect the evaluation of the three best performing systems more thoroughly. In addition, a formalization to measure the variety of occurring patterns of morphological alteration (used affixes/affix combinations, systematic recurrences of roots...) as a complementary measure for morphological challenges seems desirable. We will further have multiple annotators to morphologically annotate the named entities of the GermEval reference, in order to estimate the confidence of our observation by measuring inter-annotator agreement.

Acknowledgment. These research activities were funded by grants from the H2020 EU projects ALIGNED (GA-644055) and FREME (GA-644771) and the Smart Data Web BMWi project (GA-01MD15010B).

Appendix

Table 4. Distribution of the morphological complexities in the annotated subsets

Compl.	TPi	FNi	FN ExB	Example SNE	Example TNE
$\mathcal{C}_0\mathcal{D}_0$	910 (94.20%)	442 (69.28%)	1149 (74.47%)	Mozart	Mozart
$\mathcal{C}_0\mathcal{D}_0f$	27 (2.80%)	47 (7.37%)	98 (6.35%)	Mozarts	Mozart
$\mathcal{C}_1\mathcal{D}_0$	0 (0.00%)	62 (9.72%)	101 (6.55%)	Mozart-Konzert	Mozart
$\mathcal{C}_1\mathcal{D}_0f$	0 (0.00%)	15 (2.35%)	24 (1.56%)	Mozart-Konzerten	Mozart
$\mathcal{C}_1\mathcal{D}_0m$	0 (0.00%)	3 (0.47%)	5 (0.32%)	Pieterskirche	Pieter
$\mathcal{C}_1\mathcal{D}_0mf$	0 (0.00%)	3 (0.47%)	4 (0.26%)	Reichstagsabgeordneten	Reichstag
$\mathcal{C}_0\mathcal{D}_1$	0 (0.00%)	9 (1.41%)	20 (1.30%)	Donaldismus	Donald
$\mathcal{C}_0\mathcal{D}_1f$	0 (0.00%)	1 (0.16%)	4 (0.26%)	Donaldismusses	Donald
$\mathcal{C}_0\mathcal{D}_1m$	0 (0.00%)	7 (1.10%)	10 (0.65%)	Nestorianismus	Nestorius
$\mathcal{C}_0\mathcal{D}_1mf$	0 (0.00%)	1 (0.16%)	2 (0.13%)	Spartiaten	Sparta
$\mathcal{C}_0\mathcal{D}_1c$	5 (0.52%)	16 (2.51%)	61 (3.95%)	Japanisch	Japan
$\mathcal{C}_0\mathcal{D}_1cf$	9 (0.93%)	8 (1.25%)	14 (0.91%)	Japanischen	Japan
$\mathcal{C}_0\mathcal{D}_1cm$	1 (0.10%)	1 (0.16%)	6 (0.39%)	Europäisch	Europa
$\mathcal{C}_0\mathcal{D}_1cmf$	10 (1.04%)	8 (1.25%)	19 (1.23%)	Europäischen	Europa
$\mathcal{C}_2\mathcal{D}_0$	0 (0.00%)	3 (0.47%)	5 (0.32%)	Bibelforscherfrage	Bibel
$\mathcal{C}_2\mathcal{D}_0mf$	0 (0.00%)	1 (0.16%)	1 (0.06%)	Erderkundungssatelliten	Erde
$\mathcal{C}_1\mathcal{D}_1$	0 (0.00%)	1 (0.16%)	2 (0.13%)	Benediktinerstift	Benedikt
$\mathcal{C}_1\mathcal{D}_1f$	0 (0.00%)	2 (0.31%)	2 (0.13%)	Transatlantikflüge	Atlantik
$\mathcal{C}_1\mathcal{D}_1m$	0 (0.00%)	1 (0.16%)	2 (0.13%)	Römerstrasse	Rom
$\mathcal{C}_0\mathcal{D}_2$	0 (0.00%)	1 (0.16%)	2 (0.13%)	Geismarerin	Geismar
$\mathcal{C}_0\mathcal{D}_2f$	0 (0.00%)	1 (0.16%)	2 (0.13%)	Hüttenbergerinnen	Hüttenberg
$\mathcal{C}_0\mathcal{D}_2m$	0 (0.00%)	0 (0.00%)	1 (0.06%)	Rheinländerin	Rheinland
$\mathcal{C}_0\mathcal{D}_2cf$	0 (0.00%)	1 (0.16%)	1 (0.06%)	Austropolnischen	Polen
$\mathcal{C}_0\mathcal{D}_2cmf$	4 (0.41%)	0 (0.00%)	3 (0.19%)	Transatlantischen	Atlantik
$\mathcal{C}_3\mathcal{D}_0$	0 (0.00%)	1 (0.16%)	1 (0.06%)	25-US-Dollar-Marke	US
$\mathcal{C}_1\mathcal{D}_2cf$	0 (0.00%)	2 (0.31%)	2 (0.13%)	Gesamtschweizerischen	Schweiz
$\mathcal{C}_1\mathcal{D}_2cmf$	0 (0.00%)	1 (0.16%)	2 (0.13%)	Skialpinisten	Alpen
Total	966	638	1543		

References

Benikova, D., Biemann, C., Kisselew, M., Padó, S.: GermEval 2014 named entity recognition shared task: companion paper. In: Workshop Proceedings of the 12th Edition of the KONVENS Conference, Hildesheim, Germany, pp. 104–112 (2014a)

Benikova, D., Biemann, C., Reznicek, M.: NoSta-D named entity annotation for German: guidelines and dataset. In: Calzolari, N., Choukri, K., Declerck, T., Loftsson, H., Maegaard, B., Mariani, J., Moreno, A., Odijk, J., Piperidis, S. (eds.) Proceedings of the Ninth International Conference on Language Resources and Evaluation (LREC 2014), Reykjavik, Iceland, pp. 2524–2531. European Language Resources Association (ELRA) (2014b)

Benikova, D., Muhie, S., Prabhakaran, Y., Biemann, S.C.: GermaNER: free open German named entity recognition tool. In: Proceedings of the International Conference of the German Society for Computational Linguistics and Language Technology, Duisburg-Essen, Germany, pp. 31–38. German Society for Computational Linguistics and Language Technology (2015)

Capsamun, R., Palchik, D., Gontar, I., Sedinkina, M., Zhekova, D.: DRIM: named entity recognition for German using support vector machines. In: Proceedings of the KONVENS GermEval Shared Task on Named Entity Recognition, Hildesheim, Germany, pp. 129–133 (2014)

Chrupała, G., Klakow, D.: A named entity labeler for German: exploiting Wikipedia and distributional clusters. In: Calzolari, N., Choukri, K., Maegaard, B., Mariani, J., Odijk, J., Piperidis, S., Rosner, M., Tapias, D. (eds.) Proceedings of the Seventh International Conference on Language Resources and Evaluation (LREC 2010), Valletta, Malta. European Language Resources Association (ELRA) (2010). ISBN 2-9517408-6-7

Derczynski, L., Maynard, D., Rizzo, G., van Erp, M., Gorrell, G., Troncy, R., Petrak, J., Bontcheva, K.: Analysis of named entity recognition and linking for tweets. Inf. Process. Manag. **51**(2), 32–49 (2015)

Faruqui, M., Padó, S.: Training and evaluating a German named entity recognizer with semantic generalization. In: Proceedings of KONVENS 2010, Saarbrücken, Germany (2010)

Finkel, J.R., Manning, C.D.: Nested named entity recognition. In: Proceedings of the 2009 Conference on Empirical Methods in Natural Language Processing, pp. 141–150. Association for Computational Linguistics (2009)

Hänig, C., Bordag, S., Thomas, S.: Modular classifier ensemble architecture for named entity recognition on low resource systems. In: Workshop Proceedings of the 12th Edition of the KONVENS Conference, Hildesheim, Germany, pp. 113–116 (2014)

Helmers, L.A.: Eigennamenerkennung in Web-Korpora des Deutschen. Eine Herausforderung für die (Computer)linguistik. Bachelor thesis, Humboldt-Universität zu Berlin (2013)

Reimers, N., Eckle-Kohler, J., Schnober, C., Gurevych, I.: GermEval-2014: nested named entity recognition with neural networks. In: Proceedings of the KONVENS GermEval Shared Task on Named Entity Recognition, Hildesheim, Germany, pp. 117–120 (2014)

Schäfer, R., Bildhauer, F.: Building large corpora from the web using a new efficient tool chain. In: Calzolari, N., Choukri, K., Declerck, T., Doğan, M.U., Maegaard, B., Mariani, J., Moreno, A., Odijk, J., Piperidis, S. (eds.) Proceedings of the Eight International Conference on Language Resources and Evaluation (LREC 2012), Istanbul, Turkey, pp. 486–493. European Language Resources Association (ELRA) (2012)

Schmid, H.: Improvements in part-of-speech tagging with an application to German. In: Proceedings of the ACL SIGDAT-Workshop, pp. 47–50 (1995)

Schüller, P.: MoSTNER: morphology-aware split-tag German NER with Factorie. In Workshop Proceedings of the 12th Edition of the KONVENS Conference, Hildesheim, Germany, pp. 121–124 (2014)

Detecting Named Entities and Relations
in German Clinical Reports

Roland Roller[1]([✉]), Nils Rethmeier[1], Philippe Thomas[1], Marc Hübner[1],
Hans Uszkoreit[1], Oliver Staeck[2], Klemens Budde[2], Fabian Halleck[2],
and Danilo Schmidt[2]

[1] Language Technology Lab, DFKI, Berlin, Germany
{roland.roller,nils.rethmeier,philippe.thomas,marc.huebner,
hans.uszkoreit}@dfki.de
[2] Charité Universitätsmedizin, Berlin, Germany
{oliver.staeck,klemens.budde,fabian.halleck,danilo.schmidt}@charite.de

Abstract. Clinical notes and discharge summaries are commonly used
in the clinical routine and contain patient related information such as
well-being, findings and treatments. Information is often described in
text form and presented in a semi-structured way. This makes it difficult
to access the highly valuable information for patient support or clinical
studies. Information extraction can help clinicians to access this information. However, most methods in the clinical domain focus on English
data. This work aims at information extraction from German nephrology reports. We present on-going work in the context of detecting named
entities and relations. Underlying to this work is a currently generated
corpus annotation which includes a large set of different medical concepts, attributes and relations. At the current stage we apply a number
of classification techniques to the existing dataset and achieve promising
results for most of the frequent concepts and relations.

1 Introduction

Within the clinical routine many patient related information are recorded
in unstructured or semi-structured text documents and are stored in large
databases. These documents contain valuable information for clinicians which
can be used to, e.g., improve/support the treatment of long-term patients or
clinical studies. Even today information access is often manual, which is cumbersome and time-consuming. This creates a demand for efficient and easy tools
to access relevant information. Information extraction (IE) can support this process by detecting particular medical concepts and the relations between them to
gather the context. Such structured information can be used to improve use-cases
such as the generation of cohort groups or clinical decision support.

Generally, IE can be addressed in many different ways. If sufficient amounts
of training instances are available, supervised learning is often the technique of
choice, as it directly models expert knowledge. In context of detecting medical

© The Author(s) 2018
G. Rehm and T. Declerck (Eds.): GSCL 2017, LNAI 10713, pp. 146–154, 2018.
https://doi.org/10.1007/978-3-319-73706-5_12

concepts (named entity recognition; NER) and their relations (relation extraction; RE) conditional random fields (CRF) (Lafferty et al. 2001) and support vector machines (SVM) (Joachims 1999) have been very popular supervised methods that were frequently used for the last decade. In recent years neural network based supervised learning has gained popularity (see, e.g., Nguyen and Grishman (2015); Sahu et al. (2016); Zeng et al. (2014)).

In context of IE from German clinical data not much work has been done so far. One reason for that is the unavailability of clinical data resources in German language, as discussed in Starlinger et al. (2016). Only a few publications address the topic of NER and RE from German clinical data. Hahn et al. (2002) focus on the extraction of medical information from German pathology reports in order to acquire medical domain knowledge semi-automatically, while Bretschneider et al. (2013) presents a method to detect sentences which express pathological and non-pathological findings in German radiology reports. Krieger et al. (2014) present first attempts to analyzing German patient records. The authors focus on parsing and RE, namely: symptom-body-part and disease-body-part relations. Toepfer et al. (2015) present an ontology-driven information extraction approach which was validated and automatically refined by a domain expert. Their system aims to find objects, attributes and values from German transthoracic echocardiography reports.

Instead, we focus on detecting medical concepts (also referred to as NE) and their relations from German nephrology reports. For both tasks, NER and RE, two different learning methods are tested: first a well established method (CRF, SVM) and later a neural method for comparison. However, the paper describes on-going work, both in terms of corpus annotations and classification methods. The goal of this paper is to present first results for our use case and target domain.

2 Data and Methods

The following section overviews our corpus annotations and the models we use. Note that, due to the (short) format of the paper, method descriptions are brief. We refer the reader to the corresponding publications for details.

2.1 Annotated Data

Our annotation schema includes a wide range of different concepts and (binary) relations. The most frequent concepts used in the experiments are listed in Tables 1 and 2, including a brief explanation. The ongoing annotations (corpus generation) include German discharge summaries and clinical notes from a kidney transplant department. An example of our annotations is presented in Fig. 1. Both types of documents are generally written by medical doctors, but have apparent differences. For more details on corpus generation please see Roller et al. (2016).

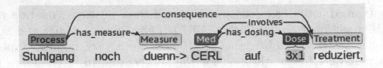

Fig. 1. Annotated sentence

For the following experiments 626 clinical notes are used for training and evaluation. Clinical notes are rather short and written during or shortly after a patient visit. Currently, only 267 of those documents contain annotated relations. The overall frequency of named entities and relations is included with the results in the experimental section (see Tables 3 and 4).

Table 1. Annotated concepts

Label	Explanation
Body_Part	Body parts; organs
Med_Con	Medical_Condition: symptom, diagnosis and observation
Process	Body's own biological processes
Health_State	Positive, wanted finding; contrary to *Med_Con*
Treatment	Therapeutic procedures, treatments
Medication	Drugs, medicine
Med_Spec	Medical_Specification: closer definition; describing lexemes, often adjectives
Local_Spec	Local_Specification: anatomical descriptions of position and direction

2.2 Machine Learning Methods

NER – Conditional Random Field (CRF). Conditional random fields have been used for many biomedical and clinical named entity recognition tasks, such as gene name recognition (Leaman and Gonzalez 2008), chemical compound recognition (Rocktäschel et al. 2012), or disorder names (Li et al. 2008). One disadvantage of CRFs is that the right selection of features can be crucial to achieving optimal results. However, for a different domain or language important features might change. In this work we are not interested in exhaustive feature engineering. Instead, we intend to re-use an existing feature setup as described by Jiang et al. (2015) who use word-level and part-of-speech information around the target concept. For our experiment we use the CRF++[1] implementation.

NER – Character-level Neural Network (CharNER NN). In addition to the well established CRF for NER we also use a neural CRF implementation[2] as

---

[1] https://taku910.github.io/crfpp/.

[2] https://github.com/ozanarkancan/char-ner.

Table 2. Annotated relations

Label	Explanation
hasState	Describes the state of health (positive and negative) of different entities (e. g., *Process*, *Med_Con*, *Body_Part*)
involves	Describes a relation between *Treatment* and *Medication*: e. g., to use or to discontinue a medication
has_measure	Links a *Measurement* to a corresponding concept
is_located	Links a positional information (*Body_part*, *Local_spec*) to concepts such as *Med_con* or *Process*
is_specified	Links *Medical_Spec* to a corresponding concept (e. g., *Med_Con*, *Process*

introduced by Kuru et al. (2016). The model uses a character-level Bidirectional-LSTM with a CRF objective. Using character level inputs has the advantage of reducing the unknown vocabulary word problem, as the vocabulary size and hence the feature sparsity are reduced compared to words allowing character models to compensate for words unseen during training, which helps on smaller datasets.

RE – Support Vector Machine (SVM). SVMs are often the method of choice in context of supervised relation extraction (Tikk et al. 2010). Besides their advantages, SVMs also suffer from the issue of optimal feature/kernel selection. Other problems are related to the bias of positive and negative instances in training and test data which can significantly influence the classification results (Weiss and Provost 2001). Again, feature selection is not in our interest. For this reason we use the Java Simple Relation Extraction[3] (jSRE) (Giuliano et al. 2006) which uses a shallow linguistic kernel and bases on LibSVM (Chang and Lin 2011). jSRE provides reliable classification results and has been shown to achieve state-of-the-art results for various tasks, such as protein-protein extraction (Tikk et al. 2010), drug-drug extraction (Thomas et al. 2013) and extraction of neuroanatomical connectivity statements (French et al. 2012).

RE – Convolutional Neural Network (CNN). Besides SVM, we also use a convolutional neural network for relation extraction. We employ a Keras[4] implementation of the model described by Nguyen and Grishman (2015) using a TensorFlow[5] backend and a modified activation layer. The architecture consists of four main layers: (a) lookup tables to encode words and argument positions into vectors, (b) a convolutional layer to create n-gram features, (c) a max-pooling layer to select the most relevant features and (d) a sigmoid activation layer for binary classification.

---

[3] https://hlt.fbk.eu/technologies/jsre.
[4] https://keras.io.
[5] https://www.tensorflow.org.

3   Experiment

In this section named entity recognition and relation extraction on German nephrology reports are carried out. Given a sentence (token sequence), the task of NER is to assign the correct named entity label to the given tokens in the test data. Relation extraction takes a sentence including the different named entity labels as input and determines for each entity pair whether one of our target relations exists. Both classification tasks are evaluated based on precision, recall and F1-Score. Note, due to space reasons, not all relations of the example in Fig. 1 are used for the experiment.

Table 3. Concept classification results

Label	Freq.	CRF			CharNER NN		
		Prec	Rec	F1	Prec	Rec	F1
Medical_Condition	2453	95.17	75.16	83.98	89.12	82.15	**84.93**
Treatment	1680	85.79	69.63	76.81	80.46	76.37	**78.33**
State_of_Health	1451	86.68	76.35	81.18	83.55	80.80	**82.14**
Medication	1214	92.28	68.56	78.55	90.37	82.39	**86.17**
Process	1145	90.53	60.56	72.57	84.74	66.29	**74.02**
Body_part	840	96.96	65.23	**77.90**	89.15	68.78	77.53
Medical_Specification	764	78.76	48.82	**60.20**	65.32	53.04	58.21
Local_Specification	189	95.83	31.94	45.87	81.84	49.77	**61.05**

Table 4. Relation classification results

Label	Freq.	SVM			CNN		
		Prec	Rec	F1	Prec	Rec	F1
hasState	388	86.86	86.86	**86.86**	81.96	88.10	84.64
involves	370	88.96	78.38	83.33	81.51	90.42	**85.58**
has_measure	427	80.25	88.52	**84.19**	81.47	76.61	78.97
is_located	162	46.96	85.80	60.70	65.14	64.12	**63.48**
is_specified	112	94.85	82.14	**88.04**	76.34	83.83	79.89

3.1   Preprocessing

To carry out the experiment text documents are processed by a sentence splitter, a tokenizer, stemmer and Part-of-Speech (POS) tagger. The sentence splitting and tokenization are essential to split documents into single sentences and single word tokens. We use JPOS (Hellrich et al. 2015), to tag Part-of-Speech information, since the tool is specialized for German clinical data. POS tags are used for both the CRF and SVM. Additionally, we stem words for jSRE using the German Snowball stemmer in NLTK. CharNER and CNN do not require additional linguistic features as input.

For both NER and RE the experiments are carried out multiple times – for each single named entity type and each single relation for two reasons: Firstly, in context of named entities tokens might be assigned to multiple labels which our classifiers can not handle directly. Secondly, jSRE does not handle multi-class classification. Hence, we use a One-vs. rest (OvR) classification to train separate models for each NER/RE type.

3.2   Named Entity Recognition

Setup. NER type evaluation uses the OvR setup to train a single classifier (CRF or CharNER) per class. The experiment run as a reduced 10-fold cross-validation on 3 out of 10 stratified dataset splits, since the CharNER model took a very long time to compute, despite using a GPU. Specifically, each split has a 80% training, a 10% validation and a 10% test part. To further save time, we determined the CharNERs optimal parameters on only one splits' validation part for only one out of eight entity types (body part). Afterwards, we applied the found parameters to the other entity types and splits to produce average test part scores. Thus, the parameter settings may not be optimal for all entity types. In practice, the CRF trained in hours compared to days for the Bi-LSTM. Both models were evaluated using the 3-fold setup for comparison.

Results. The results for named entity recognition are shown in Table 3. Even though classifiers are not necessarily optimal (e.g., no feature engineering), the results are promising. All concepts with a frequency above 800 have an F1-Score above 70. Moreover, all concepts can be detected at a high level of precision. Both classifiers produce comparable results, with better F1 for the CharNER and a focus on precision for the CRF.

3.3   Relation Extraction

Setup. Our relation extraction task considers only entity pairs within the same sentence. While positive relation instances can be directly taken from the annotations, negative relation instances (used for training and testing) are generated by creating new (unseen) combinations between entities. The relation extraction experiment is then carried out within a 5-fold cross-validation using NE gold labels as input.

Due to the comparably smaller size of the dataset, hyperparameters of the CNN have been slightly modified in comparison to (Nguyen and Grishman 2015). As before, we used one relation type from one fold to find the optimal parameters and then applied those parameters to the other folds and types. This resulted in a reduced position embeddings dimensionality (from 50 to 5) compared to the original model. We also used pre-trained German word embeddings[6].

---

[6] https://devmount.github.io/GermanWordEmbeddings/.

Results. The relation extraction results are presented in Table 4. Most relations can be detected at an F1-Score of 80. Only the relation *is_located* produces a surprisingly low precision which results in a reduced F1. Overall, the results are very promising and leave space for further improvements using improved classification models.

4   Conclusion and Future Work

This work presented first results in context of detecting various named entities and their relations from German nephrology reports. For each task two different methods have been tested. Even though preliminary classification methods have been used (no feature engineering, sub-optimal tuning) and the relatively small size of training and evaluation data, the results are already very encouraging. Generally, the results indicate, that the classification of such information is not too complex. However, a more detailed analysis is necessary to support this assumption.

Future work will focus on increasing the corpus size, and extending/improving our classification models (e.g., elaborate hyperparameter search and selection of pre-trained embeddings). Then those models will be used for further use-cases such as general information access of clinical documents and cohort group generation.

Acknowledgements. This research was supported by the German Federal Ministry of Economics and Energy (BMWi) through the project MACSS (01MD16011F).

References

Bretschneider, C., Zillner, S., Hammon, M.: Identifying pathological findings in german radiology reports using a syntacto-semantic parsing approach. In: Proceedings of the 2013 Workshop on Biomedical Natural Language Processing, Sofia, Bulgaria, pp. 27–35. Association for Computational Linguistics, August 2013

Chang, C.-C., Lin, C.-J.: LIBSVM: a library for support vector machines. ACM Trans. Intel. Syst. Technol. **2**, 1:27–27:27 (2011)

French, L., Lane, S., Xu, L., Siu, C., Kwok, C., Chen, Y., Krebs, C., Pavlidis, P.: Application and evaluation of automated methods to extract neuroanatomical connectivity statements from free text. Bioinformatics **28**(22), 2963–2970 (2012). https://doi.org/10.1093/bioinformatics/bts542

Giuliano, C., Lavelli, A., Romano, L.: Exploiting shallow linguistic information for relation extraction from biomedical literature. In: Proceedings of the 11th Conference of the European Chapter of the Association for Computational Linguistics (EACL 2006), Trento, Italy (2006)

Hahn, U., Romacker, M., Schulz, S.: medSynDiKATe - a natural language system for the extraction of medical information from findings reports. Int. J. Med. Inform. **67**(1), 63–74 (2002)

Hellrich, J., Matthies, F., Faessler, E., Hahn, U.: Sharing models and tools for processing German clinical texts. Stud. Health Technol. Inf. **210**, 734–738 (2015)

Jiang, J., Guan, Y., Zhao, C.: WI-ENRE in CLEF eHealth evaluation lab 2015: clinical named entity recognition based on CRF. In: Working Notes of CLEF 2015 - Conference and Labs of the Evaluation forum, Toulouse, France, 8–11 September 2015 (2015)

Joachims, T.: Making large-scale support vector machine learning practical. In: Schölkopf, B., Burges, C.J.C., Smola, A.J. (eds.) Advances in Kernel Methods, pp. 169–184. MIT Press, Cambridge (1999). ISBN 0-262-19416-3

Krieger, H.-U., Spurk, C., Uszkoreit, H., Xu, F., Zhang, Y., Müller, F., Tolxdorff, T.: Information extraction from german patient records via hybrid parsing and relation extraction strategies. In: Calzolari, N., Choukri, K., Declerck, T., Loftsson, H., Maegaard, B., Mariani, J., Moreno, A., Odijk, J., Piperidis, S. (eds) Proceedings of the Ninth International Conference on Language Resources and Evaluation (LREC 2014), Reykjavik, Iceland, May 2014. European Language Resources Association (ELRA) (2014). ISBN 978-2-9517408-8-4

Kuru, O., Can, O.A., Yuret, D.: CharNER: character-level named entity recognition. In: Proceedings of COLING 2016, the 26th International Conference on Computational Linguistics: Technical Papers, Osaka, Japan, December 2016, pp. 911–921. The COLING 2016 Organizing Committee (2016)

Lafferty, J., McCallum, A., Pereira, F., et al.: Conditional random fields: probabilistic models for segmenting and labeling sequence data. In: Proceedings of the Eighteenth International Conference on Machine Learning, ICML, vol. 1, pp. 282–289 (2001)

Leaman, R., Gonzalez, G.: BANNER: an executable survey of advances in biomedical named entity recognition. In: Proceedings of the Pacific Symposium on Biocomputing 2008, Kohala Coast, Hawaii, USA, 4–8 January 2008, pp. 652–663 (2008)

Li, D., Kipper-Schuler, K., Savova, G.: Conditional random fields and support vector machines for disorder named entity recognition in clinical texts. In: Proceedings of the Workshop on Current Trends in Biomedical Natural Language Processing, BioNLP 2008, Stroudsburg, PA, USA, pp. 94–95. Association for Computational Linguistics (2008). http://dl.acm.org/citation.cfm?id=1572306.1572326. ISBN 978-1-932432-11-4

Nguyen, T.H., Grishman, R.: Relation extraction: perspective from convolutional neural networks. In: Proceedings of the 1st Workshop on Vector Space Modeling for Natural Language Processing, pp. 39–48, Denver, Colorado, June 2015. Association for Computational Linguistics (2015)

Rocktäschel, T., Weidlich, M., Leser, U.: ChemSpot: a hybrid system for chemical named entity recognition. Bioinformatics 28(12), 1633–1640 (2012)

Roller, R., Uszkoreit, H., Xu, F., Seiffe, L., Mikhailov, M., Staeck, O., Budde, K., Halleck, F., Schmidt, D.: A fine-grained corpus annotation schema of German nephrology records. In: Proceedings of the Clinical Natural Language Processing Workshop, vol. 28, no. 1, pp. 69–77 (2016)

Sahu, S.K., Anand, A., Oruganty, K., Gattu, M.: Relation extraction from clinical texts using domain invariant convolutional neural network. In: Proceedings of the 15th Workshop on Biomedical Natural Language Processing, Berlin, Germany. Association for Computational Linguistics (2016)

Starlinger, J., Kittner, M., Blankenstein, O., Leser, U.: How to improve information extraction from German medical records. it-Inf. Technol. 59(4), 171–179 (2016)

Thomas, P., Neves, M., Rocktäschel, T., Leser, U.: WBI-DDI: drug-drug interaction extraction using majority voting. In: Proceedings of the 7th International Workshop on Semantic Evaluation (SemEval) (2013)

Tikk, D., Thomas, P., Palaga, P., Hakenberg, J., Leser, U.: A comprehensive benchmark of kernel methods to extract protein-protein interactions from literature. PLoS
Comput. Biol. **6**, e1000837 (2010). https://doi.org/10.1371/journal.pcbi.1000837

Toepfer, M., Corovic, H., Fette, G., Klügl, P., Störk, S., Puppe, F.: Fine-grained
information extraction from German transthoracic echocardiography reports. BMC
Med. Inform. Decis. Making **15**(1), 1–16 (2015). https://doi.org/10.1186/s12911-
015-0215-x. ISSN 1472-6947

Weiss, G., Provost, F.: The effect of class distribution on classifier learning: an empirical
study. Technical report, Rutgers Univ. (2001)

Zeng, D., Liu, K., Lai, S., Zhou, G., Zhao, J.: Relation classification via convolutional
deep neural network. In: Proceedings of COLING 2014, the 25th International Conference on Computational Linguistics: Technical Papers, Dublin, Ireland, August
2014, pp. 2335–2344. Dublin City University and Association for Computational
Linguistics (2014)

In-Memory Distributed Training of Linear-Chain Conditional Random Fields with an Application to Fine-Grained Named Entity Recognition

Robert Schwarzenberg[✉], Leonhard Hennig, and Holmer Hemsen

DFKI GmbH, Alt-Moabit 91c, 10559 Berlin, Germany
{robert.schwarzenberg,leonhard.hennig,holmer.hemsen}@dfki.de

Abstract. Recognizing fine-grained named entities, i.e., *street* and *city* instead of just the coarse type *location*, has been shown to increase task performance in several contexts. Fine-grained types, however, amplify the problem of data sparsity during training, which is why larger amounts of training data are needed. In this contribution we address scalability issues caused by the larger training sets. We distribute and parallelize feature extraction and parameter estimation in linear-chain conditional random fields, which are a popular choice for sequence labeling tasks such as named entity recognition (NER) and part of speech (POS) tagging. To this end, we employ the parallel stream processing framework Apache Flink which supports in-memory distributed iterations. Due to this feature, contrary to prior approaches, our system becomes iteration-aware during gradient descent. We experimentally demonstrate the scalability of our approach and also validate the parameters learned during distributed training in a fine-grained NER task.

1 Introduction

Fine-grained named entity recognition and typing has recently attracted much interest, as NLP applications increasingly require domain- and topic-specific entity recognition beyond standard, coarse types such as persons, organizations and locations (Ling and Weld 2012; Del Corro et al. 2015; Abhishek et al. 2017). In NLP tasks such as relation extraction or question answering, using fine-grained types for entities can significantly increase task performance (Ling and Weld 2012; Koch et al. 2014; Dong et al. 2015). At the same time, freely-available, large-scale knowledge bases, such as Freebase (Bollacker et al. 2008), DBpedia (Auer et al. 2007) and Microsoft's Concept Graph (Wang et al. 2015) provide rich entity type taxonomies for labeling entities. However, training models for fine-grained NER requires large amounts of training data in order to overcome data sparsity issues (e.g., for low-frequency categories or features), as well as labeling noise, e.g., as introduced by training datasets created with distant supervision (Plank et al. 2014; Abhishek et al. 2017). Furthermore, the diversity of entity type taxonomies and application scenarios often requires the frequent adaptation or re-training of models. The speed and efficiency with which we

© The Author(s) 2018
G. Rehm and T. Declerck (Eds.): GSCL 2017, LNAI 10713, pp. 155–167, 2018.
https://doi.org/10.1007/978-3-319-73706-5_13

can (re-)train models thus becomes a major criterion for selecting learning algorithms, if we want to fully make use of these larger datasets and richer type taxonomies.

Linear-chain CRFs (Lafferty et al. 2001) are a very popular approach to solve sequence labeling tasks such as NER (Strauss et al. 2016). Parameter estimation in CRFs is typically performed in a supervised manner. Training, however, is time-consuming with larger datasets and many features or labels. For instance, it took more than three days to train a part-of-speech tagging model (45 labels, around 500k parameters) with less than 1 million training tokens on a 2.4 GHz Intel Xeon machine, Sutton and McCallum (2011) report. This is due to the fact that during training, linear-chain CRFs require to perform inference for each training sequence at each iteration.

Fortunately, linear-chain CRFs hold potential for parallelization. During gradient descent optimization it is possible to compute local gradients on subsets of the training data which then need to be accumulated into a global gradient. Li et al. (2015) recently demonstrated this approach by parallelizing model training within the MapReduce framework (Dean and Ghemawat 2008). The authors distributed subsets of the training data among the mappers of their cluster, which computed local gradients in a *map* phase. The local gradients were then accumulated into a global gradient in a subsequent *reduce* step. The *map* and *reduce* steps can be repeated until convergence, using the global gradient to update the model parameters at each iteration step. For large data sets, NER experiments showed that their approach improves performance in terms of run times. However, for each learning step, their system invokes a new Hadoop job, which is very time-consuming due to JVM startup times and disk IO for re-reading the training data. As the authors themselves point out, in-memory strategies would be much more efficient.

In this paper, we employ a very similar parallelization approach as Li et al., but implement the training within an efficient, iteration-aware distributed processing framework. The framework we choose allows us to efficiently store model parameters and other pre-computed data in memory, in order to keep the de/serialization overhead across iterations to a minimum (Alexandrov et al. 2014; Ewen et al. 2013).

Our contributions in this paper are:

- a proof-of-concept implementation of a distributed, iteration-aware linear-chain CRF training (Sect. 3),
- the experimental verification of the scalability of our approach, including an analysis of the communication overhead trade-offs (Sects. 4 and 5), and
- the experimental validation of the parameters learned during distributed training in a fine-grained NER and typing task for German geo-locations (Sects. 6 and 7).

In what follows, we first define linear-chain CRFs more formally and explain in detail how parameter estimation can be parallelized. We then discuss the details of our implementation, followed by several experimental evaluations.

2   Parallelization of Conditional Random Fields

This section closely follows Sutton and McCallum (2011) and Li et al. (2015). Assume $O = o_1 \ldots o_T$ is a sequence of observations (i.e., tokens) and $L = l_1 \ldots l_T$ is a sequence of labels (i.e., NE tags). Formally, a linear-chain CRF can then be defined as

$$p(L|O) = \frac{1}{Z(O)} \prod_{t=1}^{T} \exp \left(\sum_{k}^{K} \theta_k f_k(l_{t-1}, l_t, o_t) \right) \tag{1}$$

where f_k denotes one of K binary indicator – or feature – functions, each weighted by $\theta_k \in \mathbb{R}$, and Z is a normalization term, which iterates over all possible assignments

$$Z(O) = \sum_{L'} \prod_{t=1}^{T} \exp \left(\sum_{k}^{K} \theta_k f_k(l'_{t-1}, l'_t, o_t) \right). \tag{2}$$

The parameters θ_k are estimated in a way such that the conditional log-likelihood of the label sequences in the training data, denoted by $\mathbf{L}$ in the following, is maximized. This can be achieved with gradient descent routines.

Partially deriving $\mathbf{L}$ by θ_k yields

$$\frac{\partial \mathbf{L}}{\partial \theta_k} = \mathbf{E}(f_k) - \mathbf{E}_\theta(f_k) \tag{3}$$

where

$$\mathbf{E}(f_k) = \sum_{i=1}^{N} \sum_{t=1}^{T} f_k(l_{t-1}^{(i)}, l_t^{(i)}, o_t^{(i)}) \tag{4}$$

is the expected value of feature k in the training data $D = \{O^{(i)}, L^{(i)}\}_{i=1}^{N}$, and

$$\mathbf{E}_\theta(f_k) = \sum_{i=1}^{N} \sum_{t=1}^{T} \sum_{l,l'} f_k(l, l', o_t^{(i)}) p(l, l'|O^{(i)}; \theta) \tag{5}$$

is the expected value of the feature according to the model with parameter tensor θ. The inconvenience with Eq. 5 is that it requires us to perform marginal inference at each iteration, for each training sequence.

Fortunately, according to Eqs. 4 and 5, Eq. 3 can be computed in a data parallel fashion since

$$\frac{\partial \mathbf{L}}{\partial \theta_k} = \sum_{i=1}^{N} \left(\sum_{t=1}^{T} f_k(l_{t-1}^{(i)}, l_t^{(i)}, o_t^{(i)}) \right.$$
$$\left. - \sum_{t=1}^{T} \sum_{l,l'} f_k(l, l', o_t^{(i)}) p(l, l'|O^{(i)}; \theta) \right) \tag{6}$$

The next section explains how we distributed and parallelized the training phase.

3   Implementation

We partitioned the data into disjoint chunks of size p which we distributed among the mappers in a Flink cluster. Each mapper computed a local gradient on the chunk it received. In a subsequent *reduce* job, the local gradients were accumulated into a global one:

$$\left.\begin{matrix} \sum_{i=1}^{p}(E^{(i)}(f_k) - E_\theta^{(i)}(f_k))\big\} \ \text{map} \\[6pt] \sum_{i=p+1}^{2p}(E^{(i)}(f_k) - E_\theta^{(i)}(f_k))\big\} \ \text{map} \\[6pt] \vdots \end{matrix}\right\} (+) \ \text{reduce} \tag{7}$$

We used the global gradient to update the current model parameters at each iteration. The information flow is depicted in Fig. 1. As can be seen, before the first iteration, we also distributed feature extraction among the mappers.

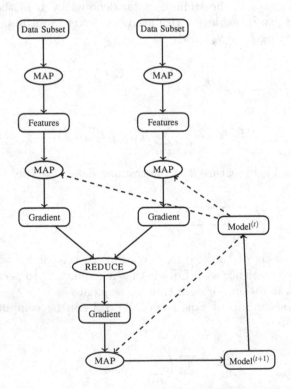

Fig. 1. Distributed iteration step. The dashed lines represent Flink broadcasts.

Our system marries two powerful tools, the probabilistic modeling library FACTORIE[1] (McCallum et al. 2009) and the parallel processing engine Apache Flink[2] (Alexandrov et al. 2014). It inherits features and functions from both tools.

The authors of FACTORIE convincingly promote it as a tool which preserves the 'traditional, declarative, statistical semantics of factor graphs while allowing imperative definitions of the model structure and operation.' Furthermore, Passos et al. (2013) compared FACTORIE's performance with established libraries such as scikit-learn, MALLET and CRFSuite and found that it is competitive in terms of accuracy and efficiency.

We distributed the model we implemented in FACTORIE with the help of Apache Flink. Flink provides primitives for massively parallel iterations and when compiling a distributed program which contains iterations, it analyses the data flow, identifies iteration-invariant parts and caches them to prevent unnecessary recomputations, Ewen et al. (2013) explain. Thus, contrary to prior approaches, due to Flink, our distributed system becomes 'iteration-aware'.

FACTORIE already supported local thread-level parallelism as well as distributed hyper-parameter optimization. Nonetheless, we had to overcome several obstacles when we ported the library into the cluster. For instance, in FACTORIE, object hash identities are used to map gradient tensors onto corresponding weight tensors during training. These identities get lost when an object is serialized in one JVM and deserialized in another JVM. To preserve identities throughout de/serialization among the virtual machines within the cluster, we cached relevant object hashes. We thus ended up using a slightly modified library.

4   Scalability Experiments

We tested our system on a NER task with seven types (including the default type). We compared our distributed parallel system with a local sequential counterpart in which we removed all Flink directives. In both versions our model consisted of a label-label factor and an observation-label factor[3]. During training, we used a likelihood objective, a belief propagation inference method which was tailored to linear chains and a constant step-size optimizer; all of which FACTORIE's modular design allows to plug in easily.

To evaluate performance, we varied three values

1. the level of parallelism,
2. the amount of training instances, and
3. the number of parameters, K.

Points (2)–(3) were varied for both the local version and the distributed version. When we tested the local version we kept the number of participating

---

[1] Version 1.2 (modified).

[2] Version 1.0.

[3] We refer to a token's features as *observations*.

computational nodes constant at one. In particular, no local thread parallelism was allowed, which is why point one does not apply to the local version.

All distributed experiments were conducted on an Apache Hadoop YARN cluster consisting of four computers (+1 master node). The local experiments were carried out using the master node. Two of the computers were running Intel Xeon CPUs E5-2630L v3 @ 1.80 GHz with 8 cores, 16 threads and 20 MB cache (as was the master node), while the other two computers were running Intel Xeon CPUs E5-2630 v3 @ 2.40 GHz again with 8 cores, 16 threads and 20 MB cache.

Each yarn task manager was assigned 8 GB of memory, of which a fraction of 30% was reserved for Flink's internal memory management. We used the Xmx option on the master node (with a total of 32 GB RAM). The nodes in the cluster thus had slightly less RAM available for the actual task than the master node. However, as a general purpose computer, the master node was also carrying out other tasks. We observed a maximal fluctuation of 1.7% (470 s vs. 478 s) for the same task carried out on different days. Loading the data from local files and balancing it between the mappers in the cluster was considered part of the training, as was feature extraction.

5   Scalability Evaluation

We first performed several sanity checks. For example, we made sure that multiple physical machines were involved during parameter estimation and that disjoint chunks of the training data reached the different machines. We also checked that the gradients computed by the mappers differed and that they were accumulated correctly during the reduce phase.

The most convincing fact that led us to believe that we correctly distributed feature extraction and parameter estimation was that after we trained the local version and the distributed version using the same training set – with just a few parameters and for just a few iterations – extremely similar parameters were in place. Consequently, the two models predicted identical labels on the same test set containing 5k tokens. The parameters diverge the more features are used and the more training steps are taken. We suspect that this is due to floating point imprecisions that result in different gradients at some point.

The first two experiments we conducted addressed the scalability of our distributed implementation. The results are summarized in Figs. 2 and 3. Figure 2 shows that our distributed implementation managed to outperform its sequential counterpart after a certain level of parallelism was reached. The level of parallelism required to beat the local version increased with the amount of parameters.

Figure 3 shows to what extent we were able to counterbalance an increase in training size with an increase in parallelism. The results suggest that our model was indeed able to dampen the effect of increasing amounts of training examples. The average rate of change in execution times was higher when we kept the number of nodes constant. As we doubled the level of parallelism along with the training size, the rate of change reduced significantly. We also compared the distributed implementation with the local implementation in Fig. 3. As can be

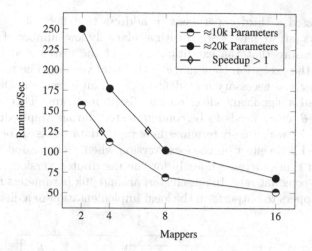

Fig. 2. Execution times for increasing numbers of mappers (M). Each training involved around 100k tokens (numbers rounded for better readability) and 25 iterations. The diamonds mark the points from which on the distributed version needed less time than its local counterpart. The sequential version needed 125 s for around 10k parameters and 126 s for twice as many parameters.

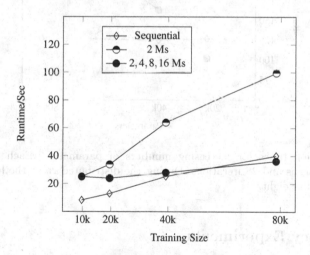

Fig. 3. Scalability of the distributed model. The figure offers a comparison between the execution times required by the local version and the distributed version to process an increasing (doubling) amount of training data. The distributed version was tested with a fixed number of mappers (M) and with an increasing (doubling) number of mappers (starting with two mappers at around 10k training instances). For each run, around 20k parameters were considered and the number of iterations was fixed at ten.

seen, the average rate of change is higher for the local version than for the distributed version with an increasing level of parallelism. However, it is still much lower when compared to the distributed runs with a fixed level of parallelism.

We conducted a third experiment to address the effect of communication overhead. Thus far, we have worked with a relatively low number of parameters. This was to ensure that the execution times of the distributed version were falling within the execution time range of the local version. The reason for why the low number was necessary is evident in Fig. 4: an increase in the number of parameters had a significant effect on the distributed runs. This is due to the fact that it is θ which needs to be communicated during MapReduce and it is also the size of θ which co-determines how much data needs to be cached. By contrast, it had little effect on the local version when we increased the size of θ. The execution times increase linearly for the distributed version, while locally they stay at a constant rate. In our cluster, around 40k parameters require more than eight mappers to outperform the local implementation in a distributed run.

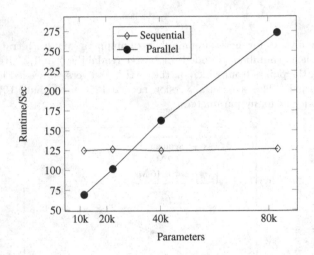

Fig. 4. Execution times for increasing numbers of parameters. Each run involved around 100k tokens and 25 iterations. During the distributed runs, the level of parallelism was fixed at eight.

6   Accuracy Experiments

The sections above address the scalability of our approach. In this section we report on experiments which demonstrate that our distributed linear-chain CRF learns meaningful parameters. We tested our model in a NER task.

The task was to recognize six fine-grained geospatial concepts in German texts, namely streets ('Berliner Straße'), cities ('Berlin'), public transportation hubs ('Berliner Straße'), routes ('U6'), distances ('5 km') and the super-type location ('Germany'). The task involves the typical challenges of NER, such as disambiguation. Furthermore, the training sets (which were annotated by trained linguists) contained user-generated content, which is why noise was also an issue.

Table 1. Datasets. Size and noise. We refer to noise as the percentage of tokens that the Enchant v 1.6.0 Myspell de_DE dictionary did not recognize.

Dataset	Tokens	Noise
RSS	20152	35.6%
Twitter	12606	45.3%

Table 1 characterizes the two datasets and explains what we refer to as *noise*. The RSS dataset consists of a sample of traffic reports crawled from more than 100 RSS feeds that provide traffic and transportation information about road blocks, construction sites, traffic jams, or rail replacement services. Feed sources include federal and state police, radio stations, Deutsche Bahn, and air travel sources. Traffic reports are typically very brief, may be semi-structured (e.g., location, cause and length of a traffic jam), and often contain telegraph-style sentences or phrases. The Twitter dataset consists of a sample of German-language tweets that were retrieved via the Twitter search API using a list of approximately 150 domain-relevant users/channels and 300 search terms. Channels include, e.g., airline companies, traffic information sources, and railway companies. Search terms comprise event-related keywords such as "traffic jam" or "roadworks", but also major highway names, railway route identifiers, and airport codes. Both datasets therefore consist of documents which contain traffic- and mobility-related information that refer to the fine-grained location types defined previously.

Besides the well-established features in NER (e.g., word shape, affixes) our application ('Locator') also considered task specific features and took measures towards text normalization. In the end, a larger number of parameters (100k–150k) was in place than during the scalability experiments.

We again used the FACTORIE components listed in Sect. 4, such as the BP method for chains and a constant step size optimizer. FACTORIE provides more sophisticated optimizers such as LBFGS or Adagrad. In our current system, however, only the model parameters survive a Flink-iteration step but methods like LBFGS and Adagrad need further information about past update steps.

We conducted a ten-fold cross-validation on the datasets. Feature extraction and parameter estimation were performed in parallel in the way described above. The level of parallelism was fixed at four, for all experiments. After training, the models were serialized and saved for the test phase. The test runs took place on a single machine.

To put the performance of our model into perspective, we also conducted a ten-fold cross-validation using the Stanford NER (v. 3.6.0) in its standard configuration[4]. The Stanford NER used the same tokenizer as our system.

---

[4] The configuration file we used can be found in the appendix.

7   Accuracy Evaluation

The results of our accuracy experiments are summarized in Table 2. The F-scores achieved on the Twitter dataset and the scores achieved on the RSS dataset reveal similar trends for both systems: In both cases, the RSS-score is higher than the Twitter-score.

Our distributed model slightly outperforms the Stanford NER on the Twitter dataset but is beaten on the RSS dataset. Since the Twitter dataset is noisier than the RSS dataset, we suspect that the task-specific features and text normalization methods of our system have a greater impact in this case.

Overall, we conclude that the experiments provide sufficient proof that during distributed training our system indeed learns meaningful parameters. It achieves comparable scores.

Table 2. Results of 10-fold NER experiments. Classification performance was evaluated on token level so that multiple-token spans resulted in multiple true positives or false negatives, for instance. To compensate class imbalances, for each fold, we weighted the fine-grained scores (i.e., precision (P), recall (R) and F1-score (F1) of the entity 'street') by the support of the entity in the test set and averaged over all fine-grained scores. The listed scores are averages over the ten fold scores.

System	Dataset	P	R	F1
Locator	RSS	80.7	75.8	75.2
Stanford	RSS	82.8	78.8	**80.5**
Locator	Twitter	57.0	50.4	**51.7**
Stanford	Twitter	79.0	35.9	47.2

8   Discussion and Conclusion

We distributed and parallelized feature extraction and parameter estimation in linear-chain CRFs. The sequence labeling experiments we conducted suggest that our system learns meaningful parameters and is able to counterbalance growing amounts of training data with an increase in the level of parallelism (see Table 2 and Figs. 2 and 3). We reached speedups greater than one and F-scores comparable to the ones produced by a state-of-the-art approach.

To achieve this, we combined the parallel processing engine Apache Flink with the probabilistic modeling library FACTORIE. Our proof-of-concept implementation now inherits functions and features from both tools.

Contrary to prior approaches, for instance, it is iteration-aware during distributed gradient descent. In addition, our system also benefits from FACTORIE's modular design and rich pool of functions. With little programming effort it is possible to plug in alternative choices for an optimizer or an inference method.

There is, however, room for improvement. The choice for an optimizer, for instance, is restricted by the fact that currently, only the model parameters survive an iteration. But some optimization procedures that FACTORIE provides, like LBFGS, require additional information about past updates. Enhancing the system to provide this feature remains future work.

Furthermore, the increase in runtime in Fig. 4 seems disproportionate. Working with sparse vectors to reduce the amount of data that needs to be cached will most likely reduce runtime. There might also be a serialization bottleneck. Registering customized serializers for FACTORIE's types with Flink may thus also improve performance. Fortunately, the number of features is typically fixed at some point in most settings. At this point the amount of available training data and the number of mappers in the cluster determine from when on our approach pays off.

Acknowledgments. This research was partially supported by the German Federal Ministry of Economics and Energy (BMWi) through the projects SD4M (01MD15007B) and SDW (01MD15010A), and by the German Federal Ministry of Education and Research (BMBF) through the project BBDC (01IS14013E).

Appendix A. Stanford NER Properties File

```
trainFile=path/to/training_file
serializeTo=path/to/model
map=word=0,answer=1

useClassFeature=true
useWord=true
useNGrams=true
noMidNGrams=true
maxNGramLeng=6
usePrev=true
useNext=true
useSequences=true
usePrevSequences=true
maxLeft=1
useTypeSeqs=true
useTypeSeqs2=true
useTypeySequences=true
wordShape=chris2useLC
useDisjunctive=true
```

References

Abhishek, A., Anand, A., Awekar, A.: Fine-grained entity type classification by jointly learning representations and label embeddings. In: Proceedings of the 15th Conference of the European Chapter of the Association for Computational Linguistics, vol. 1: Long Papers, pp. 797–807. Association for Computational Linguistics, Valencia, April 2017. http://www.aclweb.org/anthology/E17-1075

Alexandrov, A., Bergmann, R., Ewen, S., Freytag, J.-C., Hueske, F., Heise, A., Kao, O., Leich, M., Leser, U., Markl, V., et al.: The stratosphere platform for big data analytics. VLDB J. **23**(6), 939–964 (2014)

Auer, S., Bizer, C., Kobilarov, G., Lehmann, J., Cyganiak, R., Ives, Z.: DBpedia: a nucleus for a web of open data. In: Aberer, K., et al. (eds.) ASWC/ISWC -2007. LNCS, vol. 4825, pp. 722–735. Springer, Heidelberg (2007). https://doi.org/10.1007/978-3-540-76298-0_52

Bollacker, K., Evans, C., Paritosh, P., Sturge, T., Taylor, J.: Freebase: a collaboratively created graph database for structuring human knowledge. In: Proceedings of the 2008 ACM SIGMOD International Conference on Management of Data, SIGMOD 2008, pp. 1247–1250. ACM, New York (2008). https://doi.org/10.1145/1376616.1376746. http://147.46.216.176/w/images/9/98/SC17.pdf. ISBN 978-1-60558-102-6

Dean, J., Ghemawat, S.: MapReduce: simplified data processing on large clusters. Commun. ACM **51**(1), 107–113 (2008)

Del Corro, L., Abujabal, A., Gemulla, R., Weikum, G.: Finet: context-aware fine-grained named entity typing. In: Proceedings of the 2015 Conference on Empirical Methods in Natural Language Processing, pp. 868–878. Association for Computational Linguistics, Lisbon, September 2015. http://aclweb.org/anthology/D15-1103

Dong, L., Wei, F., Sun, H., Zhou, M., Xu, K.: A hybrid neural model for type classification of entity mentions. In: Proceedings of the 24th International Conference on Artificial Intelligence, IJCAI 2015, pp. 1243–1249. AAAI Press, Buenos Aires (2015). http://dl.acm.org/citation.cfm?id=2832415.2832422. ISBN 978-1-57735-738-4

Ewen, S., Schelter, S., Tzoumas, K., Warneke, D., Markl, V.: Iterative parallel data processing with stratosphere: an inside look. In: Proceedings of the 2013 ACM SIGMOD International Conference on Management of Data, pp. 1053–1056. ACM (2013)

Koch, M., Gilmer, J., Soderland, S., Weld, D.S.: Type-aware distantly supervised relation extraction with linked arguments. In: Proceedings of the 2014 Conference on Empirical Methods in Natural Language Processing (EMNLP), pp. 1891–1901. Association for Computational Linguistics, Doha, October 2014. http://www.aclweb.org/anthology/D14-1203

Lafferty, J., McCallum, A., Pereira, F.: Conditional random fields: probabilistic models for segmenting and labeling sequence data. In: Proceedings of the ICML, vol. 1, pp. 282–289 (2001)

Li, K., Ai, W., Tang, Z., Zhang, F., Jiang, L., Li, K., Hwang, K.: Hadoop recognition of biomedical named entity using conditional random fields. IEEE Trans. Parallel Distrib. Syst. **26**(11), 3040–3051 (2015)

Ling, X., Weld, D.: Fine-grained entity recognition. In: Proceedings of AAAI 2012 (2012). http://www.aaai.org/ocs/index.php/AAAI/AAAI12/paper/view/5152

McCallum, A., Schultz, K., Singh, S.: Factorie: probabilistic programming via imperatively defined factor graphs. In: Advances in Neural Information Processing Systems, pp. 1249–1257 (2009)

Passos, A., Vilnis, L., McCallum, A.: Optimization and learning in FACTORIE. In: NIPS Workshop on Optimization for Machine Learning (OPT) (2013)

Plank, B., Hovy, D., McDonald, R.T., Søgaard, A.: Adapting taggers to Twitter with not-so-distant supervision. In: COLING, pp. 1783–1792 (2014)

Strauss, B., Toma, B.E., Ritter, A., de Marneffe, M.-C., Xu, W.: Results of the WNUT16 named entity recognition shared task. In: Proceedings of the 2nd Workshop on Noisy User-Generated Text, pp. 138–144 (2016)

Sutton, C., McCallum, A.: An introduction to conditional random fields. Found. Trends Mach. Learn. 4(4), 267–373 (2011)

Wang, Z., Wang, H., Wen, J.-R., Xiao, Y.: An inference approach to basic level of categorization. In: Proceedings of the 24th ACM International on Conference on Information and Knowledge Management, CIKM 2015, pp. 653–662. ACM, New York (2015). https://doi.org/10.1145/2806416.2806533. ISBN 978-1-4503-3794-6

Online-Media and Online-Content

What Does This Imply? Examining the Impact of Implicitness on the Perception of Hate Speech

Darina Benikova$^{(\boxtimes)}$, Michael Wojatzki, and Torsten Zesch

Language Technology Lab, University of Duisburg-Essen, Duisburg, Germany
{darina.benikova,michael.wojatzki,torsten.zesch}@uni-due.de

Abstract. We analyze whether implicitness affects human perception of hate speech. To do so, we use Tweets from an existing hate speech corpus and paraphrase them with rules to make the hate speech they contain more explicit. Comparing the judgment on the original and the paraphrased Tweets, our study indicates that implicitness is a factor in human and automatic hate speech detection. Hence, our study suggests that current automatic hate speech detection needs features that are more sensitive to implicitness.

1 Introduction

With the rise of social media, hate speech (HS) has moved into the focus of public attention. However, as its perception depends on linguistic, contextual, and social factors (Stefanowitsch 2014), there is no consensus on what constitutes HS. We examine a specific dimension of this challenge – whether implicitness affects HS perception. Consider the following Tweets:

Im. Everything was quite ominous with the train accident. Would like to know whether the train drivers were called Hassan, Ali or Mohammed #Refugee Crisis

Ex. Everything [...] - The train drivers were Muslims. #RefugeeCrisis

One could argue that the first Tweet is more offensive, since it evokes racist stereotypes by using allegedly prototypical Muslim first names as an implicit way of blaming Muslims in general. However, one could counter-argue that the second Tweet is more offensive, as it explicitly accuses Muslims of being involved in a train accident. Additionally, the first Tweet is hedged by *Would like to know whether*, whereas it is implied that the second statement is rather factual. It remains unresolved whether implicit or explicit HS is perceived as more offensive and what the role of hedging is (Sanchez and Vogel 2013).

In addition to the influence on the perception of HS, implicitness is a challenge for automatic HS detection, as most approaches rely on lists of abusive terms or phrases (Waseem and Hovy 2016).

Or in terms of the above example, the classifier learns that it is HS to agitate against *Muslims*, but fails to learn the connection to *Hassan*.

© The Author(s) 2018
G. Rehm and T. Declerck (Eds.): GSCL 2017, LNAI 10713, pp. 171–179, 2018.
https://doi.org/10.1007/978-3-319-73706-5_14

To shed light on the influence of implicitness on the perception of HS, we construct a dataset[1] in which we can experimentally control for implicitness. We select implicit HS instances from the German Hate Speech Twitter Corpus (Ross et al. 2016) and create explicit paraphrased counterparts[2]. We then conduct a user study, wherein we ask participants to rate the offensiveness of either implicit or explicit Tweets. We also show that a supervised classifier is unable to detect HS on both datasets.

We hypothesize that there is a measurable difference in the perception of implicit and explicit statements in both human and automatic performance. However, we cannot estimate the direction of the difference.

2   Theoretical Grounding

Our work is grounded in (i) research on detecting HS, (ii) the annotation and detection of implicit opinions, and (iii) on paraphrasing.

Detecting Hate Speech. Hitherto, there has been no work on HS detection considering the issues posed by implicitness. Approaches based on n-grams or word lists, e.g., (Sood et al. 2012; Chen et al. 2012) are limited to detecting explicit insults or abusive language. Methods involving more semantics, e.g., by incorporating Brown clusters (Waseem and Hovy 2016; Warner and Hirschberg 2012) are unlikely to cope with implicitness, as the necessary inferences go beyond word-relatedness.

Implicit Opinions. If we define HS as expressing a (very) negative opinion against a target, there is a clear connection to aspect-based sentiment analysis. However, sentiment analysis usually only models explicit expressions. For instance, the popular series of SemEval tasks on detecting aspect based sentiment, intentionally exclude implicit sentiment expressions and expressions requiring co-reference resolution in their annotation guidelines (Pontiki et al. 2014, 2015, 2016). Contrarily, the definition of stance, namely being in favor or against a target (i.e., a person, a group or any other controversial issue) explicitly allows to incorporate such inferences (for annotation guidelines see Mohammad et al. (2016) or Xu et al. (2016)). Thus, HS can also be considered as expressing a hateful stance towards a target.

Consequently, we define explicit HS as expressing hateful sentiment and implicit HS as the instances which do not express hateful sentiment, but hateful stance. Therefore, this work relates to studies which use explicit opinion expressions to predict or rationalize stance (Boltužić and Šnajder 2014; Hasan and Ng 2014; Sobhani et al. 2015; Wojatzki and Zesch 2016).

---

[1] https://github.com/MeDarina/HateSpeechImplicit.

[2] All examples in this paper are extracted from this corpus and were translated to English. None of the examples reflects the opinion or political orientation of the authors.

Paraphrasing. The implicit and explicit versions of a Tweet can be seen as para-phrases, i. e., units of texts containing semantically equivalent content (Madnani and Dorr 2010). Paraphrases can be classified according to the source of differ-ence between the two texts. Incorporating implicit stances is equivalent to the paraphrase class of *Ellipsis* or the *Addition/Deletion* class.

The modification of hedges corresponds to the classes of *Quantifiers* and *General/Specific substitution* (Bhagat and Hovy 2013; Rus et al. 2014; Vila et al. 2014). To the best of our knowledge, paraphrasing techniques have not been used in the context of HS and its analysis.

3   Manufacturing Controllable Explicitness

The basis of our data set is the German Hate Speech corpus (Ross et al. 2016) that contains about 500 German Tweets annotated for expressing HS against refugees or not. We chose this corpus because it is freely available and addresses a current social problem, namely the debate on the so-called *European refugee crisis.* To construct a data set in which we can control for implicitness, we perform the following steps: (1) Restriction to Tweets which contain HS, i. e., at least one annotator flagged a Tweet as such (2) Removal of Tweets containing explicit HS markers, as described in Sect. 3.1 (3) Paraphrasing the remaining Tweets to be explicit, so that we obtain a dataset which has both an implicit and an explicit version of each Tweet.

3.1   Indicators for Explicit Hate Speech

We first identify tokens that are clear indicators for HS by retrieving words that are most strongly associated with HS.[3] We restrict ourselves to nouns, named entities, and hashtags, as we do not observe strong associations for other POS tags. We compute the collocation coefficient *Dice* (Smadja et al. 1996) for each word and inspect the end of the spectrum associated with the HS class.

We observe the – by far – strongest association for the token *#rapefugee.* Furthermore, we perceive strong association for cognates of *rape* such as *rapist* and *rapes.*

To further inspect the influence of these indicators, we compute the proba-bility of their occurrence predicting whether a Tweet is HS or not. We find a probability of 65.8% for *#rapefugee* and of even 87.5% for the group of nouns related to *rape.* When inspecting the Tweets containing those explicit HS indi-cators, we observe that they are often considered as HS regardless of whether the rest of the Tweet is protective of refugees. Because of this simple heuristic, we remove those Tweets from our data set.

---

[3] Tokenization is done with Twokenizer (Gimpel et al. 2011) and POS-tagging with Stanford POS-tagger (Toutanova et al. 2003).

3.2 Paraphrasing

To make the Tweets explicit, we paraphrase them according to a set of rules[4], which correspond to previously mentioned paraphrase classes. We apply as many rules as possible to one Tweet in order to make it as explicit as possible. As the corpus is concerned with the refugee crisis, we define *Islam*, *Muslim*, and *refugee* as the targets of HS. If a phrase does not explicitly contain them, we paraphrase it by adding this information as a new subject, object, or adjective or by co-reference resolution. An example for this rule is shown in the first explicit paraphrase:

Im. #Vendetta, #ForcedConversion, #Sharia, #ChildBrides, #Polygamy, #GenitalMutilation - don't see how it belongs to us.
Ex.1 [. . .] - don't see how **Islam** belongs to us.
Ex.2 [. . .] - It **doesn't** belongs to us.
Ex.3 [. . .] - **Islam doesn't** belongs to us.

If the message of the phrase is softened through hedges such as modals (e.g., *could*, *should*) and epistemic modality with first person singular (e.g., *I think*, *in my opinion*) these are either removed or reformulated to be more explicit. This reformulation is shown in the second explicit paraphrase in the example above. However, as we apply as many rules as possible, the Tweet would be paraphrased to its final version as shown in the third paraphrase in the example above. Rhetorical questions are paraphrased to affirmative phrases, e.g.,

– Yesterday the refugees came. Today there's burglary. Coincidence?
– Yesterday the refugees came. [. . .] **Not a coincidence!**

Furthermore, implicit generalizations are made explicit through the use of quantifiers.

– 90% of all refugees want to come to Germany, only because nobody else will give them money! Islamize in passing. #Lanz
– **All** refugees want to come to Germany, [. . .].

The paraphrasing process was performed independently by two experts, who chose the same instances of implicit stance, but produced slightly differing paraphrases.

The experts merged the two sets by choosing one of the two paraphrased versions after a discussion.

3.3 Supervised Machine Learning

To examine the influence of implicitness on automatic HS detection, we re-implement a state-of-the-art system. We adapt the systems of Waseem and Hovy (2016) and Warner and Hirschberg (2012) to German data. Thus, we rely on an

---

[4] https://github.com/MeDarina/HateSpeechImplicit.

SVM equipped with type-token-ratio, emoticon ratio, character, token, and POS uni-, bi-, and trigams features.

For our classification, we consider Tweets as HS in which at least one annotator flagged it as such since we aim at training a high-recall classifier. The resulting class distribution is 33% HS and 67% NO HS. First, we establish baselines by calculating a majority class baseline and conducting a ten-fold cross-validation. We report macro-F_1 for all conducted experiments. While the majority class baseline results in a macro-F_1 of .4, we obtain a macro-F_1 of .65 for the cross-validation.

To inspect the influence of implicitness, we conduct a train-/test-split with the selected implicit Tweets as test instances and the remaining Tweets as train instances. We achieve a macro-F_1 of only .1, regardless whether we use the explicit or implicit version of the Tweets. Although the performance is higher than the majority class baseline, the drop is dramatic compared to the cross-validation.

First, these results indicate that implicitness is a major problem in HS detection and thus should be addressed by future research. Second, as results are the same for the more explicit version, the classifier seems to be incapable of recognizing explicit paraphrases of implicit Tweets. Although this was expected since we did not add HS indicating tokens during paraphrasing, it may be highly problematic as implicitness may alter human perception of HS.

4   User Study

After the exclusion of explicit Tweets, a set of 36 implicit Tweets remained, which were paraphrased into an explicit version. To analyze the difference in their perception, we conducted an online survey using a *between-group* design with implicitness as the experimental condition. The randomly assigned participants had to make a binary decision for each Tweet on whether it is HS and rate its offensiveness on a six-point scale, in accordance with Ross et al. (2016). The participants were shown the definition of *HS* of the European ministerial committee[5].

As understanding the content of the Tweets is crucial, we filtered according to native knowledge of German which resulted in 101 participants. They reported a mean age of 27.7 years, 53.4% considering themselves female, 41.6% male and 1% other genders. 39.6% had a university entrance qualification, 58.4% a university degree, and 1% had another education level. More than 90% stated that they identify as Germans which may question the representativeness of our study. Especially, the educational and ethnic background might be factors strongly influencing the perception of HS. 55 remained in the implicit condition and 46 in the explicit condition.

---

[5] http://www.egmr.org/minkom/ch/rec1997-20.pdf.

5   Results

First, we inspect how often the Tweets are identified as HS. On average, we find that 31.6% of the Tweets are rated as HS in the explicit ($M_{explicit} = 11.3$)[6] and 40.1% in the implicit condition ($M_{implicit} = 14.4$). Interestingly, we observe a high standard deviation ($SD_{explicit} = 11.3$ and $SD_{implicit} = 14.6$) for both conditions. These findings underline how difficult it is for humans to reliably detect HS and thus align with the findings of Ross et al. (2016). A χ^2 test shows that the answer to this question is not significantly differently distributed in the two conditions, ($\chi^2_{(22,N=57)} = 4.53$, $p < .05$). Regarding intensity, encoded from 1–6, we do not find statistically significant differences between the explicit ($M = 3.9$, $SD = .94$) and the implicit ($M = 4.1$, $SD = .98$) condition according to a T-test ($t(97.4) = 1.1$, $p > .05$). To further analyze this difference, we inspect the difference for each instance, which is visualized in Fig. 1. All except one of the significantly differing instances are perceived as more hateful in the implicit version. For all cases, we observe that the implicit version is more global and less directed, which could be due to the fact that the vague and global formulation targets larger groups. Instances 6 and 10 contain rhetorical questions, which may be perceived as hidden or more accusing than the affirmative rather factual version. The one case in which the explicit form is more offensive is the only instance containing a threat of violence, which becomes more directed through making it explicit.

We also compute the change in the binary decisions between HS and NO HS on the level of individual instances using χ^2. Three of the eight significantly less offensive explicit instances on the scale are also significantly less often considered being HS in the binary decision. Similarly, instance 24, which is perceived significantly more offensive is more frequently considered as HS. Thus, we conclude that there is a relationship between the offensiveness and the HS rating

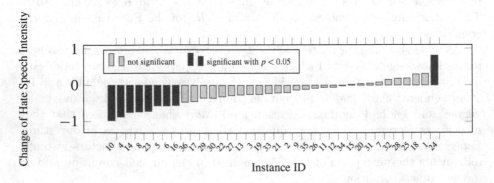

Fig. 1. Change in HS intensity between implicit and explicit versions.

---

[6] Statistical measures are reported according to the American Psychological Association (1994): M = Mean, SD = standard deviation, p = probability; N = number of participants/annotators.

and that both are equally affected by implicitness. However, the direction of this relationship, is moderated by the contentual factors (e. g., the presence of a threat) which need further investigation.

6 Conclusion and Future Work

In this study we show that there are individual instances of explicit HS which are perceived significantly different compared to their implicit counterparts. However, on average, the polarity of this deviation remains unclear and seems to be moderated by content variables.

In all cases where the implicit version is perceived as more intensely hateful, the Tweets were rather insulting than threatening. The perception change might be due to several reasons: the sly, potentially deceiving nature of implicitness might be perceived as more hateful, whereas the same content expressed clearly might be perceived as more honest and thus less hateful.

Furthermore, although implicitness has an influence on the human perception of HS, the phenomenon is invisible to automatic classifiers. This poses a severe problem for automatic HS detection, as it opens doors for more intense HS hiding behind the phenomenon of implicitness.

Since this study is based on 36 Tweets, the generalizability of the findings may be limited. Thus, in future work a larger study with more data and more fine-grained distinctions between classes such as *insulting* and *threatening content* would give more insight in the correlation between implicitness and HS perception. Additionally it would be interesting to produce implicit paraphrases of explicitly expressed HS and see the effect. Furthermore, more diverse focus groups, such as representatives of diverse religions, origins, and educational backgrounds are required.

Acknowledgements. This work is supported by the German Research Foundation (DFG) under grant No. GRK 2167, Research Training Group "User-Centred Social Media".

References

American Psychological Association: Publication Manual of the American Psychological Association. American Psychological Association, Washington (1994)

Bhagat, R., Hovy, E.: What is a paraphrase? Comput. Linguist. **39**(3), 463–472 (2013). ISSN 04194217

Boltužić, F., Šnajder, J.: Back up your stance: recognizing arguments in online discussions. In: Proceedings of the First Workshop on Argumentation Mining, Baltimore, USA, pp. 49–58 (2014)

Chen, Y., Zhou, Y., Zhu, S., Xu, H.: Detecting offensive language in social media to protect adolescent online safety. In: 2012 International Conference on Privacy, Security, Risk and Trust (PASSAT), and 2012 International Conference on Social Computing (SocialCom), Amsterdam, Netherlands, pp. 71–80. IEEE (2012)

Gimpel, K., Schneider, N., O'Connor, B., Das, D., Mills, D., Eisenstein, J., Heilman, M., Yogatama, D., Flanigan, J., Smith, N.A.: Part-of-speech tagging for Twitter: annotation, features, and experiments. In: Proceedings of the 49th Annual Meeting of the Association for Computational Linguistics: Human Language Technologies: Short Papers, Portland, USA, vol. 2, pp. 42–47 (2011)

Hasan, K.S., Ng, V.: Why are you taking this stance? Identifying and classifying reasons in ideological debates. In: Proceedings of the EMNLP, Doha, Qatar, pp. 751–762 (2014)

Madnani, N., Dorr, B.J.: Generating phrasal and sentential paraphrases: a survey of data-driven methods. Comput. Linguist. 36(3), 341–387 (2010)

Sanchez, L.M., Vogel, C.: IMHO: an exploratory study of hedging in web forums. In: Proceedings of the SIGDIAL 2013 Conference, Metz, France, pp. 309–313. Association for Computational Linguistics, August 2013

Mohammad, S.M., Kiritchenko, S., Sobhani, P., Zhu, X., Cherry, C.: Semeval-2016 task 6: detecting stance in tweets. In: Proceedings of the International Workshop on Semantic Evaluation, San Diego, USA (2016, to appear)

Pontiki, M., Galanis, D., Pavlopoulos, J., Papageorgiou, H., Androutsopoulos, I., Manandhar, S.: Semeval-2014 task 4: aspect based sentiment analysis. In: Proceedings of SemEval 2014, pp. 27–35 (2014)

Pontiki, M., Galanis, D., Papageorgiou, H., Manandhar, S., Androutsopoulos, I.: SemEval-2015 task 12: aspect based sentiment analysis. In: Proceedings of the 9th SemEval, Denver, Colorado, pp. 486–495. Association for Computational Linguistics (2015)

Pontiki, M., Galanis, D., Papageorgiou, H., Androutsopoulos, I., Manandhar, S., AL-Smadi, M., Al-Ayyoub, M., Zhao, Y., Qin, B., De Clercq, O., Hoste, V., Apidianaki, M., Tannier, X., Loukachevitch, N., Kotelnikov, E., Bel, N., Jiménez-Zafra, S.M., Eryiğit, G.: SemEval-2016 task 5: aspect based sentiment analysis. In: Proceedings of the 10th SemEval, San Diego, California, pp. 19–30. Association for Computational Linguistics (2016)

Ross, B., Rist, M., Carbonell, G., Cabrera, B., Kurowsky, N., Wojatzki, M.: Measuring the reliability of hate speech annotations: the case of the European refugee crisis. In: Beißwenger, M., Wojatzki, M., Zesch, T. (eds.) Proceedings of NLP4CMC III: 3rd Workshop on Natural Language Processing for Computer-Mediated Communication. Bochumer Linguistische Arbeitsberichte, Bochum, Germany, vol. 17, pp. 6–9 (2016)

Rus, V., Banjade, R., Lintean, M.C.: On paraphrase identification corpora. In: LREC, pp. 2422–2429 (2014)

Smadja, F., McKeown, K.R., Hatzivassiloglou, V.: Translating collocations for bilingual lexicons: a statistical approach. Comput. Linguist. 22(1), 1–38 (1996)

Sobhani, P., Inkpen, D., Matwin, S.: From argumentation mining to stance classification. In: Proceedings of the NAACL HLT 2015, Denver, USA, pp. 67–77 (2015)

Sood, S.O., Antin, J., Churchill, E.F.: Using crowdsourcing to improve profanity detection. In: AAAI Spring Symposium: Wisdom of the Crowd, vol. 12, p. 06 (2012)

Stefanowitsch, A.: Was ist überhaupt hate-speech. In: Stiftung, A.A. (ed.) Geh sterben. Umgang mit Hate-Speech und Kommentaren im Internet, pp. 11–13 (2014)

Toutanova, K., Klein, D., Manning, C.D., Singer, Y.: Feature-rich part-of-speech tagging with a cyclic dependency network. In: Proceedings of the 2003 Conference of the North American Chapter of the Association for Computational Linguistics on Human Language Technology, Sapporo, Japan, vol. 1, pp. 173–180. Association for Computational Linguistics (2003)

Vila, M.M., Martí, A., Rodríguez, H., et al.: Is this a paraphrase? What kind? Paraphrase boundaries and typology. Open J. Mod. Linguist. 4(01), 205 (2014)

Warner, W., Hirschberg, J.: Detecting hate speech on the world wide web. In: Proceedings of LSM 2012, pp. 19–26. ACL (2012)

Waseem, Z., Hovy, D.: Hateful symbols or hateful people? Predictive features for hate speech detection on Twitter. In: Proceedings of NAACL-HLT, pp. 88–93 (2016)

Wojatzki, M., Zesch, T.: Stance-based argument mining - modeling implicit argumentation using stance. In: Proceedings of the KONVENS, Bochum, Germany, pp. 313–322 (2016)

Xu, R., Zhou, Y., Wu, D., Gui, L., Du, J., Xue, Y.: Overview of NLPCC shared task 4: stance detection in Chinese microblogs. In: Lin, C.-Y., Xue, N., Zhao, D., Huang, X., Feng, Y. (eds.) ICCPOL/NLPCC -2016. LNCS, vol. 10102, pp. 907–916. Springer, Cham (2016). https://doi.org/10.1007/978-3-319-50496-4_85

Automatic Classification of Abusive Language and Personal Attacks in Various Forms of Online Communication

Peter Bourgonje, Julian Moreno-Schneider, Ankit Srivastava,
and Georg Rehm[(✉)]

Language Technology Lab, DFKI GmbH,
Alt-Moabit 91c, 10559 Berlin, Germany
georg.rehm@dfki.de

Abstract. The sheer ease with which abusive and hateful utterances can be made online – typically from the comfort of your home and the lack of any immediate negative repercussions – using today's digital communication technologies (especially social media), is responsible for their significant increase and global ubiquity. Natural Language Processing technologies can help in addressing the negative effects of this development. In this contribution we evaluate a set of classification algorithms on two types of user-generated online content (tweets and Wikipedia Talk comments) in two languages (English and German). The different sets of data we work on were classified towards aspects such as racism, sexism, hatespeech, aggression and personal attacks. While acknowledging issues with inter-annotator agreement for classification tasks using these labels, the focus of this paper is on classifying the data according to the annotated characteristics using several text classification algorithms. For some classification tasks we are able to reach f-scores of up to 81.58.

1 Introduction

Hateful conduct, abusive language and verbal aggression are by no means new phenomena. Comments and statements of this type seriously hamper a constructive private discussion or public debate. The sheer ease with which hateful utterances can be made – typically from the comfort of your home and the lack of any immediate negative repercussions – using today's digital communication technologies, is responsible for their significant increase and global ubiquity. In recent years, the topic has received an increasing amount of attention from multiple stakeholders. Among these are social scientists who want to analyse this phenomenon and reasons for abusive online behaviour and politicians who realise that major parts of public debates and social discourse are carried out online. In addition, we have seen that not only such online discussions but also the perception of concepts, politicians, elections and civil rights movements can be influenced using highly targeted social media marketing campaigns. We live in a time in which online media, including online news and online communication,

G. Rehm and T. Declerck (Eds.): GSCL 2017, LNAI 10713, pp. 180–191, 2018.
https://doi.org/10.1007/978-3-319-73706-5_15

have an unprecedented level of social, political and also economic relevance. This situation creates a plethora of challenges with regard to the key question how best to address the importance and relevance of online media and online content with technological means while at the same time not putting in place a centralised infrastructure that can be misused for the purpose of censorship or surveillance. One challenge is to separate high quality content from offensive, hateful, abusive or massively biased content. While these tasks have been mostly in the realm of journalism, they are getting more and more transferred to the end user of online content, i.e., the analysis, curation and assessment of information is no longer carried out by professional news editors or journalists exclusively – the burden of fact checking is more and more left to the reader.

In the social sciences and humanities, research on the phenomena and characteristics of Computer-Mediated Communication (CMC) has a long tradition. Initially, scholars concentrated on different types of novel communication media such as electronic mail, Internet Relay Chat (IRC), Usenet newsgroups, and different types of hypertext systems and documents, especially personal home pages, guestbooks and, later, discussion fora (Runkehl et al. 1998; Crystal 2001; Storrer 2001; Döring 2002). Early on, researchers focused upon the obvious differences between these new forms of written digital communication and the established, traditional forms, especially when it comes to linguistic phenomena that can be observed on the text surface, such as smileys and emoticons, specific acronyms and technological aspects of communication. Many authors observed that the different forms of internet-mediated communication have a certain *oral* and *spoken* style, quality and conceptualisation to them, as if produced spontaneously in a casual conversation, while, at the same time, being realised in a *written* medium (Haase et al. 1997).

If we now fast forward to 2017, a completely different picture emerges. About 40–50% of the global population has access to the Internet, most of whom also use the World Wide Web and one or more of the big social networks. The internet has become mainstream and acts like an amplifier, maybe also as an enabler, of social trends. We already mentioned some of the current challenges of this massive penetration of our lives through Internet-based forms of communication. The social, political and economic relevance of online media, online news and online communication could not be any more crucial. While early analyses and discussions of computer-mediated communication and discourse, e.g., (Reid, 1991), observed that their participants were involved in the "deconstruction of boundaries" and the "construction of social communities", today the exact opposite seems to be case: both offline and online can we observe the (disturbing) trend of increased nationalism and the exclusion of foreigners, immigrants and seemingly arbitrary minorities – boundaries are constructed, social communities deconstructed.

One last aspect is worth pointing out: up to now there has not really been any major need for automatic classification approaches of online content, with two notable exceptions. The first are online advertisements, either in the form of unsolicited spam email or in the form of online ads, either embedded in web documents or presented as pop-out windows. The second exception is sentiment analysis of social media data, driven by a clear use case: knowing what your customers or voters say or think about you as a company or politician. We are now slowly approaching a state of play, in which automatic means may be needed to classify online content or parts of online content into additional dimensions such as, for example, "hatespeech", "abusive language", maybe even "fake news" and "alternative facts". While spam mail can be classified and categorised with a fairly high accuracy (and online ads taken care of with an ad blocker), sentiment analysis already poses more difficult challenges (such as irony, sarcasm and scope issues). And it remains to be seen if abusive language can be identified accurately using automatic means and if additional analysis dimensions have to be incorporated if automatic approaches are to be used in a real application scenario.

The research presented in this paper has been carried out under the umbrella of a two-year research and technology transfer project. We collaborate with four SME partners that all face the challenge of having to process, to analyse and to make sense of large amounts of digital information. The four companies cover four different use cases and sectors (Rehm and Sasaki 2015), including journalism. For these partners we develop a platform that provides several semantic and knowledge technologies. In this article, we focus upon the application of several classification algorithms to establish the feasibility of the detection and classification of abusive language. We do this by evaluating the classification algorithms on three publicly available data sets. While the definition of abusive language, and, consequently, inter-annotator agreement of relevant corpora are crucial and far from resolved issues in this area of research, we focus on classification using linguistic features. Our goal is to establish a solid baseline for these three, publicly available corpora. The remainder of this paper is divided into four sections. Section 2 discusses related work, most notably on inter-annotator agreement, an important prerequisite for accurate classification. Section 3 describes the analysed and classified data sets. Section 4 discusses the results and Sect. 5 provides a summary and ideas for future work.

2   Related Work

Today, when it comes to the characteristics of online media and communication, several challenges are being discussed over and over again. We believe that the Natural Language Processing (NLP) and Language Technology (LT) communities can provide at least parts of the adequate social and technical solutions for, among others, hatespeech, fake news (including orchestrated disinformation campaigns), politically biased journalism, trolling, cyber-bullying and abusive language.

Hateful or aggressive conduct online has received an increasing amount of attention in recent years. For an accurate classification, however, consensus is needed on what constitutes abusive language, hate speech and aggressive conduct, and what is still acceptable within the boundaries of free speech.

While automated methods for detecting and classifying language use – such as spam vs. no spam – and sentiment – such as positive vs. negative – are typical application scenarios for NLP technologies, the question of what is acceptable and no longer acceptable within the boundaries of free speech puts this sensitive question and area of research into the intersection of different disciplines, including linguistics, sociology (Jones et al. 2013; Phillips 2015), psychology (Kowalski and Limber 2013; Dreißing et al. 2014), law (Marwick and Miller 2014; Banks 2010; Massaro 1991) and also common sense.

Many researchers approaching this topic acknowledge the difficulty in reaching a consensus. Ross et al. (2016) introduce a German corpus of hate speech on the European refugee crisis and report low inter-annotator agreement scores (Krippendorff's α between 0.18 and 0.29). Waseem (2016) investigates inter-annotator agreement comparing amateur annotations using CrowdFlower and expert annotations using precise instructions and reports a Cohen's Kappa of 0.14. Van Hee et al. (2015) work on classification of cyberbullying using a Dutch corpus and report Kappa scores between 0.19 and 0.69. Kwok and Wang (2013) investigate racist tweets and report an overall inter-annotator agreement of only 33%. Nobata et al. (2016) report a relatively high agreement for binary classification of clean vs. abusive for Yahoo! comments (Kappa = 0.843), but this number drops significantly when different subcategories for the abusive comments are introduced (such as hate, derogatory language and profanity, with Kappa decreasing to 0.456).

Another complicating issue is the fact that abusive language is often extra-linguistic in nature. Whether a particular utterance is considered abusive or not, often depends on other aspects including context, (ethnicity of the) author, (ethnicity of the) targeted person or group, etc. (Nand et al. 2016; Waseem and Hovy 2016; Warner and Hirschberg 2012). An excellent overview of NLP-based approaches towards hate speech detection is provided by Schmidt and Wiegand (2017).

In this paper we focus on the classification task and present several classification scores using multiple available data sets.

3   Data Sets

The experiments reported in this paper are conducted on three different data sets. The first one (ET, see Table 1) is provided by Waseem and Hovy (2016) and consists of English tweets. We scraped the tweets' actual content; of the 16,907 tweet IDs provided on the authors' GitHub page, we were able to retrieve 15,979 tweets (the smaller number most likely due to deleted tweets or time-outs during scraping). The tweets were classified into the classes *none* (10,939 instances), *sexism* (3,131 instances) or *racism* (1,909 instances). Despite missing out on 928 annotated tweets, the distribution over our version of the data set is the same as the one reported by Waseem and Hovy (2016), with respectively 68%, 20%, 12% of tweets being annotated as *none*, *sexist*, *racist*. For this and the other two data sets, Table 1 provides some examples.

Table 1. The three data sets – key facts and examples

Data Set: English Tweets (ET) – (Waseem and Hovy, 2016)

15,979 English language tweets (no. of words 273,805, avg.: 17)

Classes   *none* (10,939), *sexism* (3,131), *racism* (1,909)

Examples @Fixer_Turkey Why were innocent civilians in prison? *(none)*

@shaner38 Hope not. How will she pay her bills? *(sexism)*

@FalconEye123456 May Allah bless him with 72 virgin pigs. *(racism)*

Data Set: German Tweets (GT) – (Ross et al., 2016)

---

469 German language tweets (no. of words: 8,650, avg.: 18)

Classes   *hateful* (104), *non-hateful* (365)

Examples Deutsche Frauen an #rapefugees opfern. #wasistlinks *(hateful)*

Flüchten, wo andere Urlaub machen. #Idomeni #refugeesnotwelcome *(non-hateful)*

Gegen #Multikulti hab ich eigentlich nichts, gegen #Islamisierung schon. *(non-hateful)*

Data Set: Wikipedia Talk (WT) – (Wulczyn et al., 2016)

11,304 English language Wikipedia Talk comments (no. of words: 739,494, avg.: 65)

Classes   *aggression* (8,674) vs. *no aggression* (2,630) – *attack* (2,498) vs. *no attack* (8,806)

Examples You stick to your talk page, I'll d mine, right? 20: *(none)*

::::Yes, and Kudpung himself called for an admin's desysop in the section just above this one. What base hypocrisy. Perhaps he does not realize his own membership in his "anti-admin brigade", the existence of which he has never provided a shred of evidence for. *(attack)*

== Thomas W == : Don't bother telling him anything. He'll cry to his butt buddy Bishonen, who happens to have admin powers. *(aggression)*

== Suck it! == If you can't understand this common American idiom then perhaps you shouldn't be editing Wikipedia. At any rate, why are you monitoring my talk page, stalker? *(aggression)*

The second data set (GT, see Table 1) is provided by Ross et al. (2016) and consists of German tweets. With only 469 tweets, this data set is considerably smaller. They were annotated by two expert annotators who indicated a tweet to be either *hateful* or *not hateful*. In addition, the second annotator also scored the tweet on a scale of 1 (not offensive at all) to 6 (very offensive). The distribution of *hateful* vs. *non-hateful* tweets for annotator 1 was 110–359 and for annotator 2 it was 98–371.

The third data set (WT, see Table 1) is described by Wulczyn et al. (2016) and consists of user comments on Wikipedia Talk pages. This corpus is annotated for toxicity, aggression and personal attacks on users; the annotations are obtained through crowd-sourcing. Due to the sheer size of the complete data set, we only downloaded part of it (user comments from 2013 to 2015)[1] as well as the annotations for aggression and personal attacks.[2] This resulted in 11,304 annotated comments, 8.806 were annotated as cases of an *attack* and 2,498 as cases of *no attack*. 2,630 comments were annotated as containing *aggression* and 8.676 as *no aggression*. In the case of aggression, a rating was annotated as well. On a scale of -3 (very aggressive) to 3 (very friendly),[3] the distribution from -3 to 3 was as follows: -3: 772; -2: 635; -1: 1.223; 0: 7,623; 1: 717; 2: 243; 3: 91.

4 Evaluation

We applied a set of classification algorithms (Bayes, Bayes expectation maximization, C4.5 Decision Trees, Multivariate Logistic Regression, Maximum Entropy and Winnow2) on all three corpora using the Mallet Machine Learning for Language toolkit (McCallum 2002).[4] All classifiers use a Bag of Words (BOW) feature set (word unigrams). The figures in Table 2 are the result of

Table 2. Results of our classification experiments

	Bayes	Bayes exp. max.	C4.5	Logistic Regression	Maximum Entropy	Winnow2
English Tweets (ET)						
accuracy	84.61	84.01	82.95	**85.67**	83.67	76.66
precision	80.54	79.57	79.07	**83.57**	81.20	69.85
recall	**78.63**	77.97	74.37	77.45	74.37	69.62
f-score	79.10	78.34	76.17	**80.06**	77.20	69.32
German Tweets (GT) – (binary, exp. 2)						
accuracy	**80.21**	74.26	76.81	79.15	76.38	77.23
precision	72.76	73.59	72.54	**77.18**	73.62	74.65
recall	77.57	79.49	77.85	**79.74**	77.31	76.37
f-score	70.93	68.97	69.85	**75.41**	74.20	73.05
Wikipedia (WT) – *Attack* (binary)						
accuracy	**83.11**	82.70	81.08*	80.90	77.71	77.77
precision	**81.78**	81.33	79.27*	79.36	76.03	77.11
recall	**83.14**	82.83	81.31*	80.97	77.87	77.83
f-score	**81.58**	81.36	79.27*	79.74	76.65	77.28
Wikipedia (WT) – *Aggression* (rating)						
accuracy	67.13	**67.40**	66.81*	65.28	57.77	55.73
precision	57.21	56.05	54.08*	**57.42**	57.21	54.07
recall	67.27	66.94	66.42*	65.68	58.18	55.73
f-score	59.13	59.00	58.14*	**59.95**	55.26	54.53

	Bayes	Bayes exp. max.	C4.5	Logistic Regression	Maximum Entropy	Winnow2
German Tweets (GT) – (binary, exp. 1)						
accuracy	75.74	**78.93**	74.04	77.23	75.96	71.91
precision	70.65	**75.07**	69.30	74.80	72.46	72.41
recall	74.78	76.06	74.98	**76.58**	74.85	72.68
f-score	65.84	69.74	70.66	71.98	**73.02**	71.15
German Tweets (GT) – (rating)						
accuracy	36.60	35.32	**37.87**	33.40	34.89	25.53
precision	42.51	39.76	**56.22**	31.39	31.90	38.17
recall	38.53	38.19	**38.76**	36.34	35.71	25.84
f-score	27.43	27.03	23.68	30.34	**30.75**	24.06
Wikipedia (WT) – *Aggression* (binary)						
accuracy	**82.19**	82.10	79.58*	80.42	77.17	79.08
precision	**80.68**	80.60	78.13*	78.91	75.26	77.25
recall	**82.01**	81.87	80.18*	80.46	77.29	78.57
f-score	**80.60**	80.57	78.37*	79.23	75.80	77.45

---

[1] https://figshare.com/articles/Wikipedia_Talk_Corpus/4264973.

[2] https://figshare.com/projects/Wikipedia_Talk/16731.

[3] While the documentation states a range from -2 to 2, we actually found a range of -3 to 3 in the annotations (https://meta.wikimedia.org/wiki/Research:Detox/Data_Release).

[4] http://mallet.cs.umass.edu/api/.

ten-fold cross-validation[5] with a 90–10 distribution of training and test data. Note that the table incidentally contains f-scores that are lower than their corresponding precision and recall scores due to averaging over the precision, recall and f-scores for every class that exists in the data.

While Waseem and Hovy (2016) report better results for character n-grams compared to word n-grams (73.89 vs. 64.58) on their data set (ET), Mallet's logistic regression implementation, using word unigrams, outperforms the best scoring feature set in Waseem and Hovy (2016) (i. e., 80.06 vs. 73.89). The influence of using character n-grams vs. word n-grams may be language dependent. Nobata et al. (2016) report better performance with character n-grams, while Van Hee et al. (2015) report better performance with word n-grams (on a Dutch corpus) but in the above example, the same language and data is used. The type of classifier may also influence the features that are used, but Waseem and Hovy (2016) also use a logistic regression classifier. We have experimented with other features, such as word-ngrams, character-ngrams, and for the tweets cleaned the content using a set of regular expressions,[6] but the best results were obtained with BOW features, as reported in Table 2. In addition, the most informative features are shown in Fig. 1.

The data set of Ross et al. (2016) is significantly smaller (GT). The annotations were done by two expert annotators. We have trained the set of classifiers twice, using these annotations (binary, expert 1 and binary, expert 2), and have made no attempts to resolve a final or definitive label, exactly because of the problems with inter-annotator agreement. While the results for the best scoring algorithm in the case of binary classification is still reasonable, performance

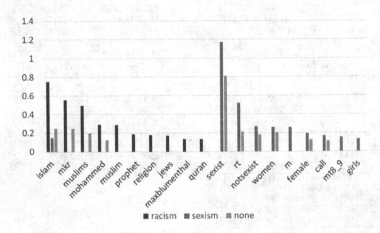

Fig. 1. The ten most informative features for classes in the (Waseem and Hovy, 2016) data set

---

[5] Except for the numbers marked with an asterisk; these are the result of three-fold cross-validation due to the large amount of time needed for training and execution.

[6] The set we used is inspired on the script available at https://nlp.stanford.edu/projects/glove/preprocess-twitter.rb.

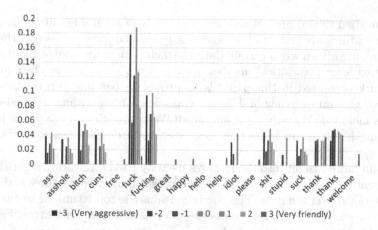

Fig. 2. The ten most informative features for classes in the Wikipedia talk aggression ratings data set

drops dramatically when using multi-label classification (six classes). The figures for the GT data set are of limited reliability due to its small size. Ross et al. (2016) do not report classification results and focus on inter-annotator agreement. We include the results in Table 2, but due to large score deviations for individual runs and for reasons of brevity, do not discuss the results for this data set.

For our subsection of the WT data set (Wulczyn et al., 2016), we see a similar pattern in the sense that binary classification scores are reasonable (81.58 and 80.60 for Bayes) but performance drops significantly when using multi-class classification (seven classes).

Wulczyn et al. (2016) do not mention any f-scores, but they do refer to experiments using their best performing personal attack classifier on comments made in 2015 (excluding "administrative comments and comments generated by bots"). The different setup in this study makes it hard to compare, as they are not performing binary classification directly, but assign a continuous score between 0 and 1. As the authors note, "even though the thresholded model-scores give good estimates of the rate of attacks over a random sample of comments, it is not given that they also give accurate estimates when partitioning comments into different groups". Using this method, however, the authors do report a precision of 0.63 and a recall of 0.63, when using a threshold value of 0.425.

In an attempt to get a better overview of which classification algorithms work best for which type of data, we found that the preferred classification algorithm is highly data-specific. For example, Ranawana and Palade (2006) provide an overview of multi-class classification algorithms and identify the most distinct features in order to combine several algorithms for one classification problem. In their experiments, they work on a data set of human DNA from the Berkeley Drosophila Genome Project website. Andreopoulos et al. (2009) describes a variety of clustering algorithms applied to the biomedical domain. Caruana and

Niculescu-Mizil (2006) present an empirical study of several learning algorithms and their features applied on 11 data sets. It remains difficult to predict the best-performing algorithm when certain data characteristics (like size of the data set, average text length, variation, number of classes, etc.) are known. The main goal of the work presented in this paper is to provide a baseline performance indication and give an overview of different classification algorithms applied on the data sets included. We only use a simple BOW approach and the job of feature engineering is left as an important next step towards classification of abusive language.

To gain more insight into the features used by the best scoring algorithm for the different data sets, we extracted the ten most informative features and report scores averaged over ten runs. The x-axis contains the top 10 unique words for all existing classes. The y-axis represents the information gain per feature. Features for the first data set are shown in Fig. 1 for the three existing classes. The only overlap between the "racism" and "sexism" features (in the top 10) is *islam*, which was apparently also a relatively frequent term in the tweets classified as "sexist" in the training set. The considerable overlap between the "none" class and the other two is likely to reflect the fact that not all tweets containing these words were annotated as either "racist" or "sexist". As also noted by Warner and Hirschberg (2012), classification of abusive language can be conceptualised as a word disambiguation task: sometimes the usage of a word is abusive, sometimes it is not. The features that are less straightforward as words represent the name of a cooking show (mkr),[7] the established abbreviation of "retweet" (rt), the twitter handle of a user ($mt8\_9$)[8] and (probably) the result of Mallet tokenising the m in *I'm* as a separate word.

The ten most informative features for the classification task of *aggression* on a scale of -3 (very aggressive) to 3 (very friendly) in the WT data set for the best performing algorithm (Bayes) are shown in Fig. 2.[9] Remarkably, the top 10 most informative features (words) for the seven classes are represented by only 18 unique words. The words in this set associated with friendliness ("welcome", "please", "great" and "happy", for example) are only present in class 3. On the other end of the spectrum, class -3 only contains one word often associated with friendliness ("thanks"). Overall, there is a large degree of overlap between the classes, suggesting that the lower-ranked features also play an important role in classification. Upon manual investigation of the lower-ranked features, we found that the neutral class (0) seemed to function mostly as background noise, canceling out both extremes, as is the case for the classifier for the Waseem and Hovy (2016) data set. The negative digit classes (-1, -2 and -3) contain a large vocabulary of curse words, whereas the positive digit classes (1, 2 and 3) contain more communicative, constructive and cooperative terms like "ask", "questions", "discussion", etc.

---

[7] https://en.wikipedia.org/wiki/My_Kitchen_Rules.

[8] https://twitter.com/mt8_9.

[9] Due to the small corpus size of the GT data set, we refrain from showing the ten most informative features here.

5   Conclusion

We apply a range of classification algorithms on three data sets that differ in language (usage), size and domain/genre: A corpus of English tweets annotated for racist and sexist tweets (ET); a corpus of German tweets annotated for hate speech (GT); and a corpus of English Wikipedia user comments who, unlike tweets, have less strict length restrictions (WT). While many studies on this topic focus on inter-annotator agreement (Sect. 2), we establish a baseline for classification based on these three corpora and data sets. We describe the differences between the performance of different classification algorithms and the features used by the respective best performing algorithm. Although consensus on what needs to be detected, which is typically reflected by inter-annotator agreement, is important to construct relevant training corpora, our results indicate that automatic classification can provide reasonable results and does not have to be a bottle neck in attempts to automatically detect abusive language. Nevertheless, further research is needed to reach an agreement on definitions of abusive language, hate speech, hateful conduct, cyber-bullying and other phenomena of online communication in order to arrive at accurate and meaningful classification approaches. Additionally, the simple feature set (BOW) used in the experiments can and should be complemented with more semantically and context-aware components such as sentiment values, dependency parsing (to provide insight on scope of abusive elements), and other more sophisticated NLP techniques. Finally, we would like to emphasise that the extra-linguistic nature of abusive behaviour and the fact that, apart from language usage, accurate detection depends on the author, target audience, communicative intent and other context specifics, is not addressed in this paper and, thus, left as an important part of future work in this area.

Acknowledgments. The authors wish to thank the anonymous reviewers for their helpful feedback. The project "Digitale Kuratierungstechnologien (DKT)" is supported by the German Federal Ministry of Education and Research (BMBF), "Unternehmen Region", instrument "Wachstumskern-Potenzial" (No. 03WKP45). More information at http://www.digitale-kuratierung.de.

References

Andreopoulos, B., An, A., Wang, X., Schroeder, M.: A roadmap of clustering algorithms: finding a match for a biomedical application. Briefings in Bioinformatics 10(3), 297–314 (2009). https://doi.org/10.1093/bib/bbn058. ISSN 1477-4054

Banks, J.: Regulating hate speech online. Int. Rev. Law Comput. Technol. 24(3), 233–239 (2010)

Caruana, R., Niculescu-Mizil, A.: An empirical comparison of supervised learning algorithms. In: Proceedings of the 23rd International Conference on Machine Learning, pp. 161–168. ACM (2006)

Crystal, D.: Language and the Internet. Cambridge University Press, Cambridge (2001)

Döring, N.: Personal home pages on the web: a review of research. J. Comput.-Mediat. Commun. 7(3) (2002). http://www.ascusc.org/jcmc/

Dreißing, H., Bailer, J., Anders, A., Wagner, H., Gallas, C.: Cyberstalking in a large sample of social network users: prevalence, characteristics, and impact upon victims. Cyberpsychology Behav. Soc. Netw. **17**(2), 61–67 (2014)

Haase, M., Huber, M., Krumeich, A., Rehm, G.: Internetkommunikation und Sprachwandel. In: Weingarten, R. (ed.) Sprachwandel durch Computer, pp. 51–85. Westdeutscher Verlag, Opladen (1997). https://doi.org/10.1007/978-3-322-91416-3_3

Jones, L.M., Mitchell, K.J., Finkelhor, D.: Online harassment in context: trends from three youth internet safety surveys (2000, 2005, 2010). Psychol. Violence **3**(1), 53 (2013)

Kowalski, R.M., Limber, S.P.: Psychological, physical, and academic correlates of cyberbullying and traditional bullying. J. Adolesc. Health **53**(1), S13–S20 (2013)

Kwok, I., Wang, Y.: Locate the hate: detecting tweets against blacks. In: Proceedings of the Twenty-Seventh AAAI Conference on Artificial Intelligence, AAAI 2013, pp. 1621–1622. AAAI Press (2013). http://dl.acm.org/citation.cfm?id=2891460.2891697

Marwick, A.E., Miller, R.W.: Online harassment, defamation, and hateful speech: a primer of the legal landscape. Fordham Center on Law and Information Policy Report, June 2014

Massaro, T.M.: Equality and freedom of expression: the hate speech dilemma. William Mary Law Rev. **32**(211) (1991)

McCallum, A.K.: MALLET: a machine learning for language toolkit (2002). http://www.cs.umass.edu/~mccallum/mallet

Nand, P., Perera, R., Kasture, A.: "How bullying is this message?": a psychometric thermometer for bullying. In: Proceedings of COLING 2016, the 26th International Conference on Computational Linguistics: Technical Papers, Osaka, Japan, pp. 695–706. The COLING 2016 Organizing Committee, December 2016. http://aclweb.org/anthology/C16-1067

Nobata, C., Tetreault, J., Thomas, A., Mehdad, Y., Chang, Y.: Abusive language detection in online user content. In: Proceedings of the 25th International Conference on World Wide Web, WWW 2016, Republic and Canton of Geneva, Switzerland, pp. 145–153. International World Wide Web Conferences Steering Committee (2016). https://doi.org/10.1145/2872427.2883062, ISBN 978-1-4503-4143-1

Phillips, W.: This Is Why We Can't Have Nice Things: Mapping the Relationship Between Online Trolling and Mainstream Culture. The MIT Press, Cambridge (2015)

Ranawana, R., Palade, V.: Multi-classifier systems: review and a roadmap for developers. Int. J. Hybrid Intell. Syst. **3**(1), 35–61 (2006). http://dl.acm.org/citation.cfm?id=1232855.1232859, ISSN 1448-5869

Rehm, G., Sasaki, F.: Digitale Kuratierungstechnologien - Verfahren für die effiziente Verarbeitung, Erstellung und Verteilung qualitativ hochwertiger Medieninhalte. In: Proceedings of the 2015 International Conference of the German Society for Computational Linguistics and Language Technology, GSCL 2015, pp. 138–139 (2015)

Reid, E.M.: Electropolis: communication and community on internet relay chat, Honours thesis, University of Melbourne, Department of History (1991). http://www.aluluei.com/electropolis.htm

Ross, B., Rist, M., Carbonell, G., Cabrera, B., Kurowsky, N., Wojatzki, M.: Measuring the reliability of hate speech annotations: the case of the European refugee crisis. In: Beißwenger, M., Wojatzki, M., Zesch, T. (eds.) Proceedings of NLP4CMC III: 3rd Workshop on Natural Language Processing for Computer-Mediated Communication, Bochumer Linguistische Arbeitsberichte, Bochum, vol. 17, pp. 6–9, September 2016

Runkehl, J., Schlobinski, P., Siever, T.: Sprache und Kommunikation im Internet - Überblick und Analysen. Westdeutscher Verlag, Opladen (1998)

Schmidt, A., Wiegand, M.: A survey on hate speech detection using natural language processing. In: Proceedings of the Fifth International Workshop on Natural Language Processing for Social Media, Valencia, Spain, pp. 1–10. Association for Computational Linguistics, April 2017. http://www.aclweb.org/anthology/W17-1101

Storrer, A.: Getippte Gespräche oder dialogische Texte? Zur kommunikationstheoretischen Einordnung der Chat-Kommunikation. In: Lehr, A., Kammerer, M., Konerding, K.-P., Storrer, A., Thimm, C., Wolski, W. (eds.) Sprache im Alltag. Beiträge zu neuen Perspektiven der Linguistik, pp. 439–466. de Gruyter, Berlin (2001). Herbert Ernst Wiegand zum 65. Geburtstag gewidmet

Van Hee, C., Lefever, E., Verhoeven, B., Mennes, J., Desmet, B., De Pauw, G., Daelemans, W., Hoste, V.: Detection and fine-grained classification of cyberbullying events. In: Proceedings of the International Conference Recent Advances in Natural Language Processing, pp. 672–680. INCOMA Ltd., Shoumen (2015). http://aclweb.org/anthology/R15-1086

Warner, W., Hirschberg, J.: Detecting hate speech on the world wide web. In: Proceedings of the Second Workshop on Language in Social Media, LSM 2012, Stroudsburg, PA, USA, pp. 19–26. Association for Computational Linguistics (2012). http://dl.acm.org/citation.cfm?id=2390374.2390377

Waseem, Z.: Are you a racist or am i seeing things? annotator influence on hate speech detection on Twitter. In: Proceedings of the First Workshop on NLP and Computational Social Science, Austin, Texas, pp. 138–142. Association for Computational Linguistics, November 2016. http://aclweb.org/anthology/W16-5618

Waseem, Z., Hovy, D.: Hateful symbols or hateful people? predictive features for hate speech detection on Twitter. In: Proceedings of the NAACL Student Research Workshop, San Diego, California, pp. 88–93. Association for Computational Linguistics, June 2016. http://www.aclweb.org/anthology/N16-2013

Wulczyn, E., Thain, N., Dixon, L.: Ex machina: personal attacks seen at scale. CoRR, abs/1610.08914 (2016). http://arxiv.org/abs/1610.08914

Token Level Code-Switching Detection Using Wikipedia as a Lexical Resource

Daniel Claeser[✉], Dennis Felske[✉], and Samantha Kent[✉]

Fraunhofer FKIE, Fraunhoferstraße 20, 53343 Wachtberg, Germany
{daniel.claeser,dennis.felske,samantha.kent}@fkie.fraunhofer.de

Abstract. We present a novel lexicon-based classification approach for code-switching detection on Twitter. The main aim is to develop a simple lexical look-up classifier based on frequency information retrieved from Wikipedia. We evaluate the classifier using three different language pairs: Spanish-English, Dutch-English, and German-Turkish. The results indicate that our figures for Spanish-English are competitive with current state of the art classifiers, even though the approach is simplistic and based solely on word frequency information.

1 Introduction

Code-switching (CS) or code-mixing can be defined as a linguistic phenomenon in which multilingual speakers use languages interchangeably. A distinction is made between inter-sentential CS, where the switch occurs at sentence level, and intra-sentential CS, where the switch occurs within a sentence at the phrase or word level (Bullock and Toribio 2009). In turn, intra-sentential CS can be divided into two different types. Alternation is the switching of different languages whilst keeping the grammatical structure of each language intact. Contrastingly, in insertion, lexical items from one language are included within the grammatical structure of another (Muysken 2000).

In recent years, multilingual written communication that includes these different types of CS has become more prevalent and there has been a growing interest in the automatic identification of codeswitched language on social media. This paper seeks to contribute to that growing body of work and introduce a simple lexical look-up classifier that identifies code-switching between different languages on Twitter. The focus lies on three different language pairs: Spanish-English, Dutch-English, and German-Turkish.

2 Related Work

Currently, a range of different classification approaches have been presented for a variety of different languages (see Solorio et al. (2014) and Molino et al. (2016) for a current overview). Methods vary from the use of more complex deep learning algorithms (Jaech et al. 2016) to a range of different lexicon-based approaches, those of which are most relevant to our approach are discussed in turn below.

© The Author(s) 2018
G. Rehm and T. Declerck (Eds.): GSCL 2017, LNAI 10713, pp. 192–198, 2018.
https://doi.org/10.1007/978-3-319-73706-5_16

Maharjan et al. (2015) compare a lexical lookup classification approach to current state of the art classifiers in order to identify Spanish-English code-switched Tweets. They demonstrate a very simple dictionary approach in which the classifier assigns a language based on whether the token was present in a Spanish or English dictionary. If the token is either present in both dictionaries or absent in both dictionaries, the tag associated with the majority language in the training data is assigned. The results illustrate that the most elementary binary classification approach yields an F1 score of 0.61 at Tweet level and 0.73 at token level.

Chanda et al. (2016) combine a dictionary method with the use of n-gram categorization and additional processing of Bengali suffixes in order to identify English-Bengali CS. In contrast to the previous binary classification approach, the lexical look-up in the English, Bengali, and hand-crafted slang dictionary does not solely determine the language of the token. The inclusion of the additional features yields an accuracy level of 86.27%. Notably, when predicting the languages based solely on the dictionary approach an accuracy level similar to the approach outlined above is achieved (72.54%). In a further step, the authors compare various machine learning techniques to construct the actual classifier and find that the most accurate results were achieved using an IBk algorithm, where an accuracy of 90.54% is reached in their social media corpus (compared to 91.65% in their corpus not extracted from social media).

Shirvani et al. (2016) combine 14 different features, including character n-grams, prefixes and suffixes, a Spanish-English dictionary, Spanish and English Part-Of-Speech (POS) tagging, Brown clustering, as well as a number of additional binary features. Logistic regression is used to determine the probability of the various possible labels using different combinations of these 14 features. This language classifier is more complex and contains more features compared to the previous approach. The results indicate a further increase in overall performance with a weighted F1 score of 91.3% at Tweet level and an overall accuracy of 97.3% at token level, with 93.8% and 98.4% for English and Spanish respectively.

3 Datasets

We used *corpora* that were previously collected and annotated to evaluate the classifier. The English-Spanish Twitter corpus was provided for the Shared Task Challenge for EMNLP 2016 (Molina et al. 2016), the Turkish-German corpus was created by Çetinoğlu (2016) and the Dutch-English Twitter corpus was provided by the University of Amsterdam (Dongen 2017). The first two corpora are distributed in the form of Tweet IDs that are to be downloaded using the Twitter API. However, due to Twitter's policy, we were only able to download a fraction of the original corpora, even though both were assembled in 2016. Overall, we managed to procure a total of 1028 Tweets (7133 tokens) from the English-Spanish corpus and 145 Tweets (1720 tokens) from Turkish-German corpus. The Dutch English corpus was provided in plain text and we thus managed to reproduce it completely (1284 Tweets, 16050 tokens). For each of the language pairs, we were able to use the full Twitter corpus as an evaluation set because we did not need a training corpus.

The *dictionaries* were built using the Wikipedia dumps for each of the five languages in the corpora (version: "all pages with complete edit history" on 01/03/2017). Crucially, those packages contain both the user discussion sections for each article as well as the actual article itself. The input size of the dictionaries varies for each of the languages, the influence of which will be discussed in the results section. The dictionaries were created as follows. The basic format is a token list which was obtained by parsing the Wikipedia dumps for the respective language. The raw input was stripped of all special characters before being changed to lower case, tokenised and ranked according to their frequency. Later, the dictionaries were cropped at 5 million types each, based on the idea that tokens that rank lower than 5 million are mostly hapax legomena.

Using Wikipedia as the input source to create the dictionaries has a number of different advantages. Firstly, it is freely available and distributable under the CC license, free of charge, and easy to access for any language that is present on Wikipedia. Secondly, due to the fact that the dictionaries contain text from both the articles and the comments section, we captured both formal and informal language. Consequently, the dictionaries contain a wide range of vocabulary, ranging from subject specific vocabulary to a variety of different abbreviations. This turned out to be a crucial aspect, because almost all tokens in the Tweets are present in the dictionaries even though language on social media is characteristically colloquial. Furthermore, Wikipedia tends to reflect current topics, even if there is a delay compared to social media, which ensures that the vocabulary in the dictionaries is up to date.

4   Classification

The classification process is based on a number of different assumptions. Firstly, we assume that if the dictionaries are large enough, all tokens will be present in all dictionaries, regardless of the language. Crucially however, the rank of the token will be different in each of the dictionaries, and it is likely that a word stems from the language in which the rank in the dictionary is the highest. So in the first step of the token-level classification of Tweets, the rank of the token is retrieved from the respective language dictionaries, and the language in which the rank is highest is assigned to the token. In the rare case that a token is in fact not present in the dictionaries the tag 'none' is assigned. In the final step, all tokens that are classified as 'none' are assigned to the majority language of the Tweet.

Secondly, the assumption is made that some tokens are not unique to a specific language. This is particularly true for language pairs such as Dutch-English, which share many overlapping lexical items. So in order to account for orthographically identical words that are frequently used in both languages, two further steps are introduced. In the first of these steps, tokens that are ranked very highly in both languages simultaneously, for example the word 'me' in English and Spanish, are considered to be grammatical function words that are identical in both languages and thus should be assigned to both languages. Therefore, they are initially tagged as 'ambiguous'. The rank threshold at which tokens are classified as 'ambiguous' was iteratively determined to be 702 for EN-ES, 127 for EN-NL and 112 for DE-TR. This tag 'ambiguous' is only temporary, and once the classification process has been completed it is reassigned to the majority language found in the Tweet.

In the second additional step, a context-based rule is introduced to account for tokens that are being misclassified because they are orthographically identical and frequent in both languages, but are not categorized as grammatical function words. In these cases, if the language of both the preceding and following word is the same, the token is reassigned to match that language. This step accounts for words that are borrowed from another language and have been integrated into the lexicon and should therefore not be classified as codeswitching. However, this step is only incorporated if the ranks of the particular token in the respective dictionaries are sufficiently similar. The maximum distance between the ranks was iteratively determined for each language pair and is 16.000 for EN-ES, 27.000 for EN-NL and 0 for DE-TR.

5   Results

Table 1 shows the results of the classification process described above. Note that these figures include all tokens found in the Tweets that have been labeled as either one of the languages in the language pair. This does not include Twitter handles and hashtags or emoticons, as these were classified as 'other' and excluded from further evaluation. In general, the performance of the classifier has exceeded our prior expectations, with an F1 as high as 0.963 and 0.983 for the Spanish-English Tweets. However, it is also evident that there are still some challenges to overcome and that each language pair has particular characteristics that influence the performance of the classifier.

Table 1. Evaluation of token-based classification

	Spanish (EN-ES)	English (EN-ES)	English (EN-NL)	Dutch (EN-NL)	German (DE-TR)	Turkish (DE-TR)
P	.964	.985	.453	.995	.915	.939
R	.962	.980	.618	.822	.845	.771
F1	.963	.983	.524	.900	.879	.847

Spanish-English outperformed the other language pairs with precision and recall competitive to the state of the art as presented in Shirvani et al. (2016). Accounting for a significant proportion of misclassified tokens are either words containing irregular orthography, such as 'ooooooooommmmmmmg', 'noooooo', and 'meee'. All of these tokens could be normalised in a pre-processing step in order to improve performance. Having said that, the majority of words containing deviating spelling or abbreviations, for example 'jajajaj' ('hahaha') in Spanish and 'btw' ('by the way') in English, are actually captured by the classifier. This suggests that the incorporation of both the Wikipedia article and the comment section is an important element in this classification approach. Furthermore, some words were misclassified because they are present in both languages. Examples of homonyms include the verb forms 'prove', 'embrace' and 'continue'. The status of several other tokens, such as 'ugh', 'ahh', 'pfft' and 'wey', could be contested as they do not strictly belong to either language. Interestingly, the noun 'broncos' was false-positively identified as English, probably based on the occurrences of the Denver American Football team in the EnglishWikipedia. The classifier does not have a separate named entities tagger and named entities are considered to be part of the language of origin. This means that named entities correctly identified as Spanish are for example 'san antonio', 'gloria trevi' and 'san marcos'.

Turkish-German was the second best performing language pair. The Turkish dictionary was the smallest one available, but since Turkish and German are part of very different language families and are the least similar, the pair performed with F1 scores of 0.847 and 0.879. Similar to the Spanish-English corpus, misclassified tokens consist of words with incorrect or irregular spelling, such as '*verstandnis' ('understanding'), '*nasilsn' ('how are you'), '*anlatacann' (abbr. 'you will tell') and '*seniiii gelisiniiii bekliyorum gözleee' ('looking forward to seeing from you soon'), while '*insallahhhh' ('god willing') and '*anladiiim' ('i understand') were correctly identified as Turkish. Named identities assigned to the correct language by the classifier include 'Bahar', 'Sezen Aksu', 'Ezel', 'Tolga Cigerci', 'Frau Geiger', 'Galatasaray' and 'Bochum'. The small size of the Turkish dictionary, however, had obvious influence on recall for Turkish as several inflected forms of otherwise common lemmata such as 'kıyımıza' ('to our shores') and 'domuzköpeği' (ad-hoc literal translation from German idiom 'innerer Schweinehund', 'inner temptation') were not originally recognized as Turkish by the dictionary look-up, but reassigned to Turkish by the context rule. This rule did not manage to capture all such forms evading the dictionary based assignment and failed with tokens such as '*yalmazdım' ('i would not write'), '*varediyorsun' ('you create'), 'çiğdemim' ('my crocus', 'my love'). These examples highlight a characteristic property of Turkish: as an agglutinating language, Turkish attaches a broad range of particles, for example the prepositions 'de'/'da' ('in', 'with') or possessive pronouns like 'im' (my), directly to the words in the open word classes. However, these patterns are highly regular and thus accessible for resolution through parsing. Applying a morphological parser to words not found in the base lexicon might thus greatly improve recall on inflected Turkish words or phrases and compensate for the lack of entries in the smaller dictionary. Such a parser-based dictionary compensation mechanism might additionally improve recall on intra-word level codeswitched tokens such as 'partyler' ('party' + plural morpheme), 'deutschlandda' ('germany' +

preposition 'in') and 'schatzim' ('darling' + possessive pronoun 'my'), which are all German lemmata combined with Turkish morphemes denoting plural, location and possession. Çetinoğlu (2016) gives further examples of intra-word code-switching in the corpus.

Out of the three language pairs examined in this paper, English-Dutch is the most similar. It must also be taken into account that the English-Dutch corpus was constructed to examine code mixing more generally, rather than to evaluate automatic language classifiers. This means that there is less CS in this corpus than in the other two language pairs and there are more single word inclusions as opposed to intra-sentential CS. In general, Dutch contains many words that have been borrowed from English and have either been fully integrated into the Dutch vocabulary or are used even though there are Dutch equivalents. Consequently, the Dutch dictionary contains many English words. Words such as 'arrogant', 'stress', 'weekend', and 'incident', have been tagged as English, but they should be classified as Dutch because there are no other Dutch equivalents and therefore they cannot be considered to be CS. Contrastingly, words such as 'happy', 'same', and 'highlights', are English words with Dutch equivalents that are used frequently in the Dutch language, but are incorrectly classified as Dutch. They should be classified as English and considered to be CS, but are misclassified due to the context rule. The many overlapping lexical items explain why the F1 score for English is much lower when combined with Dutch than it is with Spanish.

6 Conclusion and Future Work

We presented a simple dictionary-based classification system for the identification of CS on Twitter. The results for Spanish-English are comparable to current state of the art classifiers even though the approach taken in this paper is much more simplistic. The classifier does not need any external toolkits or additional features such as a POS tagger or suffix information as it relies solely on the frequency information of the tokens within each dictionary. The use of Wikipedia as a dictionary resource has allowed for the classification of formal language as well as the colloquial language that is characteristic of language on social media. The classifier managed to successfully identify the language of sequences such as 'jajajaj' and 'omg' automatically. Nevertheless, many irregular tokens were not identified correctly and the addition of a token simplification rule, to reduce sequences such as 'noooooo' to 'no', would improve performance. This approach can be adapted to any language as long as the resources on Wikipedia are available and of an appropriate size. The difficulty lies in finding appropriate CS material on which to train and test the classifier. Once new Twitter corpora containing CS have been created, we plan on incorporating a wider variety of languages and focusing on how to improve the classification of closely related languages.

References

Bullock, B.E., Toribio, A.J.: Themes in the study of code-switching. In: Cambridge Handbooks in Language and Linguistics, pp. 1–18. Cambridge University Press (2009). https://doi.org/10.1017/CBO9780511576331.002

Çetinoglu, Ö.: A Turkish-German code-switching corpus. In: LREC (2016)

Chanda, A., Das, D., Mazumdar, C.: Unraveling the English-Bengali code-mixing phenomenon. In: EMNLP 2016, p. 80 (2016)

Dongen, N.: Analysis and prediction of Dutch-English code-switching in Dutch social media messages. Master's thesis, Universiteit van Amsterdam (2017)

Jaech, A., Mulcaire, G., Hathi, S., Ostendorf, M., Smith, N.A.: A neural model for language identification in code-switched tweets. In: EMNLP 2016, p. 60 (2016)

Maharjan, S., Blair, E., Bethard, S., Solorio, T.: Developing language-tagged corpora for code-switching tweets. In: LAW@ NAACL-HLT, pp. 72–84 (2015)

Molina, G., AlGhamdi, F., Ghoneim, M., Hawwari, A., Rey-Villamizar, N., Diab, M., Solorio, T.: Overview for the second shared task on language identification in code-switched data. In: Proceedings of the Second Workshop on Computational Approaches to Code Switching, pp. 40–49 (2016)

Muysken, P.: Bilingual Speech: A Typology of Code-Mixing, vol. 11. Cambridge University Press, Cambridge (2000)

Shirvani, R., Piergallini, M., Gautam, G.S., Chouikha, M.: The Howard University system submission for the shared task in language identification in Spanish-English codeswitching. In: Proceedings of The Second Workshop on Computational Approaches to Code Switching, pp. 116–120 (2016)

Solorio, T., Blair, E., Maharjan, S., Bethard, S., Diab, M., Ghoneim, M., Hawwari, A., AlGhamdi, F., Hirschberg, J., Chang, A., et al.: Overview for the first shared task on language identification in code-switched data. In: Proceedings of the First Workshop on Computational Approaches to Code Switching, pp. 62–72 (2014)

How Social Media Text Analysis Can Inform Disaster Management

Sabine Gründer-Fahrer[1]([✉]), Antje Schlaf[1], and Sebastian Wustmann[2]

[1] Institute for Applied Informatics, Hainstraße 11, 04109 Leipzig, Germany
gruender@uni-leipzig.de, antje.schlaf@informatik.uni-leipzig.de
[2] CID GmbH, Gewerbepark Birkenhain 1, 63579 Freigericht, Germany
s.wustmann@cid.de

Abstract. Digitalization and the rise of social media have led disaster management to the insight that modern information technology will have to play a key role in dealing with a crisis. In this context, the paper introduces a NLP software for social media text analysis that has been developed in cooperation with disaster managers in the European project *Slandail*. The aim is to show how state-of-the-art techniques from text mining and information extraction can be applied to fulfil the requirements of the end-users. By way of example use cases the capacity of the approach will be demonstrated to make available social media as a valuable source of information for disaster management.

1 Introduction

The emerging field of *crisis informatics* (e.g., Palen et al. (2010)) is driven by the insight that, in the digital age, the ability to efficiently access and process huge amounts of unstructured data is crucial to situational awareness, knowledge building, and decision-making of organizations responsible for saving lives and property of people affected by a crisis. Disaster events like hurricane Katrina, 9/11, the Haiti earthquake, or the Central-European Flooding 2013 have demonstrated that there is urgent need to understand how information is shared during a crisis and to improve strategies and technologies for turning information into relevant insights and timely actions. Within crisis informatics, social media offer an interesting new opportunity for improvement of disaster management by providing fast, interactive communication channels and enabling participation of the public (Starbird and Palen 2011). However, social media data are *big data* in terms of volume, velocity, variety and veracity, and, accordingly, the demands and challenges with respect to the development of appropriate information technologies are especially high.

The paper presents possibilities for social media analysis that arise within a disaster management software that has been developed as part of the *Slandail* project (Slandail 2014), funded by the European community. *Slandail* deals with data in different modalities (texts and images) and languages (English, German and Italian) as well as with the integration of cross-lingual and cross-cultural

G. Rehm and T. Declerck (Eds.): GSCL 2017, LNAI 10713, pp. 199–207, 2018.
https://doi.org/10.1007/978-3-319-73706-5_17

aspects of crisis communications and has a special focus on issues related to the legal and ethical correctness of data use. End-users from Ireland, Germany, and Italy have been involved in the development of the system from design to testing.

The focus of the paper is on text analysis functionalities using NLP methods from the fields of text mining and information extraction that have been contributed by the two German partner organizations in cooperation with the disaster control authorities Landeskommando and Bezirksverbindungskommando in Saxony. The prototype of the software (Topic Analyst) has been implemented at CID and further developed in cooperation with InfAI during the course of the *Slandail* project. The software module is currently under consideration by German authorities for future use in German disaster management.

2   Approach

Computational methods from NLP offer a wide variety of possibilities for systematically and efficiently searching, filtering, sorting and analyzing huge amounts of data and thereby can enable end-users from disaster management to face the problem of information overload posed by social media. In this section, we describe how interests on the side of disaster management have guided our choice of the methods used and our way to apply them in context of our software.

2.1   End-User Requirements

Aspects: First of all, disaster managers want to find structured information on what is happening in a crisis situation ('what?'). Equally important aspects are the place ('where?') and the time of the event ('when?'). Further relevant information may concern the organizations involved in the event ('who?').

Perspectives: Beside the current state of the event with respect to all of these aspects, disaster managers are interested in current changes of state and in the development of the event over time in order to detect hot spots or trends.

Combinations of Filters: Taking into account the variety of possible circumstances and different roles disaster managers may have to play in context of a crisis, the analysis tool must allow for great flexibility in combining all aspects just mentioned (e.g., 'how did a certain aspect of the situation develop at a certain location?').

Granularity: Similarly, since disaster managers are interested in an overview as well as in special details of the situation, there has to be the possibility of zooming in and out and looking at the event with different levels of granularity.

Relevance: A special case of guiding the attention of the end-users is the filtering out of irrelevant or wrong information.

Usability: Finally, in context of an application in disaster management, efficiency and user-friendliness of software are of high importance.

2.2 Implementation of Requirements

Aspects: For the first aspect of the analysis ('what?'), we referred to topic model analysis on basis of the HDP-CRF algorithm (Teh and Jordan 2010). In an unsupervised setting, topic modelling reveals the latent thematic structure in huge collections of documents. Furthermore, we applied hashtag statistics and keyword extraction by comparison of term distributions between the target collection and a reference corpus (*differential analysis*). For keywords or hashtags, co-occurrences analysis can reveal relations between concepts or entities.

Regarding the second aspect ('where?'), we either referred to meta data information or conducted location extraction using a list of location markers from *OpenStreetMap* (OpenStreetMap 2016) together with rule-based and context-sensitive techniques. By means of related longitude latitude coordinates, locations can be projected on a map.

Temporal information ('when?'), is provided by social media meta data. Names of organizations ('who?') got extracted by an NER approach that combines machine learning, rule based and context-sensitive techniques.

Perspectives: To take into account the different possible temporal perspectives, we not only provided means to summarize but also aggregate measurements of the various aspects over time. Additionally, we enabled calculation of growth or shrink from one interval to the next for all aspects.

Combination of Filters: All aspects of analysis as well as all meta data can be used as filter criteria and can be applied separately or in combination to create different sub-collections of data as input for analysis in line with special interests.

Granularity: The software offers possibilities for zooming in and out of a situation within the dimension of each aspect. Beside this, it integrates the mentioned statistically based *distant reading* procedures for entire collections or sub-collections of text with possibilities of manual *close reading* of single documents.

Relevance: As a provisional indication of the relevance of a message, we used the number of shares or retweets it received. On the one hand, the fact that many people found a message relevant, may really prove its relevance, on the other hand, even if the shared or retweeted message was not really relevant or even wrong, it may gain relevance from the point of view of disaster management because many people read it.

Usability: Beside the performance of the software in real-time or near-real-time, its easy handling and the intuitive visualization of analysis results have been in focus of our work. The software is accessible by an interactive graphical web interface with filter panels, drag-and-drop functionality, clickable graphs and configurable dashboards. For analysis of data, there are available two main modules – monitoring (*dashboard*) and analysis (*browser*). While the dashboards in the monitoring module are supposed to give a continuous overview over some predefined fields of interest, the analysis module allows for specific ad hoc investigations and close-reading of documents.

3   Examples

In this section, we demonstrate main functionalities of our software by means of example. As test data sets we used Facebook and Twitter data that had been created during the Central European flooding in June 2013 in Germany and Austria. For the Facebook flood corpus, we collected data from public pages or groups containing the words 'Hochwasser' or 'Fluthilfe' in their names via the public API (about 36k messages). For the Twitter flood corpus (about 354k tweets), we retrieved the current version of the research corpus of the QuOIMA project (QuOIMA 2011), that had been collected from the API filtering by disaster-related hash tags as well as by names of manually chosen public accounts connected to disaster management and flood aid (ibid.).

The example use case we present will be built around the topic extraction functionality. The dashboard in Fig. 1 gives an overview of topics and topic proportions for the Facebook flood corpus for the entire period of the event.

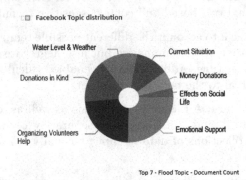

Fig. 1. Topics and topic proportions Facebook flood

By clicking on a name of a topic, it is possible to change to the analysis mode and to close-read or further analyze the messages belonging to this topic. The analysis view is shown in Fig. 2.

In the analysis mode, one could get an overview of the content of messages in a certain topic by showing typical topic words, for instance. Figure 3 includes typical words for the volunteering topic. By touching one of the words with the mouse, its co-occurring terms will get connected to it by edges to form a graph.

The dashboard in Fig. 4 changes temporal perspective and analyses the development of topics over the time of the event for the Twitter flood corpus. Again, clicking on the dots on the graph lines gives access to the messages showing the respective topic at the respective day for inspection or further analysis.

The dashboard in Fig. 5 gives an example of filtering for relevance of messages by number of their retweets. The peak around 20th June is connected to heavy rainfalls and thunderstorms that made alarm levels and subjective worries of the people rise anew but finally did not cause mayor new floodings. By only showing messages retweeted more than 6 times, a disaster manger searching for

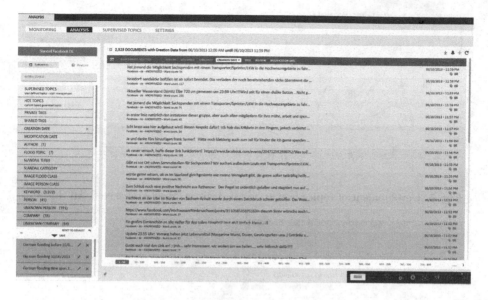

Fig. 2. Analysis modus with document view and filter panel

Fig. 3. Typical words for topic 'organizing volunteer's help'

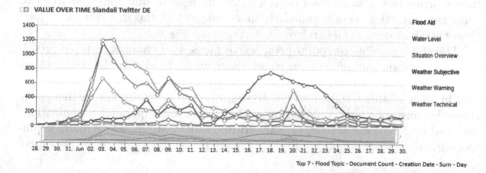

Fig. 4. Development of topics over time for Twitter

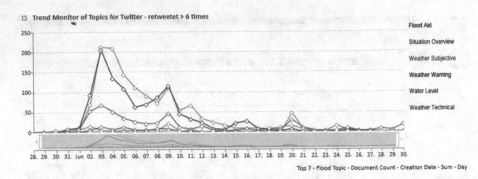

Fig. 5. Development of topics over time for Twitter tweets retweeted >6 times

Fig. 6. Significant words and topics at a day in Twitter

practically relevant information can filter out precaution and pure worries as 'noise'.

To get an idea of what is happening at the 20th June to cause these emotional reactions, one could either change to the close-reading modus for the relevant topics or extract a situational overview by the help of differential analysis to show keywords that were significantly more frequent at this day (target corpus) than they had been before (reference corpus), see Fig. 6.

The third possible temporal perspective focuses on changes in topic prominence at one day or interval in comparison with the day or interval before. Figure 7 illustrates a possible outcome of analysis for Facebook.

On basis of this insight into hot topics, one could, again, ask further questions. For instance, one could be interested in the most popular organizations involved in donations in kind, or want to find out locations where many volunteering activities are organized. Figures 8 and 9 reveal the results of the respective analyses.

While our examples were developing from the point of view of the topic aspect ('what?'), each other aspect can equally well serve as a starting point

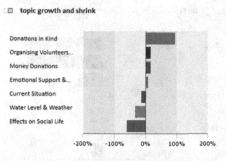

Fig. 7. Change of topic prominence for Facebook for a day

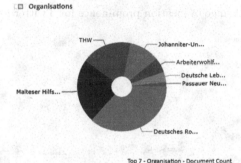

Fig. 8. Prominent organizations in topic 'donations' at a day

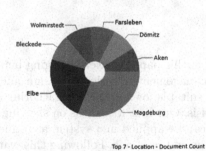

Fig. 9. Prominent locations in topic 'volunteering' at a day

for filtering and further analysis. For instance, the location aspect ('where?') is often of high interest for disaster managers. As before, the temporal perspective can either reveal an overview (as in Fig. 9), significant changes or the temporal development of the aspect. In Fig. 10, geographical hot spots are identified, while Fig. 11 shows the geographical unfolding of the flood event over its lifetime.

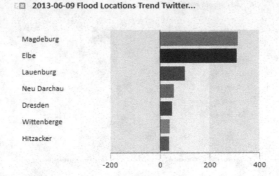

Fig. 10. Change of location prominence for Twitter for a day

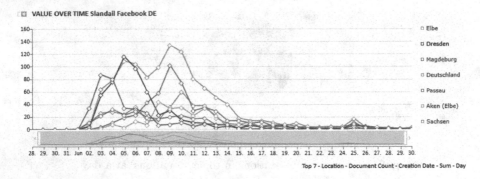

Fig. 11. Prominent locations over time for Facebook

4   Conclusion

In the work presented in this paper we were bridging between the fields of informatics and disaster management in order to design and create a social media text analysis software suitable for information gathering and knowledge acquisition in context of a crisis. Our first focus was on showing which methods can be chosen and how they can be applied in a system as to meet the special interests and requirements on the end-user side. Following this, various example use cases illustrated our general approach and demonstrated the capacity of our software to extract information useful for disaster management from huge collections of social media data. An approach along these lines can help to meet the challenges and make use of the opportunities that digitalization and the rise of social media have brought to disaster management.

Acknowledgments. The research leading to these results has received funding from the European Union's Seventh Framework Programme under grant agreement No. 607691 (SLANDAIL).

References

OpenStreetMap: OpenStreetMap - Deutschland (2016). https://www.openstreetmap. de. Accessed 08 June 2017

Palen, L., Anderson, K., Mark, G., Martin, J., Sicker, D., Palmer, M., Grunwald, D.: A vision for technology-mediated support for public participation & assistance in mass emergencies and disasters. In: Proceedings of the ACM-BCS Visions of Computer Science (2010)

QuOIMA: QuOIMA Open Source Integrated Multimedia Analysis (2011). www.kiras. at/projects. Accessed 08 June 2017

Slandail: Slandail - Security System for language and image analysis (2014). http:// slandail.eu. Accessed 08 June 2017

Starbird, K., Palen, L.: "Voluntweeters": self-organizing by digital volunteers in times of crisis. In: Proceedings of the ACM-BCS Visions of Computer Science (2011)

Teh, Y.W., Jordan, M.I.: Hierarchical Bayesian nonparametric models with applications. In: Hjort, N.L., et al. (eds.) Bayesian Nonparametrics, pp. 114–133. Cambridge University Press, Cambridge (2010)

A Comparative Study of Uncertainty Based Active Learning Strategies for General Purpose Twitter Sentiment Analysis with Deep Neural Networks

Nils Haldenwang[(⊠)], Katrin Ihler, Julian Kniephoff, and Oliver Vornberger

Institute of Computer Science, Media Computer Science Group,
University of Osnabrück, Osnabrück, Germany
{nils.haldenwang,kihler,jkniepho,oliver}@uos.de

Abstract. Active learning is a common approach when it comes to classification problems where a lot of unlabeled samples are available but the cost of manually annotating samples is high. This paper describes a study of the feasibility of uncertainty based active learning for general purpose Twitter sentiment analysis with deep neural networks. Results indicate that the approach based on active learning is able to achieve similar results to very large corpora of randomly selected samples. The method outperforms randomly selected training data when the amount of training data used for both approaches is of equal size.

1 Introduction

General purpose Twitter sentiment analysis was introduced as a new sentiment classification task by Haldenwang and Vornberger (2015). The main difference to other popular Twitter sentiment analysis tasks – such as SemEval, Nakov et al. (2016) – lies in the omission of filtering the Twitter stream with regard to certain topics or types of messages. Hence, the data set consists of a representative sample of the public Twitter stream, which is relevant for applications such as monitoring the sentiment of individuals, regions or the general, unfiltered public Twitter stream.

Systems based on deep neural networks are prevalent in the related Twitter sentiment analysis tasks (Deriu et al. 2016, Rouvier and Favre 2016, Xu et al. 2016). Therefore, it seems reasonable to investigate their feasibility for general purpose Twitter sentiment analysis.

Acquiring a sufficient amount of manually annotated data for the training of deep neural networks to perform the aforementioned task is very labor intensive. One possibility to deal with low amounts of manually annotated data is the use of *distant supervision* approaches based upon emoticons as originally introduced by Pak and Paroubek (2010). Distant supervision has already successfully been used in the training process of various deep learning architectures for Twitter sentiment analysis (Severyn and Moschitti 2015, Deriu et al. 2016, Xu et al. 2016).

G. Rehm and T. Declerck (Eds.): GSCL 2017, LNAI 10713, pp. 208–215, 2018.
https://doi.org/10.1007/978-3-319-73706-5_18

While noisy labels based on emoticons provide a good starting point for the training of a deep learning system, it is probably beneficial to use manually annotated training data for the specific task to achieve satisfying results.

A common approach to reduce the manual effort is *active learning*. Settles (2010) summarizes the idea of active learning as follow: "[. . .] a machine learning algorithm can achieve greater accuracy with fewer training labels if it is allowed to choose the data from which it learns [. . .]". Given a large corpus of unlabeled data points, the learner may choose the samples from which it hopes to gain the most insights from. The labels of the chosen data points are queried from an *oracle*, in this case a human annotator. The remainder of this paper describes a study the authors conducted to assess the feasibility of various metrics for measuring the potential information gain for unlabeled samples and then choosing the samples that are to be annotated.

2   Experimental Setup

In this section we first introduce the initial deep neural network that is the starting point for all experiments and illustrate how it was parametrized. Secondly, the active learning strategies which are evaluated are described. Finally, the experimental procedure is presented.

2.1   Initial Deep Neural Network

The classifier used in these experiments is a convolutional neural network. Its basic architecture is described in Zhang and Wallace (2015). First, the tokenized tweet is transformed into a list of dense *word embeddings*. The resulting *sentence matrix* is then convolved with a certain set of *filters* of potentially varying *region sizes*. After that, the resulting *feature maps*, which are vectors describing certain "higher order features" of the tweet, activate a 1-max-pooling layer via a possibly non-linear activation function. Lastly, this pooling layer is densely connected to the output layer using softmax activation and optional dropout regularization. In contrast to Zhang and Wallace (2015), our output layer has three neurons, reflecting the fact that we want to differentiate the three classes *positive*, *negative* and *uncertain*.[1]

All weights of the network were initialized randomly except for the embedding layer, where we used *word2vec* vectors (cf. Mikolov et al. (2013)) of dimension $d = 100$ trained on a dataset of approximately 33 million tweets collected between June 2012 and August 2013 by Neubauer (2014). After some minimal preprocessing[2], this dataset contained 624,015 unique tokens, of which we used the 200,000 most frequent ones in the network. The parameters were chosen as follows: The model used was the skip-gram model, the window size was 5 words, the subsampling threshold was $t = 10^{-5}$; negative sampling was used with $k = 5$

---

[1] See Haldenwang and Vornberger (2015) for further details.
[2] replacing @-mentions and URLs by generic tokens and removing "non-words".

"noise words" and we ran two iterations of the algorithm. Most of these values were recommended by Mikolov et al. (2013), where one can also find explanations for the parameters. The rest of the network's hyperparameters was found using a search guided by the best practices laid out in Zhang and Wallace (2015): We first evaluated networks with only one region size $r \in \{2, 3, 4, 5, 6, 8, 10\}$ and $n \in \{50, 325, 600\}$ filters. The activation function f between the convolution and pooling layers was chosen from the set $\{\text{id, tanh, RelU}^3\}$ and the dropout rate (Srivastava et al. 2014) was $p \in \{0, 0.25, 0.5\}$.

We evaluated all of these combinations based on their average *macro-F_1-score* in a tenfold cross-validation using the dataset from Haldenwang and Vornberger (2015). First, each network was trained using a *distant supervision* procedure with noisy labels based on emoticons in the dataset of Neubauer (2014). Note, that the distant super vision approach only consists of positive and negative tweets, since there is no reliable noisy label for uncertain tweets. Next, the network's parameters were further refined by using the positive and negative tweets from the datasets of the SemEval competitions (Nakov et al. 2013, Rosenthal et al. 2014, 2015) for training.[4] The networks were trained using the Adagrad (Duchi et al. 2011) algorithm. Both datasets were presented once (one *epoch*) in a batch size of 50 tweets.

The best configuration turned out to be $r = 2$, $n = 50$, $f = \tanh$ and $p = 0.25$ with an average F-score of $F_1 \approx 0.56$. We also tried adding bigger filters to this configuration in multiple ways, but none of the resulting configurations could significantly surpass the above, so we do not go into further details of this process here. For the following experiments with regard to active learning, we used the version of this network that was only trained on the noisy labels, to properly reflect one of the constraints of this approach: not to have a big supply of manually labeled tweets in advance.

2.2 Investigated Active Learning Strategies

As a strategy to query the best suited tweets to label for the network, we decided to investigate *uncertainty sampling*, a strategy originally devised by Lewis and Gale (1994) which is both easy to implement and understand and thus commonly used. With this strategy, each tweet is assigned an uncertainty value which defines how uncertain the network is in finding the correct label for the tweet. The most uncertain tweets are then chosen to be labeled.

For a problem with three (or more) classes such as ours, there are different metrics available to calculate uncertainty. These metrics differ in how many of the class probabilities they take into account. In the following a short description for each of the metrics provided. A more thorough introduction and comparison can be found in the literature survey of Settles (2010).

---

[3] Mahendran and Vedaldi (2015).
[4] The neutral class does not match with the desired uncertain class and hence is ommitted here.

The *confidence* metric can be used to choose the tweet x^*_{LC} whose label the network is *least confident* about:

$$x^*_{LC} = \operatorname*{argmin}_{x} P_\theta(\hat{y}|x)$$

The confidence is defined as the probability that the class label $\hat{y}$ chosen by the network θ is correct as considered by the network itself (and as such is the highest of the three probabilities for the three class labels).

The *margin* metric also takes the second highest probability into account by calculating the difference between the probabilities of the two class labels $\hat{y}_1$ and $\hat{y}_2$ the network believes to be most likely correct:

$$x^*_M = \operatorname*{argmin}_{x} P_\theta(\hat{y}_1|x) - P_\theta(\hat{y}_2|x)$$

A tweet with a smaller margin would be considered more uncertain since the network has difficulties choosing between the labels $\hat{y}_1$ and $\hat{y}_2$.

Finally, the *entropy* metric considers the probability for all class labels $\hat{y}_i$ to calculate the amount of informativity each tweet has to offer to the network:

$$x^*_H = \operatorname*{argmax}_{x} - \sum_i P_\theta(\hat{y}_i|x) \log P_\theta(\hat{y}_i|x)$$

In our experiment we compare the effect of these metrics to find out which is most helpful for our use case.

To speed up the labeling process, we query and label the tweets in batches of 20. However, since the uncertainty values are not recalculated after picking a tweet for a batch, this could lead to the tweets in the batch being very similar to one another since they all occupy the same uncertain region of the feature space. To avoid this, we introduce diversity as a second criterion to our querying process as described in (Patra and Bruzzone 2012):

First, we choose the 60 most uncertain tweets which we then reduce to 20 both uncertain and diverse tweets by clustering them with kernel k-means into 20 clusters and picking the most uncertain tweet from each cluster.

2.3 Experimental Procedure and Data Usage

For each of the uncertainty metrics described above, the experiment is initialized with a copy of the initial deep neural network that was pretrained with the aforementioned distantly supervised data only. The corpus of unlabled tweets to chose from consisted of 100,000 tweets that were randomly sampled from the 33 million dataset of Neubauer (2014). First, all tweets in the unlabeled corpus are classified by the network and then 20 tweets are chosen to be annotated using the previously mentioned strategy. Next, after the 20 tweets are labeled by the human annotator, 10 training iterations are performed with the newly annotated tweets. This procedure is then repeated until 1,000 tweets are annotated for each uncertainty metric.

Additionally, we generated a random baseline by training a copy of the initial neural network with randomly selected, manually annotated tweets in batches of 20 with 10 training iterations.

Each generated network was then evaluated using the reliable general purpose Twitter sentiment analysis data set from Haldenwang and Vornberger (2015) as a test set. The resulting macro F_1-score is reported.

3   Results

Figure 1 shows a visualization of the experimental results. A notable observation is the effectiveness of just labeling 100 tweets, the classification performance almost doubles for all metrics. This drastic increase in performance is a strong indication that even small amounts of manually annotated data are very beneficial in addition to the noisy labeled training data. Note, that the initial score is rather low, because the network was just pretrained with positive and negative data and, hence, missclassified all uncertain samples. When measuring the score for just the positive and negative classes after pretraining, it was $F_1 \approx 0.637$. Hence, pretraining with the distantly supervised data provides a useful basis for the network's parameters.

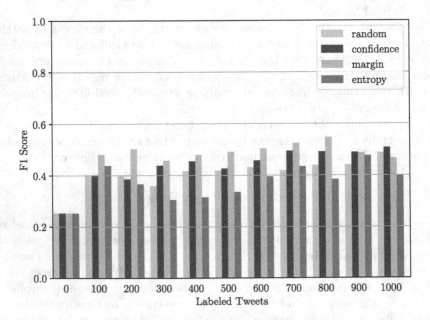

Fig. 1. Experimental results showing the macro F_1-score of the investigated metrics in steps of 100 manually annotated tweets.

The random baseline yields solid results but seems to always be outperformed by either the confidence or margin metric. The entropy metric performs worse than random in almost all cases. Moreover, it seems to be the most unstable with the strongest fluctuations in performance.

While the margin metric takes the lead for the first 800 annotated tweets, its effectiveness drastically drops at 900 and 1,000. Below 800 the confidence metric performed consistently worse than the margin metric but does not seem to suffer as severe a performance drop and at 1,000 labeled tweets takes the lead.

Overall, the best performance achieved was $F_1 \approx 0.55$ by the margin metric at 800 manually annotated tweets. The differences in classification behaviour when compared to the other metrics and the random baseline were significant. Moreover, the result is on par with training the same initial network with about 25,000 manually annotated tweets from a related domain (SemEval) and about 8,000 manually annotated tweets for the problem at hand (Haldenwang and Vornberger 2015), as was presented in Sect. 2.1, while only using a fraction of the training data.

4   Conclusions and Outlook

The results indicate that two out of three investigated uncertainty based active learning strategies consistently seem to surpass random sample selection for the investigated task.

Overall, the performance of the investigated strategies seems to be fluctuating a lot. After a certain point (more than 800 labeled tweets) the performance of all three active learning strategies seems to deteriorate or converge with the random baseline. In future work the study has to be extended to verify the aforementioned trend.

Moreover, a problem that can occur with purely uncertainty based metrics lies in their affinity to favor outliers since those are often of high uncertainty (Settles and Craven 2008). This selection of outliers may be what causes the deterioration at the last steps, since the outliers probably do not add any useful information for the correct classification of the non-outliers and may be harmful for the overall generalization of the system. In future work we plan on investigating active learning strategies which do not purely rely on the uncertainty but also take the *density weight* into account, as was suggested by Settles and Craven (2008). The basic idea is to not only select uncertain samples but also take into account the density of samples in the surrounding area to select data points which are representative for as many other uncertain samples as possible. Hopefully, this strategy can prevent pure outliers from being selected, increase the information gain and reduce the fluctuations.

Combining deep convolutional neural networks with active learning based on uncertainty sampling seems to be a promising approach for general purpose Twitter sentiment analysis which can drastically reduce the amount of manual annotation that is needed to achieve sufficient results.

References

Deriu, J., Gonzenbach, M., Uzdilli, F., Lucchi, A., De Luca, V., Jaggi, M.: Swisscheese at semeval-2016 task 4: sentiment classification using an ensemble of convolutional neural networks with distant supervision. In: Proceedings of the 10th International Workshop on Semantic Evaluation, SemEval 2016, San Diego, California, pp. 1124–1128, June 2016. Association for Computational Linguistics. http://www.aclweb.org/anthology/S16-1173

Duchi, J., Hazan, E., Singer, Y.: Adaptive subgradient methods for online learning and stochastic optimization. J. Mach. Learn. Res. **12**, 2121–2159 (2011)

Haldenwang, N., Vornberger, O.: Sentiment uncertainty and spam in Twitter streams and its implications for general purpose realtime sentiment analysis. In: Proceedings of the International Conference of the German Society for Computational Linguistics and Language Technology, GSCL 2015, pp. 157–159 (2015)

Lewis, D.D., Gale, W.A.: A sequential algorithm for training text classifiers. In: Croft, B.W., van Rijsbergen, C.J. (eds.) SIGIR 1994, pp. 3–12. Springer, London (1994). https://doi.org/10.1007/978-1-4471-2099-5_1

Mahendran, A., Vedaldi, A.: Understanding deep image representations by inverting them. In: 2015 IEEE Conference on Computer Vision and Pattern Recognition (CVPR), pp. 5188–5196. IEEE (2015)

Mikolov, T., Sutskever, I., Chen, K., Corrado, G.S., Dean, J.: Distributed representations of words and phrases and their compositionality. In: Advances in Neural Information Processing Systems, pp. 3111–3119 (2013)

Nakov, P., Rosenthal, S., Ritter, A., Wilson, T.: SemEval-2013 task 2 : sentiment analysis in Twitter. In: Proceedings of the 7th International Workshop on Semantic Evaluation (SemEval 2013), in conjunction with the 2nd Joint Conference on Lexical and Computational Semantics, *SEM 2013, vol. 2, pp. 312–320 (2013)

Nakov, P., Ritter,A., Rosenthal, S., Sebastiani, F., Stoyanov, V.: Semeval-2016 task 4: sentiment analysis in Twitter. In: Proceedings of the 10th International Workshop on Semantic Evaluation (SemEval-2016), San Diego, California, pp. 1–18. Association for Computational Linguistics, June 2016. http://www.aclweb.org/anthology/S16-1001

Neubauer, N.: Semantik und Sentiment: Konzepte, Verfahren und Anwendungen von Text-Mining. Dissertation, Universität Osnabrück (2014). https://repositorium.uni-osnabrueck.de/handle/urn:nbn:de:gbv:700--2014060612524

Pak, A., Paroubek, P.: Twitter as a corpus for sentiment analysis and opinion mining. In: LREC (2010)

Patra, S., Bruzzone, L.: A cluster-assumption based batch mode active learning technique. Pattern Recogn. Lett. **33**(9), 1042–1048 (2012)

Rosenthal, S., Ritter, A., Nakov, P., Stoyanov, V.: Semeval-2014 task 9: sentiment analysis in Twitter. In: Proceedings of the 8th International Workshop on Semantic Evaluation, SemEval 2014, Dublin, Ireland, pp. 73–80. Association for Computational Linguistics and Dublin City University, August 2014

Rosenthal, S., Nakov, P., Kiritchenko, S., Mohammad, S., Ritter, A., Stoyanov, V.: Semeval-2015 task 10: sentiment analysis in Twitter. In: Proceedings of the 9th International Workshop on Semantic Evaluation, SemEval 2015, Denver, Colorado, pp. 451–463. Association for Computational Linguistics, June 2015

Rouvier, M., Favre, B.: Sensei-lif at semeval-2016 task 4: polarity embedding fusion for robust sentiment analysis. In: Proceedings of the 10th International Workshop on Semantic Evaluation, SemEval 2016, San Diego, California, pp. 202–208. Association for Computational Linguistics, June 2016. http://www.aclweb.org/anthology/S16-1030

Settles, B.: Active learning literature survey. University of Wisconsin, Madison (2010)

Settles, B., Craven, M.: An analysis of active learning strategies for sequence labeling tasks. In: Proceedings of the Conference on Empirical Methods in Natural Language Processing, pp. 1070–1079. Association for Computational Linguistics (2008)

Severyn, A., Moschitti, A.: Unitn: training deep convolutional neural network for Twitter sentiment classification. In: Proceedings of the 9th International Workshop on Semantic Evaluation, SemEval 2015, Denver, Colorado, pp. 464–469. Association for Computational Linguistics, June 2015

Srivastava, N., Hinton, G., Krizhevsky, A., Sutskever, I., Salakhutdinov, R.: Dropout: a simple way to prevent neural networks from overfitting. J. Mach. Learn. Res. **15**(1), 1929–1958 (2014)

Xu, S., Liang, H., Baldwin, T.: Unimelb at semeval-2016 tasks 4a and 4b: an ensemble of neural networks and a word2vec based model for sentiment classification. In: Proceedings of the 10th International Workshop on Semantic Evaluation, SemEval 2016, San Diego, California, pp. 183–189. Association for Computational Linguistics, June 2016. http://www.aclweb.org/anthology/S16-1027

Zhang, Y., Wallace, B.: A sensitivity analysis of (and practitioners' guide to) convolutional neural networks for sentence classification. arXiv preprint arXiv:1510.03820 (2015)

An Infrastructure for Empowering Internet Users to Handle Fake News and Other Online Media Phenomena

Georg Rehm[(✉)]

DFKI GmbH, Language Technology Lab,
Alt-Moabit 91c, 10559 Berlin, Germany
georg.rehm@dfki.de

Abstract. Online media and digital communication technologies have an unprecedented, even increasing level of social, political and also economic relevance. This article proposes an infrastructure to address phenomena of modern online media production, circulation and manipulation by establishing a distributed architecture for automatic processing and human feedback.

1 Introduction

The umbrella term "fake news" is often used to refer to a number of different phenomena around online media production, circulation, reception and manipulation that emerged in recent years and that have been receiving a lot of attention from multiple stakeholders including politicians, journalists, researchers, non-governmental organisations, industry and civil society. In addition to the challenge of dealing with "fake news", "alternative facts" as well as "post-truth politics", there is an increasing amount of hate speech, abusive language and cyber bullying taking place online.

Among the interested stakeholders are politicians who have begun to realise that, increasingly, major parts of public debates and social discourse are carried out online, on a small number of social networks. We have witnessed that not only online discussions but also the perception of trends, ideas, theories, political parties, individual politicians, elections and societal challenges can be subtly influenced and significantly rigged using targeted social media campaigns, devised at manipulating opinions to create long-term sustainable mindsets on the side of the recipients. We live in a time in which online media, online news and online communication have an unprecedented level of social, political and economic relevance.

Due to the intrinsic danger of successful large-scale manipulations the topic is of utmost importance. Many researchers from the Social Sciences and Computer Science currently work on the topic. An idea often mentioned is to design, develop and deploy technologies to improve the situation, maybe even to solve it altogether, thanks to recent breakthroughs in AI (Metz 2016; Gershgorn 2016;

G. Rehm and T. Declerck (Eds.): GSCL 2017, LNAI 10713, pp. 216–231, 2018.
https://doi.org/10.1007/978-3-319-73706-5_19

Martinez-Alvarez 2017; Chan 2017), while at the same time *not* putting in place a centralised infrastructure, which could be misused for censorship, manipulation or mass surveillance.[1]

This article addresses key challenges of the digital age (Sect. 2) by introducing and proposing the vision of a technological infrastructure (Sect. 3); the concept has been devised in a research and technology transfer project, in which smart technologies for curating large amounts of digital content are being developed and applied by companies that cover different sectors including journalism (Rehm and Sasaki 2015; Bourgonje et al. 2016a,b; Rehm et al. 2017). Among others, we currently develop services aimed at the detection and classification of abusive language (Bourgonje et al. 2017a) and clickbait content (Bourgonje et al. 2017b). The proposed hybrid infrastructure combines automatic language technology components and user-generated annotations and is meant to empower internet users better to handle the modern online media phenomena mentioned above.

2   Modern Online Media Phenomena

The World Wide Web makes it possible for everybody to create content, to write an article on a certain topic. Until a few years ago the key challenge was to optimise the HTML code, linking and metadata to get highly ranked by the relevant search engines. Nowadays content is no longer predominantly discovered through search engines but through social media platforms: users see interesting content, which is then shared to their own connections. Many users only read a headline, identify a certain relevance to their own lives and then share the content. When in doubt, users estimate the trustworthiness of the source: potentially dubious stories about which they are skeptical are shared anyway if the friend through whom the story was discovered is considered reliable or if the number of views is rather high, which, to many users, indicates legitimacy.

There is a tendency for provocative, aggressive, one-sided, allegedly "authentic" (Marchi 2012) content. The idea is to make it as easy as possible to identify the stance of the article so that the reader's own world view is validated, implicitly urging the user to share the content. The publisher's goal is for a story to go viral, that it is shared rapidly by many users and spread through the networks to establish a reach of millions. One sub-category of this type of content is "clickbait", articles with dubious factual content, presented with misleading headlines, designed for the simple goal of generating many views. The more extreme the virality, the higher the reach, the higher the click numbers, the higher the advertisement revenue. The term "clickbait" can also refer to articles spreading political mis- or disinformation.

Content is typically discovered through a small number of social networks. While search engines and online news portals used to be the central points of

---

[1] An indicator for the relevance of the topic is the increasing number of "how to identify fake news" articles published online (Mantzarlis 2015; Bazzaz 2016; Rogers and Bromwich 2016; Wardle 2017; Walbrühl 2017).

information until a few years ago, the role of the centralised hub – and gate-keeper – is now played by social networks that help content to be discovered and go viral (Barthel et al. 2016). All social networks have as their key feature a news feed or timeline, i.e., posts, news, ads, tweets, photos presented to the user, starting with the most recent one. Nearly all social networks use machine learning algorithms to determine which content to present to a certain user. They are continuously trained through interactions with the network, i.e., "liking" a post boosts the respective topic, visiting the profile of a "friend" boosts this connection. Some networks use more fine-grained sentiments in addition to the simple "like" (see, e.g., Facebook's reactions "love", "haha", "wow", "sad", "angry"). Through "likes" of topics, connections to friends and interactions with the site, social networks create, and continuously update, for every single user, a model of their interests, which is used to select content for the user's timeline. The algorithms are designed to favour content liked or shared by those friends the user interacts with the most. This is the origin of the filter bubble phenomenon: users are predominantly exposed to content that can also be described as "safe" and "non-controversial" – content shared by friends they know and like is considered content that matches a user's interests. Content that contradicts a user's world view or that challenges their beliefs is *not* presented.

Additionally, we are faced with the challenge that more and more content is produced and spread with the sole purpose of manipulating the readers' beliefs and opinions by appealing to their emotions instead of informing them objectively. Rather, this type of opinionated, emotional, biased, often aggressive and far-right content is spread to accomplish specific goals, for example, to create support for controversial ideas or to intensify the division between two social groups. These coordinated campaigns are carried out by experts with in-depth knowledge of the underlying technologies and processes. They involve large numbers of bots and fake accounts as amplifiers (Weedon et al. 2017) as well as large budgets for online advertisements in social media, clearly targeted at very specific demographic groups the originators want to influence and then to flip to reach a specific statistical threshold. The way news are nowadays spread, circulated, consumed and shared – with less and less critical thinking or fact checking – enables this type of content to gather a large number of readers (and sharers) quickly. The filter bubble acts like an echo chamber that can amplify any type of content, from genuine, factual news to emotionally charged, politically biased news, to false news to orchestrated disinformation campaigns, created with the specific purpose of large-scale manipulation. Content of the last two categories can be hard or very hard to identify even for human experts.

A key challenge for users and machines alike is to separate objective, balanced content, be it journalistic or user-generated, from hateful, abusive or biased content, maybe produced with a hidden agenda. Even if fundamentally different in nature, both types of content share the same potential level of visibility, reach and exposure through the equalisation mechanisms of the social web, which is prone to manipulation. In the past the prerequisite tasks of fact checking, critical thinking and uncovering hidden agendas have been in the realm of (investigative) journalism – in the digital age they are more and more transferred to the actual reader of online content. The analysis and assessment of content is no longer carried out by professional journalists or news editors – the burden of fact checking and content verification is left to the reader. This aspect is getting even more crucial because the number of people who state that social networks are their *only* source of news is growing steadily (Marchi 2012). The most prominent example from recent history is the ongoing debate whether highly targeted social media ads influenced the 2016 US presidential election (Barthel et al. 2016; Rogers and Bromwich 2016; Marwick and Lewis 2017). It must be noted that a large number of fact-checking initiatives is active all over the world (Mantzarlis 2017) but they mostly rely on human expertise and, thus, do not scale (Martinez-Alvarez 2017; Dale 2017). The small number of automated fact checking initiatives are fragmented (Babakar and Moy 2016).

Table 1. Characteristics and intentions associated with different types of false news – adapted from (Wardle 2017; Walbrühl 2017; Rubin et al. 2015; Holan 2016; Weedon et al. 2017)

	Satire or parody	False connection	Misleading content	False context	Imposter content	Manipulated content	Fabricated content
Clickbait		X	X	?		?	?
Disinformation			X	X		X	X
Politically biased		?	X	?		?	X
Poor journalism		X	X	X			
To parody	X				?		X
To provoke					X	X	X
To profit	?	X			X		X
To deceive		X	X	X	X	X	X
To influence politics			X	X		X	X
To influence opinions			X	X	X	X	X

Several types of online content are often grouped together under the label "fake news". For example, Holan (2016) defines fake news as "made-up stuff, masterfully manipulated to look like credible journalistic reports that are easily spread online to large audiences willing to believe the fictions and spread the word." In reality, the situation is much more complex. Initially based on the classification suggested by Wardle (2017), Table 1 shows an attempt at bringing together the different types of false news including selected characteristics and associated intentions. The table shows the complexity of the situation and that a more fine-grained terminology is needed to discuss the topic properly, especially when it comes to designing technological solutions that are meant to address one or more of these types of content.

An additional challenge is the proliferation of hateful comments and abusive language, often used in the comments and feedback sections on social media posts. The effects can be devastating for the affected individual. Many hateful comments on repeated postings by the same person, say, a pupil, are akin to cyberbullying and cybermobbing. There is also a clear tendency to aggressive comments on, for example, the social media pages of traditional news outlets, who have to ask the users more and more to behave in a civilised way.

3 Technology Framework: Approach

Technically, online content is predominantly consumed through two possible channels, both of which rely substantially on World Wide Web technology and established web standards. Users either read and interact with content directly on the web (mobile or desktop versions of websites) or through dedicated mobile apps; this can be considered using the web implicitly as many apps make heavy use of HTML5 and other web technologies. The World Wide Web itself still is and, for the foreseeable future, will continue to be the main transport medium for online content. The infrastructure suggested by this article is, hence, designed as an additional layer on top of the World Wide Web. The scope and ambition of the challenge is immense because the infrastructure needs to be able to cope with millions of users, arbitrary content types, hundreds of languages and massive amounts of data. Its goal is to empower users by enabling them to balance out network and filter bubble effects and to provide mechanisms to filter for abusive content.

3.1 Services of the Infrastructure

The burden of analysing and fact checking online content is often shifted to the reader (Sect. 2), which is why corresponding analysis and curation services need to be made available in an efficient and ubiquitous way. The same tools to be used by *content consumers* can and should also be applied by *content creators*, e.g., journalists and bloggers. Those readers who are interested to know more about what they are currently reading should be able to get the additional information as easily as possible, the same applies to those journalists who are interested in fact-checking the content they are researching for the production of new content.

Readers of online content are users of the World Wide Web. They need, first and foremost, web-based tools and services with which they can process any type of content to get additional information on a specific piece, be it one small comment on a page, the main content component of a page (for example, an article) or even a set of interconnected pages (one article spread over multiple pages), for which an assessment is sought.

The services need to be designed to operate in and with the web stack of technologies, they need to support users in their task of reading and curating content within the browser in a smarter and, eventually, more balanced way. This can be accomplished by providing additional, also alternative opinions and view points, by presenting other, indepedent assessments, or by indicating if content is dangerous, abusive, factual or problematic in any way. Fully automatic technologies (Rubin et al. 2015; Schmidt and Wiegand 2017; Horne and Adal

2017; Martinez-Alvarez 2017) can take over a subset of these tasks but, given the current state of the art, not all, which is why the approach needs to be based both on simple and complex automatic filters and watchdogs as well as human intelligence and feedback.[2]

The tools and services should be available to every web user without the need of installing any additional third-party software. This is why the services, ideally, should be integrated into the browser on the same level as bookmarks, the URL field or the navigation bar, i.e., without relying on the installation of a plugin. The curation services should be thought of as an inherent technology component of the World Wide Web, for which intuitive and globally acknowl-edged user-interface conventions can be established, such as, for example, traffic light indicators for false news content (green: no issues found; yellow: medium issues found and referenced; red: very likely false news). Table 2 shows a first list of tools and services that could be embedded into such a system.[3] Some of these can be conceptualised and implemented as automatic tools (Horne and Adal 2017), while others need a hybrid approach that involves crowd-sourced data and opinions. In addition to displaying the output of these services, the browser interface needs to be able to gather, from the user, comments, feedback, opinions and sentiments on the current piece of content, further to feed the crowd-sourced data set. The user-generated data includes both user-generated annotations (UGA) and also user-generated metadata (UGM). Automatically generated metadata are considered machine-generated metadata (MGM).

Table 2. Selected tools and services to be provided through the infrastructure

Tool or service	Description	Approach
Political bias indicator	Indicates the political bias (Martinez-Alvarez 2017) of a piece of content, e.g., from far left to far right	Automatic
Hate speech indicator	Indicates the level of hate speech a certain piece of content contains	Automatic
Reputation indicator	Indicates the reputation, credibility (Martinez-Alvarez 2017), trustworthiness, quality (Filloux 2017) of a certain news outlet or individual author of content	Crowd, automatic
Fact checker	Checks if claims are backed up by references, evidence, established scientific results and links claims to the respective evidence (Babakar and Moy 2016)	Automatic
Fake news indicator	Indicates if a piece of content contains non-factual statements or dubious claims (Horne and Adal 2017; Martinez-Alvarez 2017)	Crowd, automatic
Opinion inspector	Inspect opinions and sentiments that other users have with regard to this content (or topic) – not just the users commenting on one specific site but all of them	Crowd, automatic

---

[2] A fully automatic solution would work only for a very limited set of cases. A purely human-based solution would work but required large amounts of experts and, hence, would not scale. This is why we favour, for now, a hybrid solution.

[3] This list is meant to be indicative rather than complete. For example, services for getting background information on images are not included (Gupta et al. 2013). Such tools could help pointing out image manipulations or that an old image was used, out of context, to illustrate a new piece of news.

3.2 Characteristics of the Infrastructure

In order for these tools and services to work effectively, efficiently and reliably, they need to have several key characteristics, which are critical for the success of the approach.

Like the Internet and the World Wide Web, the proposed infrastructure must be operated in a federated, i.e., de-centralised setup – a centralised approach would be too vulnerable for attacks or misuse. Any organisation, company, research centre or NGO should be able to set up, operate and offer services (Sect. 3.1) and pieces of the infrastructure. The internal design of the algorithms and tools may differ but their output should comply to a standardised metadata format (MGM). It is rather likely that political biases in different processing models meant to serve the same purpose cannot be avoided, which is especially likely for models based on large amounts of data, which, in turn, may inherently include a political bias. This is why users must be enabled to configure their own personalised set of tools and services to get an aggregated value, for example, with regard to the level of hate speech in content or its political bias. Services and tools must be combinable, i.e., they need to comply to standardised input and output formats (Babakar and Moy 2016). They also need to be transparent (Martinez-Alvarez 2017). Only transparent, i.e., fully documented, checked, ideally also audited approaches can be trustworthy.

Access to the infrastructure should be universal and available everywhere, i.e., in any browser, which essentially means that, ideally, the infrastructure should be embedded into the technical architecture of the World Wide Web. As a consequence, access meachnisms should be available in every browser, on every platform, as native elements of the GUI. These functions should be designed in such a way that they support users without distracting them from the content. Only if the tools are available virtually anywhere, can the required scale be reached.

The user should be able to configure and to combine multiple services, operated in a de-centralised way, for a clearly defined purpose in order to get an aggregated value. There is a danger that this approach could result in a replication and shift of the filter bubble effect (Sect. 2) onto a different level but users would at least be empowered actively to configure their own personal set of filters to escape from any resulting bubble. The same transparency criterion also applies to the algorithm that aggregates multiple values.

3.3 Building Blocks of the Proposed Infrastructure

Research in Language Technology and NLP currently concentrates on smaller components, especially watchdogs, filters and classifiers (see Sect. 4) that could be applied under the umbrella of a larger architecture to tackle current online media phenomena (Sect. 2). While this research is both important and crucial, even if fragmented and somewhat constrained by the respective training data sets (Rubin et al. 2015; Conroy et al. 2015; Schmidt and Wiegand 2017) and limited use cases, we also need to come to a shared understanding *how* such components can be deployed and made available. The proposed infrastructure consists of several building blocks (see Fig. 1).

Building Block: Natively embedded into the World Wide Web – An approach that is able to address modern online media and communication phenomena adequately needs to operate on a web-scale level. It should natively support cross-lingual processing and be technically and conceptually embedded into the architecture of the World Wide Web itself. It should be standardised, endorsed and supported not only by all browser vendors but also by all content and media providers, especially the big social networks and content hubs. Only if *all* users have *immediate* access to the tools and services suggested in this proposal can they reach its full potential. The services must be unobtrusive and cooperative, possess intuitive usability, their recommendations and warnings must be immediately understandable, it must be simple to provide general feedback (UGM) and assessments on specific pieces of content (UGA).

Building Block: Web Annotations – Several pieces of the proposed infrastructure are already in place. One key component are Web Annotations, standardised by the World Wide Web Consortium (W3C) in early 2017 (Sanderson et al. 2017a,b; Sanderson 2017). They enable users to annotate arbitrary pieces of web content, essentially creating an additional and independent layer on top of the regular web. Already now Web Annotations are used for multiple individual projects in research, education, scholarly publishing, administration and investigative journalism.[4] Web Annotations are *the* natural mechanism to enable users and readers interactively to work with content, to include feedback and assessments, to ask the author or their peers for references or to provide criticism. The natural language content of Web Annotations (UGA) can be automatically mined using methods such as sentiment analysis or opinion mining – in order to accomplish this across multiple languages, cross-lingual methods need to be applied (Rehm et al. 2016). However, there are still limitations. Content providers need to enable Web Annotations by referencing a corresponding JavaScript library. Federated sets of annotation stores or repositories are not yet foreseen, neither are native controls in the browser that provide aggregated feedback, based on automatic (MGM) or manual content assessments (UGM, UGA). Another barrier for the widespread use and adoption of Web Annotations are proprietary commenting systems, as used by all major social networks. Nevertheless, services such as Hypothes.is enable Web Annotations on any web page, but native browser support, ideally across all platforms, is still lacking. A corresponding browser feature needs to enable both free-text annotations of arbitrary content pieces (UGA) but also very simple flagging of problematic content, for example, "content pretends to be factual but is of dubious quality" (UGM). Multiple UGA, UGM or MGM annotations could be aggregated and presented to new readers of the content to provide guidance and indicate issues.

---

[4] See, for example, the projects presented at I Annotate 2015 (http://iannotate.org/2015/), 2016 (http://iannotate.org/2016/) and 2017 (http://iannotate.org/2017/).

Building Block: Metadata Standards – Another needed piece of the architecture is an agreed upon metadata schema, i.e., a controlled vocabulary, (Babakar and Moy 2016) to be used both in manual annotation scenarios (UGM) and also by automatic tools (MGM). Its complexity should be as low as possible so that key characteristics of a piece of content can be adequately captured and described by humans or machines. With regard to this requirement, W3C published several standards to represent the provenance of digital objects (Groth and Moreau 2013; Belhajjame et al. 2013a). These can be thought of as descriptions of the entities or activities involved in producing or delivering a piece of content to understand how data was collected, to determine ownership and rights or to make judgements about information to determine whether to trust content (Belhajjame et al. 2013b). An alternative approach is for publishers to use Schema.org's ClaimReview[5] markup after specific facts have been checked. The needed metadata schema can be based on the W3C provenance ontology and/or Schema.org. Additional metadata fields are likely to be needed.

Building Block: Tools and Services – Web Annotations can be used by readers of online content to provide comments or to include the results of researched facts (UGA, UGM). Automatic tools and services that act as filters or watchdogs can make use of the same mechanisms (MGM, see Sect. 3.1). These could be functionally limited classifiers, for example, regarding abusive language, or sophisticated NLU components that attempt to check certain statements against one or more knowledge graphs. Regardless of the complexity and approach, the results can be made available as globally accessible Web Annotations (that can even, in turn, be annotated themselves). Services and tools need to operate in a decentralised way, i.e., users must be able to choose from a wide variety of automatic helpers. These could, for example, support users to position content on the political spectrum, either based on crowd-sourced annotations, automatic tools, or both (see Table 2).

Building Block: Decentralised Repositories and Tools – The setup of the infrastructure must be federated and decentralised to prevent abuse by political or industrial forces. Data, especially annotations, must be stored in decentral repositories, from which browsers retrieve, through secure connections, data to be aggregated and displayed (UGM, UGA, MGM, i.e., annotations, opinions, automatic processing results etc.). In the medium to long term, in addition to annotations, repositories will also include more complex data, information and knowledge that tools and services will make use of, for example, for fact checking. In parallel to the initiative introduced in this article, crowd-sourced knowledge graphs such as Wikidata or DBpedia will continue to grow, the same is true for semantic databases such as BabelNet and many other data sets, usually available and linkable as Linked Open Data. Already now we can foresee more sophisticated methods of validating and fact-checking arbitrary pieces of content using systems that make heavy use of knowledge graphs, for example, through automatic entity recognition and linking, relation extraction, event extraction and

---

[5] https://schema.org/ClaimReview.

mapping etc. One of the key knowledge bases missing, in that regard, is a Web Annotation-friendly event-centric knowledge graph, against which fact-checking algorithms can operate.[6] Basing algorithms that are supposed to determine the truth of an arbitrary statement on automatically extracted and formally represented knowledge creates both practical and philosophical questions, among others, who checks these automatically extracted knowledge structures for correctness? How do we represent conflicting view points and how do algorithms handle conflicting view points when determining the validity of a statement? How do we keep the balance between multiple subjective opinions and an objective and scientific ground-truth?

Building Block: Aggregation of Annotations – The final key building block of the proposed infrastructure relates to the aggregation of automatic and manual annotations, created in a de-centralised and highly distributed way by human users and automatic services (UGA, UGM, MGM). Already now we can foresee very large numbers of annotations so that the aggregation and consolidation will be a non-trivial technical challenge. This is also true for those human annotations that are not based on shared metadata vocabularies but that are free text – for these free and flexible annotations, robust and also multilingual annotation mining methods need to be developed.

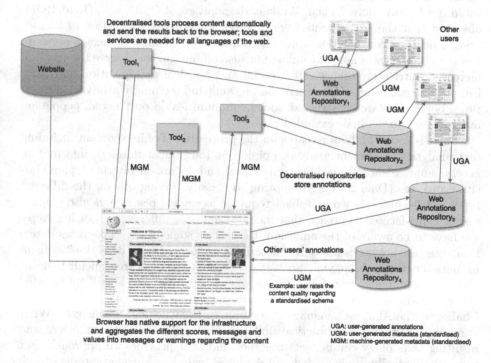

Fig. 1. Simplified architecture of the proposed infrastructure

---

[6] Promising candidates are GDELT (http://www.gdeltproject.org) and EventRegistry (http://eventregistry.org).

4   Related Work

Research on Computer-Mediated Communication (CMC) has a long tradition. Scholars initially concentrated on different types of communication media such as e-mail, IRC, Usenet newsgroups, and different hypertext systems and document types, especially personal home pages, guestbooks and, later, discussion fora (Runkehl et al. 1998; Crystal 2001; Rehm 2002). Early on, researchers focused upon the (obvious) differences between these new forms of digital communication and traditional forms, especially when it comes to linguistic phenomena that can be observed on the text surface (smileys, emoticons, acronyms etc.). Several authors pointed out that the different forms of CMC have a certain oral and spoken style, quality and conceptualisation to them, as if produced spontaneously in a casual conversation, while being realised in a written medium (Haase et al. 1997).

If we now fast forward to 2017, a vastly different picture emerges. About half of the global population has access to the internet, most of whom also use the World Wide Web and big social networks. Nowadays the internet acts like an amplifier and enabler of social trends. It continues to penetrate and to disrupt our lives and social structures, especially our traditions of social and political debates. The relevance of online media, online news and online communication could not be any more crucial. While early analyses of CMC, e.g., (Reid 1991), observed that the participants were involved in the "deconstruction of boundaries" and the "construction of social communities", today the exact opposite seems to be the case: not only online but also offline can we observe the trend of increased, intricately orchestrated, social and political manipulation, nationalism and the exclusion of foreigners, immigrants and seemingly arbitrary minorities – boundaries are constructed, social communities deconstructed, people are manipulated, individuals excluded.

There is a vast body of research on the processing of online content including text analytics (sentiment analysis, opinion and argument mining), information access (summarisation, machine translation) and document filtering (spam classification), see (Dale 2017). Attempting to classify, among others, the different types of false news shown in Table 1 requires, as several researchers also emphasise, a multi-faceted approach that includes multiple different processing steps. We have to be aware of the ambition, though, as some of the "fake news detection" use case scenarios are better described as "propaganda detection", "disinformation detection", maybe also "satire detection". These are difficult tasks at which even humans often fail. Current research in this area is still fragmented and concentrates on very specific sub-problems, see, for example, the Fake News Challenge, the Abusive Language Workshop, or the Clickbait Challenge.[7] What is missing, however, is a practical umbrella that pulls the different pieces and resulting technology components together and that provides an approach that can be realistically implemented and deployed including automatic tools as well as human annotations.

---

[7] See http://www.fakenewschallenge.org, http://www.clickbait-challenge.org, https://sites.google.com/site/abusivelanguageworkshop2017/.

5 Summary and Conclusions

Humanity is transitioning into becoming a digital society, or at least a "digital first" society, i.e., news, media, facts, rumours (Zubiaga et al. 2016; Derczynski et al. 2017; Srivastava et al. 2017), information are created, circulated and consumed online. Already now the right social media strategy can make or break an election or influence if a smaller or larger societal or demographic group (city, region, country, continent) is in favour or against constructively solving a certain societal challenge. Social media and online communication technologies can be an extremely powerful tool to bridge barriers, inform people and enable global communication and a constructive dialogue. When abused, misused or infiltrated, they are a dangerous weapon.

Computational Linguistics, Language Technology and Artificial Intelligence should actively contribute solutions to this key challenge of the digital age. If not, there is a concrete danger that stakeholders with bad intentions are able to influence parts of the society to their liking, only constrained by their political or commercial interests. Technologies need to be developed to enable every user of online media to break out of their filter bubbles and to inform themselves in a balanced way, taking all perspectives into account. Nevertheless, there is, as Dale (2017) points out, the danger that technologies developed to *detect* false news can also be used to *create* false news.

After dumb digital content, smart content and semantic content enrichment we now need to concentrate on content curation tools that enable *contextualised content*, i.e., content that can be, ideally, automatically cross-referenced and fact-checked, and for which background information can be retrieved in a robust way. This can involve assessing the validity of claims as well as retrieving related texts, facts and statements, both in favour and against a certain piece of content.

In this article a hybrid technology infrastructure that provides user- and machine-generated annotations on top of the whole World Wide Web is proposed with the ultimate goal of empowering internet users to handle false news and other online media phenomena by providing both automatic assessments of content and also by including alternative opinions into the process of media consumption. However, part of the solution could be provided by the small number of social networks which currently connect a vast majority of the online population and whose features and mechanisms are responsible for and also amplify the phenomena discussed in this article. It can be argued that these social networks have an obligation to act, for example, by modifying their algorithms to enable users to break out of their filter bubbles, by making the algorithms more transparent, or by using data analytics to detect potential manipulations. It is likely that regulatory steps will be taken on national and international levels.

Future work includes presenting this proposal in various different fora and communities, among others, researchers and technologists, standards-developing organisations (Babakar and Moy 2016) and national as well as international political bodies. At the same time, research needs to be continued and prototypes of the architecture as well as individual services developed, enabling organisations to build and to deploy decentralised tools early. While a universal, globally accessible, balanced and well maintained knowledge graph containing up-to-date information about entities and events would be convenient to have, it is out of scope with regard to the initiative proposed in this article; however, it is safe to assume that such a knowledge repository will be developed in parallel in the next couple of years. The proposed infrastructure can be used to link online content against such a knowledge graph and to measure the directions of online debates.

The proposal introduced in this article is ambitious in its scope and implications, prevention of misuse and establishing trust will play a hugely important role. How can we make sure that a certain piece of technology is only used with good intentions? Recently it has been shown that a user's social media data can reliably predict if the user is suffering from alcohol or drug abuse (Ding et al. 2017). Will this technology be used to help people or to stigmatise them? Will an infrastructure, as briefly sketched in this paper, be used to empower users to make up their own minds by providing additional information about online content or will it be used to spy on them and to manipulate them with commercial or political intentions?

Acknowledgments. The author would like to thank the reviewers for their insightful comments and suggestions. The project "Digitale Kuratierungstechnologien" (DKT) is supported by the German Federal Ministry of Education and Research (BMBF), "Unternehmen Region", instrument Wachstumskern-Potenzial (no. 03WKP45). More information: http://www.digitale-kuratierung.de.

References

Babakar, M., Moy, W.: The State of Automated Factchecking - How to make factchecking dramatically more effective with technology we have now. Full Fact (2016). https://fullfact.org/media/uploads/full_fact-the_state_of_automated_factchecking_aug_2016.pdf

Barthel, M., Mitchell, A., Holcomb, J.: Many Americans Believe Fake News Is Sowing Confusion. Pew Research Center (2016). http://www.journalism.org/2016/12/15/many-americans-believe-fake-news-is-sowing-confusion/

Bazzaz, D.: News you can use: infographic walks you through 10 questions to detect fake news. The Seattle Times (2016). http://www.seattletimes.com/education-lab/infographic-walks-students-through-10-questions-to-help-them-spot-fake-news/

Belhajjame, K., Cheney, J., Corsark, D., Garijo, D., Soiland-Reyes, S., Zednik, S., Zhao, J.: PROV-O: The PROV Ontology. W3C Recommendation, World Wide Web Consortium (W3C), April 2013a. https://www.w3.org/TR/2013/REC-prov-o-20130430/

Belhajjame, K., Deus, H., Garijo, D., Klyne, G., Missier, P., Soiland-Reyes, S., Zednik, S.: PROV Model Primer. W3C Working GroupNote, World Wide Web Consortium (W3C), April 2013b. https://www.w3.org/TR/2013/NOTE-prov-primer-20130430/

Bourgonje, P., Moreno-Schneider, J., Nehring, J., Rehm, G., Sasaki, F., Srivastava, A.: Towards a platform for curation technologies: enriching text collections with a semantic-web layer. In: Sack, H., Rizzo, G., Steinmetz, N., Mladenić, D., Auer, S., Lange, C. (eds.) ESWC 2016. LNCS, vol. 9989, pp. 65–68. Springer, Cham (2016a). https://doi.org/10.1007/978-3-319-47602-5_14

Bourgonje, P., Schneider, J.M., Rehm, G., Sasaki, F.: Processing document collections to automatically extract linked data: semantic storytelling technologies for smart curation workflows. In: Gangemi, A., Gardent, C. (eds.) Proceedings of the 2nd International Workshop on Natural Language Generation and the Semantic Web (WebNLG 2016), Edinburgh, UK, pp. 13–16. The Association for Computational Linguistics, September 2016b

Bourgonje, P., Schneider, J.M., Rehm, G.: Automatic classification of abusive language and personal attacks in various forms of online communication. In: Rehm, G., Declerck, T. (eds.) GSCL 2017. LNAI, vol. 10713, pp. 180–191. Springer, Heidelberg (2017a)

Bourgonje, P., Schneider, J.M., Rehm, G.: From clickbait to fake news detection: an approach based on detecting the stance of headlines to articles. In: Popescu, O., Strapparava, C. (eds.) Proceedings of the Second Workshop on Natural Language Processing meets Journalism - EMNLP 2017 Workshop (NLPMJ 2017), Copenhagen, Denmark, pp. 84–89, 7 September 2017b

Chan, R.: Artificial Intelligence is Going to Destroy Fake News - But A.I. can also cause the volume of fake news to explode. Inverse Innovation (2017). https://www.inverse.com/article/27723-artificial-intelligence-will-destroy-fake-news

Conroy, N.J., Rubin, V.L., Che, Y.: Automatic deception detection: methods for finding fake news. Proc. Assoc. Inf. Sci. Technol. **52**(1), 1–4 (2015)

Crystal, D.: Language and the Internet. Cambridge University Press, Cambridge (2001)

Dale, R.: NLP in a post-truth world. Nat. Lang. Eng. **23**(2), 319–324 (2017)

Derczynski, L., Bontcheva, K., Liakata, M., Procter, R., Hoi, G.W.S., Zubiaga, A.: SemEval-2017 task 8: RumourEval: determining rumour veracity and support for rumours. In: Proceedings of the 11th International Workshop on Semantic Evaluation (SemEval-2017), Vancouver, Canada, pp. 69–76. Association for Computational Linguistics, August 2017

Ding, T., Bickel, W.K., Pan, S.: Social Media-based Substance Use Prediction, May 2017. https://arxiv.org/abs/1705.05633

Filloux, F.: Quality for news is mostly about solving the reputation issue (2017). https://mondaynote.com/quality-for-news-is-mostly-about-solving-the-reputation-issue-fdebd0dcc9c2

Gershgorn, D.: In the fight against fake news, artificial intelligence is waging a battle it cannot win. Quartz (2016). https://qz.com/843110/can-artificial-intelligence-solve-facebooks-fake-news-problem/

Groth, P., Moreau, L.: PROV-overview: an overview of the PROV family of documents. W3C Working Group Note, World Wide Web Consortium (W3C), April 2013. https://www.w3.org/TR/prov-overview/

Gupta, A., Lamba, H., Kumaraguru, P., Joshi, A.: Faking Sandy: characterizing and identifying fake images on Twitter during Hurricane Sandy. In: Proceedings of the 22nd International Conference on World Wide Web, Rio de Janeiro, pp. 729–736 (2013)

Haase, M., Huber, M., Krumeich, A., Rehm, G.: Internetkommunikation und Sprachwandel. In: Weingarten, R. (ed.) Sprachwandel durch Computer, pp. 51–85. Westdeutscher Verlag, Opladen (1997)

Holan, A.D.: 2016 lie of the year: fake news. PolitiFact (2016). http://www.politifact.com/truth-o-meter/article/2016/dec/13/2016-lie-year-fake-news/

Horne, B.D., Adal, S.: This just in: fake news packs a lot in title, uses simpler, repetitive content in text body, more similar to satire than real news. In: Proceedings of the 2nd International Workshop on News and Public Opinion at ICWSM, March 2017

Mantzarlis, A.: 6 tips to debunk fake news stories by yourself. Poynter (2015). http://www.poynter.org/2015/6-tips-to-debunk-fake-news-stories-by-yourself/385625/

Mantzarlis, A.: There are now 114 fact-checking initiatives in 47 countries. Poynter (2017). https://www.poynter.org/2017/there-are-now-114-fact-checking-initiatives-in-47-countries/450477/

Marchi, R.: With Facebook, blogs, and fake news, teens reject journalistic "Objectivity". J. Commun. Inq. **36**(3), 246–262 (2012)

Martinez-Alvarez, M.: How can machine learning and AI help solving the fake news problem? (2017). https://miguelmalvarez.com/2017/03/23/how-can-machine-learning-and-ai-help-solving-the-fake-news-problem/

Marwick, A., Lewis, R.: Media Manipulation and Disinformation Online. Data and Society Research Institute, May 2017. https://datasociety.net/output/media-manipulation-and-disinfo-online/

Metz, C.: The Bittersweet Sweepstakes to Build an AI that Destroys Fake News. Wired (2016). https://www.wired.com/2016/12/bittersweet-sweepstakes-build-ai-destroys-fake-news/

Rehm, G.: Towards automatic web genre identification - a corpus-based approach in the domain of academia by example of the academic's personal homepage. In: Proceedings of the 35th Hawaii International Conference on System Sciences (HICSS-35), Big Island, Hawaii. IEEE Computer Society, January 2002

Rehm, G., Sasaki, F.: Digitale Kuratierungstechnologien - Verfahren für die effiziente Verarbeitung, Erstellung und Verteilung qualitativ hochwertiger Medieninhalte. In: Proceedings of the 2015 International Conference of the German Society for Computational Linguistics and Language Technology, GSCL 2015, pp. 138–139 (2015)

Rehm, G., Sasaki, F., Burchardt, A.: Web Annotations - A Game Changer for Language Technologies? Presentation given at I Annotate 2016, Berlin, Germany, 19/20 May 2016. http://www.slideshare.net/georgrehm/web-annotations-a-game-changer-for-language-technology, http://iannotate.org/2016/

Rehm, G., Schneider, J.M., Bourgonje, P., Srivastava, A., Nehring, J., Berger, A., König, L., Räuchle, S., Gerth, J.: Event detection and semantic storytelling: generating a travelogue from a large collection of personal letters. In: Caselli, T., Miller, B., van Erp, M., Vossen, P., Palmer, M., Hovy, E., Mitamura, T. (eds.) Proceedings of the Events and Stories in the News Workshop, Vancouver, Canada. Association for Computational Linguistics. Co-located with ACL 2017, August 2017, in print

Reid, E.M.: Electropolis: communication and community on internet relay chat. Honours thesis, University of Melbourne, Department of History (1991). http://www.aluluei.com/electropolis.htm

Rogers, K., Bromwich, J.E.: The Hoaxes, Fake News and Misinformation We Saw on Election Day. New York Times (2016). https://www.nytimes.com/2016/11/09/us/politics/debunk-fake-news-election-day.html

Rubin, V.L., Chen, Y., Conroy, N.J.: Deception detection for news: three types of fakes. Proc. Assoc. Inf. Sci Technol. **52**(1), 1–4 (2015). https://doi.org/10.1002/pra2.2015.145052010083/full

Runkehl, J., Schlobinski, P., Siever, T.: Sprache und Kommunikation im Internet - Überblick und Analysen. Westdeutscher Verlag, Opladen, Wiesbaden (1998)

Sanderson, R.: Web Annotation Protocol. W3C Recommendation, World Wide Web Consortium (W3C), February 2017. https://www.w3.org/TR/2017/REC-annotation-protocol-20170223/

Sanderson, R., Ciccarese, P., Young, B.: Web Annotation Data Model. W3C Recommendation, World Wide Web Consortium (W3C), February 2017a. https://www.w3.org/TR/2017/REC-annotation-model-20170223/

Sanderson, R., Ciccarese, P., Young, B.: Web Annotation Vocabulary. W3C Recommendation, World Wide Web Consortium (W3C), February 2017b. https://www.w3.org/TR/2017/REC-annotation-vocab-20170223/

Schmidt, A., Wiegand, M.: A survey on hate speech detection using natural language processing. In: Proceedings of the Fifth International Workshop on Natural Language Processing for Social Media, Valencia, Spain (2017)

Srivastava, A., Rehm, G., Schneider, J.M.: DFKI-DKT at SemEval-2017 Task 8: rumour detection and classification using cascading heuristics. In: Proceedings of the 11th International Workshop on Semantic Evaluation (SemEval-2017), Vancouver, Canada, pp. 477–481. Association for Computational Linguistics, August 2017

Walbrühl, D.: Das musst du wissen, um Fake News zu verstehen. Perspective Daily (2017). https://perspective-daily.de/article/213/AhopoOEF

Wardle, C.: Fake news. It's complicated. First Draft News (2017). https://firstdraftnews.com/fake-news-complicated/

Weedon, J., Nuland, W., Stamos, A.: Information Operations and Facebook, April 2017. Version 1.0 https://fbnewsroomus.files.wordpress.com/2017/04/facebook-and-information-operations-v1.pdf

Zubiaga, A., Liakata, M., Procter, R., Hoi, G.W.S., Tolmie, P.: Analysing how people orient to and spread rumours in social media by looking at conversational threads. PLoS ONE **11**(3), e0150989 (2016). https://doi.org/10.1371/journal.pone.0150989

Different Types of Automated and Semi-automated Semantic Storytelling: Curation Technologies for Different Sectors

Georg Rehm[1]([⊠]), Julián Moreno-Schneider[1], Peter Bourgonje[1],
Ankit Srivastava[1], Rolf Fricke[2], Jan Thomsen[2], Jing He[3], Joachim Quantz[3],
Armin Berger[4], Luca König[4], Sören Räuchle[4], Jens Gerth[4],
and David Wabnitz[5]

[1] Language Technology Lab, DFKI GmbH, Alt-Moabit 91c, 10559 Berlin, Germany
georg.rehm@dfki.de
[2] Condat GmbH, Alt-Moabit 91c, 10559 Berlin, Germany
[3] 3pc GmbH Neue Kommunikation, Prinzessinnenstraße 1, 10969 Berlin, Germany
[4] ART+COM AG, Kleiststraße 23-26, 10787 Berlin, Germany
[5] Kreuzwerker GmbH, Ritterstraße 12-14, 10969 Berlin, Germany
http://digitale-kuratierung.de

Abstract. Many industries face an increasing need for smart systems that support the processing and generation of digital content. This is both due to an ever increasing amount of incoming content that needs to be processed faster and more efficiently, but also due to an ever increasing pressure of publishing new content in cycles that are getting shorter and shorter. In a research and technology transfer project we develop a platform that provides content curation services that can be integrated into Content Management Systems, among others. In the project we develop curation services, which comprise semantic text and document analytics processes as well as knowledge technologies that can be applied to document collections. The key objective is to support digital curators in their daily work, i.e., to (semi-)automate processes that the human experts are normally required to carry out intellectually and, typically, without tool support. The goal is to enable knowledge workers to become more efficient and more effective as well as to produce high-quality content. In this article we focus on the current state of development with regard to semantic storytelling in our four use cases.

1 Introduction

Digital content and online media have reached an unprecedented level of relevance and importance, especially with regard to commercial but also political and societal aspects. One of the many technological challenges refers to better support and smarter technologies for digital content curators, i.e., persons, who work primarily at and with a computer, who are facing an ever increasing incoming stream of heterogeneous information and who create, in a general

G. Rehm and T. Declerck (Eds.): GSCL 2017, LNAI 10713, pp. 232–247, 2018.
https://doi.org/10.1007/978-3-319-73706-5_20

sense, new content based on the requirements, demands, expectations and conventions of the sector they work in. For example, experts in a digital agency build websites or mobile apps for clients who provide documents, data, pictures, videos and other assets that are processed, sorted, augmented, arranged, designed, packaged and then deployed. Knowledge workers in a library digitise a specific archive, add metadata and critical edition information and publish the archive online. Journalists need to stay on top of the news stream including blogs, microblogs, newswires etc. in order to produce a new article on a breaking topic. A multitude of examples exist in multiple sectors and branches of media (television, radio, blogs, journalism etc.). All these different professional environments can benefit immensely from semantic technologies that support knowledge workers, who typically work under high time pressure, in their activities: finding relevant information, highlighting important concepts, sorting incoming documents, translating articles in foreign languages, suggesting interesting topics etc. We call these different semantic services, that can be applied flexibly in different professional environments that all have to do with the processing, analysis, translation, evaluation, contextualisation, verification, synthesis and production of digital information, *Curation Technologies*.

The activities reported in this paper are carried out in the context of a two-year research and technology transfer project, Digital Curation Technologies[1] in which DFKI collaborates with four SME companies that operate in four sectors (3pc: public archives; Kreuzwerker: print journalism; Condat: television and media; ART+COM: museum and exhibition design). We develop, in prototypically implemented use cases, a flexible platform that provides generic curation services such as, e.g., summarisation, named entity recognition, entity linking and machine translation (Bourgonje et al. 2016a,b). These are integrated into the partners' in-house systems and customised to their domains so that the content curators who use these systems can do their jobs more efficiently, more easily and with higher quality. Their tasks involve processing, analysing, skimming, sorting, summarising, evaluating and making sense of large amounts of digital content, out of which a new piece of digital content is created, e.g., an exhibition catalogue, a news article or an investigative report.

We mainly work with self-contained document collections but our tools can also be applied to news, search results, blog posts etc. The key objective is to shorten the time it takes digital curators to familiarise themselves with a large collection by extracting relevant data and presenting the data in a way that enables the user to be more efficient, especially when they are not domain experts.

We develop modular language and knowledge technology components that can be arranged in workflows. Based on their output, a semantic layer is generated on top of a document collection. It contains various types of metadata as annotations that can be made use of in further processing steps, visualisations or user interfaces. Our approach bundles a flexible set of semantic services for the *production of digital content*, e.g., to recommend or to highlight interesting

---

[1] http://digitale-kuratierung.de.

and unforeseen storylines or relations between entities to human experts. We call this approach *Semantic Storytelling*.

In this article we concentrate on the collaboration between the research partner and the four SME companies. For each use case we present a prototype application, all of which are currently in experimental use in these companies.

2   Curation Technologies

The curation services are made available through a shared platform and RESTful APIs (Bourgonje et al. 2016a; Moreno-Schneider et al. 2017a; Bourgonje et al. 2016b; Srivastava et al. 2016). They comprise modules that either work on their own or that can be arranged as workflows.[2] The various modules analyse documents and extract information to be used in content curation scenarios. Interoperability between the modules is achieved through the NLP Interchange Format (NIF) (Sasaki et al. 2015). NIF allows for the combination of web services in a decentralised way, without hard-wiring specific pipelines. In the following we briefly present selected curation services.

2.1   Named Entity Recognition and Named Entity Linking

First we convert every document to NIF and then perform Named Entity Recognition (NER). NER consists of two different approaches that allow training with annotated data and/or to use dictionaries. Afterwards the service attempts to look up any named entity in its (language-specific) DBpedia page using DBpedia Spotlight (2016) to extract additional information using SPARQL.

2.2   Geographical Localisation Module and Map Visualisations

The geographical location module uses SPARQL and the Geonames ontology (Wick 2015) to retrieve the latitude and longitude of a location as specified in its DBpedia entry. The module also computes the mean and standard deviation value for latitude and longitude of all identified locations in a document. With this information we can position a document on a map visualisation.

2.3   Temporal Expression Analysis and Timelining

The temporal expression analyser consists of two approaches that can process German and English natural language text, i.e. a regular expression grammar and a modified implementation of HeidelTime (Strötgen and Gertz 2013). After identification, temporal expressions are normalised to a shared format and added to the NIF representation to enable reasoning over temporal expressions and also for archiving purposes. The platform adds document-level statistics based on normalised temporal values. These can be used to position a document on a timeline.

---

[2] Moreno-Schneider et al. (2017a) describes the Semantic Storytelling curation service and provides more technical details. The platform itself is based on the FREME infrastructure (Sasaki et al. 2015).

2.4   Text Classification and Document Clustering

We provide a generic classification service, which is based on Mallet (McCallum 2002). It assigns topics or domains such as "politics" or "sports" to documents when labeled training data is available. Annotated topics are stored in the NIF representation as RDF. Unsupervised document clustering is performed using the Gensim toolkit (Řehůřek and Sojka 2010). For the purpose of this paper we performed experiments with a bag-of-words approach and with tf/idf transformations for the Latent Semantic Indexing (LSI) (Halko et al. 2011), Latent Dirichlet Allocation (LDA) (Hoffman et al. 2010) and Hierarchical Dirichlet Process (HDP) (Wang et al. 2011) algorithms.

2.5   Coreference Resolution

For the correct interpretation and representation of events and their arguments and components, the resolution of mentions referring to entities that are not identified by the NER component (because they are realised by a pronoun or alternative formulation) is essential. For these cases we implemented a coreference resolution mechanism based on CoreNLP for English (Raghunathan et al. 2010). For German language documents we replicated this multi-sieve approach (Srivastava et al. 2017). This component increases the coverage of the NER and event detection modules.

2.6   Monolingual and Cross-Lingual Event Detection

We implemented a state-of-the-art event detection system based on Yang and Mitchell (2016) to pinpoint words or phrases in a sentence that refer to events involving participants and locations, affected by other events and spatio-temporal aspects. The module is trained on the ACE 2005 data (Doddington et al. 2004), consisting of 529 documents from a variety of sources. We apply the tool to extract generic events from the various datasets in our curation scenarios. We also implemented a cross-lingual event detection system, i.e., we translate non-English documents to English through Moses SMT (Koehn et al. 2007) and detect events in the translated documents using the system described above.

2.7   Single and Multi-document Summarisation

Automatic summarisation refers to reducing input text (from one or more documents) into a shorter version by keeping its main content intact while still conveying the actual desired meaning (Ou et al. 2008; Mani and Maybury 1999). This task typically involves identifying, extracting and reordering the most important sentences from a document (collection) into a summary. We offer three different approaches: centroid-based summarisation (Radev et al. 2000), lexical page ranking (Erkan and Radev 2004), and cluster-based link analysis (Wan and Yang 2008).

2.8   User Interaction in the Curation Technologies Prototypes

Our primary goal is to support knowledge workers by automating some of their typical processes. This is why all implemented user interfaces are inherently interactive. By providing feedback to, for example, the output of certain semantic services, knowledge workers have some amount of control over the workflow. They are also able to upload existing resources to adapt individual services. For example, we allow users to identify errors in the output (e.g., incorrectly identified entities) and provide feedback to the algorithm; NER allows users to supply dictionaries for entity linking; Event Detection allows users to supply lists of entities for the identification of agents for events.

3   Semantic Storytelling: Four Sector-Specific Use Cases

Generic Semantic Storytelling involves processing a coherent and self-contained collection of documents in order to identify and to suggest, to the human curator, on a rather abstract level, one or more potential story paths, i.e., specific relationships between entities that can then be used for the process of structuring a new piece of content. It was a conscious decision not to artificially restrict the approach (for example, to certain text types) but to keep it broad and extensible so that we can apply it to the specific needs and requirements of different sectors. In one sector a single surprising, hitherto unknown relation between two entities may be enough to construct an actual story while in others we may try to generate the base skeleton of a storyline semi-automatically (Moreno-Schneider et al. 2016). One concrete example are millions of leaked documents, in which an investigative journalist wants to find the most interesting nuggets of information, i.e., surprising relations between different entities, say, politicians and offshore banks. Our services do not necessarily have to exhibit perfect performance because humans are always in the loop in our application scenario. We want to provide robust technologies with broad coverage. For some services this goal can be fulfilled while for others, it is a bit more ambitious.

3.1   Sector: Museums and Exhibitions

The company ART+COM AG is specialised in the design of museums, exhibitions and showrooms. Their creative staff needs to be able to familiarise themselves with new topics quickly to participate in pitches or during the execution of projects. We implemented a graphical user interface (GUI) that supports the knowledge workers' storytelling capabilities, e.g., for arranging exhibits in a room or for arranging the rooms themselves, by supporting and improving the task of curating incoming content. The GUI enables the effective interaction with the content and the semantic analysis layer. Users can get a quick overview of a specific topic or drill down into the semantic knowledge base to explore deeper relationships.

Initial user research provided valuable insights into the needs of the knowledge workers in this specific use case, especially regarding the kinds of tools and environments each user is familiar with as well as extrapolating their usage patterns (Rehm et al. 2017a). Incoming content materials, provided by clients, include large heterogeneous document collections, e.g., books, images, scientific papers etc. We subdivide the curation process into the phases search, evaluate, organise.

The prototype is a web application (Fig. 1). Users can import documents, such as briefing materials from the client, or perform explorative web searches. Content is automatically analysed by the curation services. The application performs a lookup on the extracted information, e.g., named entities, on Wikidata in order to enrich the entities with useful additional information. Entities are further enriched with top-level ontology labels in order to provide an overview of the distribution of information in categories, for instance, person, organisation, and location. Intuitive visualisation of extracted information is a focus of this prototype. We realised several approaches including a network overview, semantic clustering, timelining and maps. In an evaluation the knowledge workers concluded that the implemented interfaces provides a good overview of the subject and that they would use the tool at the beginning of a project, particularly when confronted with massive amounts of text (Rehm et al. 2017a).

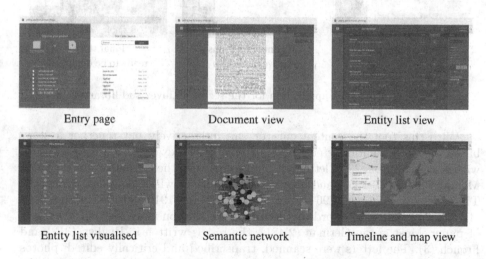

Entry page   Document view   Entity list view

Entity list visualised   Semantic network   Timeline and map view

Fig. 1. Prototype application for the sector museums and exhibitions

3.2   Sector: Public Archives, Libraries, Digital Humanities

For the company 3pc GmbH we developed an authoring environment, enabled by the curation technology platform. Many of 3pc's projects involve a client, e.g., a company, an actress or a political party, that provides a set of digital content and a rough idea how to structure and visualise these assets in the form of a website or app. A tool that can semantically process such a document collection to enable the efficient authoring of flexible, professional, convincing, visually appealing content products that provide engaging stories and that can be played out in different formats (e.g., web app, iOS or Android app, ebook etc.) would significantly reduce the effort on the side of the agency and improve their flexibility. Several screens of the authoring environment's GUI are shown in Fig. 2. It was a conscious design decision to move beyond the typical notion of a "web page" that is broken up into different "modules" using templates. The focus of this prototype are engaging stories told through informative content.

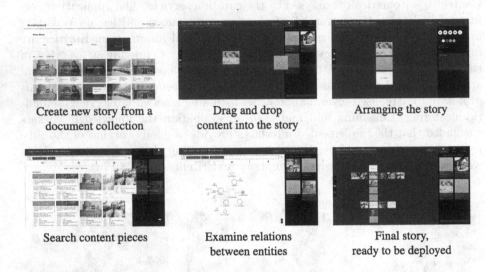

Create new story from a   Drag and drop   Arranging the story
document collection   content into the story

Search content pieces   Examine relations   Final story,
                           between entities   ready to be deployed

Fig. 2. Prototype application for the sector archives and libraries

With this tool the content curator can interactively put together a story based on the semantically enriched content. In the example use case we work with a set of approx. 2,800 letters exchanged between the German architect Erich Mendelsohn (1887–1953) and his wife Luise, both of whom travelled frequently. The collection contains 2,796 letters, written between 1910 and 1953, with a total of 1,002,742 words (359 words per letter on average) on more than 11,000 sheets of paper. Most are in German (2,481), the rest is written in English (312) and French (3). The letters were scanned, transcribed and critically edited; photos and metadata are available; this research was carried out in a project that the authors of the present paper are not affiliated with (Bienert and de Wit 2014). In the letters the Mendelsohns discuss their private and professional lives, their relationship, meetings with friends and business partners, and also their travels.

We decided to focus upon identifying all movement action events (MAE), i.e., all trips undertaken by a subject (usually the letter's author) from location A to location B with a specific departure and arrival time, using a specific mode of transport. This way we want to transform, ideally automatically, a set of interconnected *letters* into a *travelogue* that provides an engaging story to the reader and that also enables additional modes of access, e.g., through map-based or timeline-based visualisations. We wanted to explore to what extent it is possible to automate the production of an online version of such a collection. A complete description can be found in (Rehm et al. 2017b); here we only present a few examples of extracted MAEs to demonstrate the functionality (Table 1). An MAE consists of the six-tuple MAE = $<P, L_O, L_D, t_d, t_a, m>$ with P a reference to the participant (E. or L. Mendelsohn), L_O and L_D references to the origin and destination locations (named locations, GPS coordinates), t_d and t_a the time of departure and arrival and m the mode of transport. Each component is optional as long as the MAE contains at least one participant and a destination.

Table 1. Automatically extracted movement action events (MAEs)

Letter text	Extracted MAEs
Another train stopped [...] this would be the train with which Eric had to leave Cleveland	Eric, Cleveland, [], [], [], train
Because I have to leave on the 13th for Chicago	I (Erich), Croton on Hudson, NY, Chicago, 13th Dec. 1945, [], []
April 5th 48 Sweetheart - Here I am - just arrived in Palm Springs [...]	I (Erich), [], Palm Springs, [], 5th April 1948, []
Thompsons are leaving for a week - [...] at the Beverly Hills on Thursday night!!	Thompsons, [], Beverly Hills, 8th July, [], []

3.3   Sector: Journalism

Journalists write news articles based on information collected from different sources (news agencies, media streams, other news articles, sources, etc.). Research is needed on the topic and domain at hand to produce a high-quality piece. Facts have to be checked, different view points considered, information from multiple sources combined in a sensible way. The resulting piece usually combines new, relevant and surprising information regarding the event reported upon. While the amount of available information is increasing on a daily basis, the journalist's ability to go through all the data is decreasing, which is why smart technology support is needed. We want to enable journalists interactively to put together a story based on semantic content enrichment. In our various use cases, different parts of the content function as atomic building blocks (sentences, paragraphs, documents). For this use case we focus, for now, upon document-level building blocks for generating stories, i.e., documents can be rearranged, included and deleted from a storyline.[3]

---

[3] In a follow-up project we plan to use smaller content components with which we will experiment towards the generation of articles based on multiple story paths, automatically generated with the help of semantic annotations.

For the company Kreuzwerker GmbH we developed an extension for the open source newsroom software Superdesk (https://www.superdesk.org). This production environment specialises on the creation of content, i.e., the actual play-out and rendering of the content is taken care of by other parts of a larger system. The plug-in allows the semantic processing of incoming news streams to enable smart features, e.g., keyword alerts, content exploration, identifying related content, summarisation and machine translation. It also allows for the visualisation and annotation of news documents using additional databases and knowledge graphs (e.g., Linked Open Data) to enable faceted search scenarios so that the journalist has fine-grained mechanisms to locate the needle in a potentially very large digital haystack. Faceted search includes entities, topics, sentiment values and genres, complemented with semantic information from external sources (DBpedia, WikiData etc.). Menus show the annotated entities and their frequencies next to a set of related documents. Example screens of this newsroom content curation dashboard are shown in Fig. 3. The plug-in mainly operates on the (1) ingest view and the (2) authoring view. The first view allows to ingest content channels into the production environment; the semantic tools (see Sect. 2) can automatically analyse the content using, e.g., topic detection, classification (e.g., IPTC topics) and others. In the second view, the curation tools support the authoring process, to add or modify annotations and to recommend related content. A thorough description of the use case and the developed prototype can be found in (Moreno-Schneider et al. 2017b).

Initial search and filtering   Annotated entities   Arranging the storyline

Fig. 3. Prototype application for the sector journalism

The company Condat AG develops software that is used in television stations, supporting journalists to put together, e.g., news programmes, by providing access to databases of metadata-rich archives that consist of media fragments. We developed recommendation, metadata extraction and multi-document summarisation services that enable editors to process large amounts of data, to find updates of stories about events already seen and to identify relevant, but rarely used, media, to provide a certain level of surprise in the storytelling.

Current news exploration systems such as, e.g., Google News, rely on the extraction of entities and analytics processes that operate on frequencies and timestamps to populate categories that resemble traditional newspaper sections (National, World, Politics, Sports, etc.). It also includes "highlight news", which consists of named entities. At the time of writing, "London" was part of the "highlight news" but, as a label, it is not helpful – "Brexit" or "Grenfell Tower fire" would have been more appropriate. Based on simple entity frequencies we cannot distinguish between news about these independent events. We attempt to group documents from a newsstream, based on their topics, in order to generate a summary of the main topics for which we also offer a timelined view.

Annotated news items   Timeline-based view   Map visualisation

Fig. 4. Prototype application for the sector television

We group incoming news through document clustering. First we perform topic modeling using a bag-of-words representation with the vectors based on tf/idf values (Sect. 2.4). The clusters are fed to the multi-document summarisation service, summarising on a per-topic-basis and then to the timelining and multi-document summarisation service (fixed summary length of 200 words). Given that the same number of documents was clustered into topics using six different models (bag of words and tf/idf for HDP, LDA, and LSI each) and that the length of the summary for each topic was fixed at a maximum of 200 words, we discovered that the bag of words approach yields lengthier summaries than tf/idf. Additionally, in 90% of the cases, cluster-based link analysis outperformed all other approaches in terms of summary length.

The approach resembles Google News in that it focuses on entities whereas we want to move to an event-based representation of storylines. Recognised named entities focus on locations, persons or temporal expressions. To arrive at building blocks for storylines and also to identify modified news regarding stories already seen, we want to focus on the actual *events* mentioned in the document collection. We repeated the experiments by using the events extracted from the documents as features for clustering and applied the same algorithms (LSI, LDA, HDP) to the text associated with the extracted event. The clusters were summarised and timelined. While the individual services can be improved using more domain-specific training data, the goal is to present the processing results in a way that speeds up (human) access and understanding and that supports the journalist telling an interesting news story. In a previous experiment (Moreno-Schneider et al. 2016) we used biographical data and corresponding templates, which were presented as content pieces; in the present paper, events serve the same purpose. Figure 4 shows the current prototypes, i.e., entity and temporal expression annotation layers to visualise persons, locations and organisations, a timeline-based view and a map visualisation. The processing generates potential storylines for the editor who can then use them to compose stories based on these automatically extracted key facts (collated from multiple documents). The GUI provides interactive browsing and exploring of the processed news collection. Figure 5 shows an example generated from our collection of local news data.

a) [...] the Federal Office for migration and refugees counts for this year, obviously with 600-thousand newly arriving asylum seekers [...]

b) [...] Potsdam is now the authority of 600-thousand newly arriving refugees – 100-thousand more so than predicted.

c) [...] the cabinet today announced the opportunities for young asylum seekers to work better.

Fig. 5. Example: Story on refugees, composed from 18 topics and 151 documents

4   Related Work

Semantic Storytelling can be defined as the generation of stories, identification of story paths or recommendation of storylines based on a certain set of content using a concrete narrative style or voice. Thus, automatic storytelling consists of two components: a semantic representation of story structure, and the ability to automatically visualise or generate a story from this semantic representation using some form of Natural Language Generation (NLG) (Rishes et al. 2013). In NLG, notable related work is described, among others, by (Jorge et al. 2013; Dionisio et al. 2016; Mazeika 2016; Farrell and Ware 2016). While an interesting discipline that is essential to applying any system aimed at automatically generating stories, especially regarding surface realisation, we primarily focus on the generation of the semantic structure of the story.

Bowden et al. (2016) describe what a story is and how to convert it into a dialogue story, i.e., a system capable of telling a story and then retelling it in different settings to different audiences. They define a story as a set of events, characters, and properties of the story, as well as relations among them, including reactions of characters to story events. For this they use EST (Rishes et al. 2013), a framework that produces a story annotated for the tool Scheherazade as a list of Deep Syntactic Structures, a dependency-tree structure where each node contains the lexical information for the important words in a sentence. Kybartas and Bidarra (2015) present GluNet, a flexible, open source knowledge-base that integrates a variety of lexical databases and facilitates commonsense reasoning for the definition of stories.

Similar to our approach is the work of Samuel et al. (2016). They describe a writing assistant that provides suggestions for the actions of characters. This assistant is meant to be a "playful tool", which is intended to "serve the role of a digital writing partner". We perform similar processes when extracting events and entities from a document collection but our system operates on a more general level and is meant to be applied in different professional sectors.

Several related approaches concentrate on specific domains. A few systems focus on providing content for entertainment purposes (Wood 2008), others focus on storytelling in gaming (Gervás 2013), for recipes (Cimiano et al. 2013; Dale 1989) or weather reports (Belz 2008), requiring knowledge about characters, actions, locations, events, or objects that exist in this particular domain (Riedl and Young 2010; Turner 2014). A closely related approach is the one developed by Poulakos et al. (2015), which presents "an accessible graphical platform for content creators and even end users to create their own story worlds, populate it with smart characters and objects, and define narrative events that can be used by existing tools for automated narrative synthesis".

5   Conclusions

We developed curation technologies that can be applied in the sector-specific use cases of companies active in different sectors and content curation use cases. The partner companies are in need of semantic storytelling solutions that support their own in-house or their customers' content curators putting together new content products, either museum exhibitions, interactive online versions of public archives, news articles or news programmes. The motivation is to make the curators more efficient, to delegate routine tasks to the machine and to enable curators to produce higher quality products because the machine may be able to identify interesting, novel, eye-opening relationships between two pieces of content that a human is unable to recognise. The technologies, prototypically implemented and successfully applied in four sectors, show very promising results (Rehm et al. 2017a), even though the individual implementations of the interactive storytelling approaches are quite specific.

For the museums and exhibitions case we developed a prototype that allows the interactive curation, analysis and exploration of the background material of a new exhibition, supporting the knowledge workers who design the exhibition in their storytelling capabilities by helping them to identify interesting relationships. For the public archive case we implemented a prototype that semantically enriches a collection of letters so that a human expert can more efficiently tell interesting stories about the content – in our example we help the human curator to produce a travelogue about the different trips of the Mendelsohns as an alternative "view" upon the almost 2,800 letters. For newspaper journalism, we annotate named entities to generate clusters of documents that can be used as storylines. For the television case we applied a similar approach but we cluster events instead of named entities (including timelining).

This article provides a current snapshot of the technologies and approaches developed in our project. In a planned follow-up project we will experiment with Natural Language Generation approaches in order to produce natural language text – either complete documents or draft skeletons to be checked, revised and completed by human experts – based on automatically extracted information and on external knowledge provided as Linked Open Data. For this approach we anticipate a whole new set of challenges with regard to semantic storytelling.

Acknowledgments. The authors would like to thank the reviewers for their insightful comments and suggestions. The project "Digitale Kuratierungstechnologien" (DKT) is supported by the German Federal Ministry of Education and Research (BMBF), "Unternehmen Region", instrument Wachstumskern-Potenzial (no. 03WKP45). More information: http://www.digitale-kuratierung.de.

References

Belz, A.: Automatic generation of weather forecast texts using comprehensive prob-abilistic generation-space models. Nat. Lang. Eng. **14**(4), 431–455 (2008). https://doi.org/10.1017/S1351324907004664. ISSN 1351-3249

Bienert, A., de Wit, W., (eds.): EMA - Erich Mendelsohn Archiv. Der Briefwechsel von Erich und Luise Mendelsohn 1910–1953. Staatliche Museen zu Berlin, The Getty Research Institute, Los Angeles, March 2014. http://ema.smb.museum

Bourgonje, P., Moreno-Schneider, J., Nehring, J., Rehm, G., Sasaki, F., Srivastava, A.: Towards a platform for curation technologies: enriching text collections with a semantic-web layer. In: Sack, H., Rizzo, G., Steinmetz, N., Mladenić, D., Auer, S., Lange, C. (eds.) ESWC 2016. LNCS, vol. 9989, pp. 65–68. Springer, Cham (2016a). https://doi.org/10.1007/978-3-319-47602-5_14. ISBN 978-3-319-47602-5

Bourgonje, P., Moreno-Schneider, J., Rehm, G., Sasaki, F.: Processing document collections to automatically extract linked data: semantic storytelling technologies for smart curation workflows. In: Gangemi, A., Gardent, C., (eds.) Proceedings of the 2nd International Workshop on Natural Language Generation and the Semantic Web (WebNLG 2016), Edinburgh, UK, September 2016b, pp. 13–16. The Association for Computational Linguistics (2016b)

Bowden, K.K., Lin, G.I., Reed, L.I., Fox Tree, J.E., Walker, M.A.: M2D: monolog to dialog generation for conversational story telling. In: Nack, F., Gordon, A.S. (eds.) ICIDS 2016. LNCS, vol. 10045, pp. 12–24. Springer, Cham (2016). https://doi.org/10.1007/978-3-319-48279-8_2. ISBN 978-3-319-48278-1

Cimiano, P., Lüker, J., Nagel, D., Unger, C.: Exploiting ontology lexica for generating natural language texts from RDF data. In: Proceedings of the 14th European Workshop on Natural Language Generation, Sofia, Bulgaria, August 2013, pp. 10–19. Association for Computational Linguistics (2013). http://www.aclweb.org/anthology/W13-2102

Dale, R.: Cooking up referring expressions. In: Proceedings of the 27th Annual Meeting on Association for Computational Linguistics, ACL 1989, Stroudsburg, PA, USA, pp. 68–75. Association for Computational Linguistics (1989). https://doi.org/10.3115/981623.981632

Dionisio, M., Nisi, V., Nunes, N., Bala, P.: Transmedia storytelling for exposing natural capital and promoting ecotourism. In: Nack, F., Gordon, A.S. (eds.) ICIDS 2016. LNCS, vol. 10045, pp. 351–362. Springer, Cham (2016). https://doi.org/10.1007/978-3-319-48279-8_31. ISBN 978-3-319-48279-8

Doddington, G., Mitchell, A., Przybocki, M., Ramshaw, L., Strassel, S., Weischedel, R.: The automatic content extraction (ACE) program - tasks, data, and evaluation. In: Proceedings of the Fourth International Conference on Language Resources and Evaluation (LREC 2004), Lisbon, Portugal, May 2004. ELRA (2004)

Erkan, G., Radev, D.R.: LexPageRank: prestige in multi-document text summarization. In: EMNLP, Barcelona, Spain (2004)

Farrell, R., Ware, S.G.: Predicting user choices in interactive narratives using indexter's pairwise event salience hypothesis. In: Nack, F., Gordon, A.S. (eds.) ICIDS 2016. LNCS, vol. 10045, pp. 147–155. Springer, Cham (2016). https://doi.org/10.1007/978-3-319-48279-8_13. ISBN 978-3-319-48279-8

Gervás, P.: Stories from games: content and focalization selection in narrative composition. In: Proceedings of the I Spanish Symposium on Entertainment Computing, Universidad Complutense de Madrid, Madrid, Spain, September 2013

Halko, N., Martinsson, P.G., Tropp, J.A.: Finding structure with randomness: probabilistic algorithms for constructing approximate matrix decompositions. SIAM Rev. 53(2), 217–288 (2011). https://doi.org/10.1137/090771806. ISSN 0036-1445

Hoffman, M., Bach, F.R., Blei, D.M.: Online learning for Latent Dirichlet Allocation. In: Lafferty, J.D., Williams, C.K.I., Shawe-Taylor, J., Zemel, R.S., Culotta, A. (eds.) Advances in Neural Information Processing Systems 23, pp. 856–864. Curran Associates Inc. (2010). http://papers.nips.cc/paper/3902-online-learning-for-latent-dirichlet-allocation.pdf

Jorge, C., Nisi, V., Nunes, N.J., Innella, G., Caldeira, M., Sousa, D.: Ambiguity in design: an airport split-flap display storytelling installation. In: Mackay, W.E., Brewster, S.A., Bødker, S., (eds.) 2013 ACM SIGCHI Conference on Human Factors in Computing Systems, CHI 2013, Paris, France, pp. 541–546. ACM (2013). https://doi.org/10.1145/2468356.2468452. ISBN 978-1-4503-1952-2

Koehn, P., Hoang, H., Birch, A., Callison-Burch, C., Zens, R., Federico, M., Bertoldi, N., Dyer, C., Cowan, B., Shen, W., Moran, C., Bojar, O., Constantin, A., Herbst, E.: Moses: open source toolkit for statistical machine translation. In: Proceedings of ACL 2007, Prague, Czech Republic, pp. 177–180. ACL (2007)

Kybartas, B., Bidarra, R.: A semantic foundation for mixed-initiative computational storytelling. In: Schoenau-Fog, H., Bruni, L.E., Louchart, S., Baceviciute, S. (eds.) ICIDS 2015. LNCS, vol. 9445, pp. 162–169. Springer, Cham (2015). https://doi.org/10.1007/978-3-319-27036-4_15. ISBN 978-3-319-27035-7

Mani, I., Maybury, M.T. (eds.): Advances in Automatic Text Summarization. MIT Press, Cambridge (1999)

Mazeika, J.: A rules-based system for adapting and transforming existing narratives. In: Nack, F., Gordon, A.S. (eds.) ICIDS 2016. LNCS, vol. 10045, pp. 176–183. Springer, Cham (2016). https://doi.org/10.1007/978-3-319-48279-8_16. ISBN 978-3-319-48279-8

McCallum, A.K.: MALLET: a machine learning for language toolkit (2002). http://mallet.cs.umass.edu

Moreno-Schneider, J., Bourgonje, P., Rehm, G.: Towards user interfaces for semantic storytelling. In: Yamamoto, S. (ed.) HIMI 2017, Part II. LNCS, vol. 10274, pp. 403–421. Springer, Cham (2017a). https://doi.org/10.1007/978-3-319-58524-6_32

Moreno-Schneider, J., Srivastava, A., Bourgonje, P., Wabnitz, D., Rehm, G.: Semantic storytelling, cross-lingual event detection and other semantic services for a newsroom content curation dashboard. In: Popescu, O., Strapparava, C. (eds.) Proceedings of the Second Workshop on Natural Language Processing meets Journalism - EMNLP 2017 Workshop (NLPMJ 2017), Copenhagen, Denmark, September 2017b, pp. 68–73 (2017b)

Moreno-Schneider, J., Bourgonje, P., Nehring, J., Rehm, G., Sasaki, F., Srivastava, A.: Towards semantic story telling with digital curation technologies. In: Birnbaum, L., Popescuk, O., Strapparava, C. (eds.) Proceedings of Natural Language Processing Meets Journalism - IJCAI-16 Workshop (NLPMJ 2016), New York, July 2016

Ou, S., Khoo, C.S.-G., Goh, D.H.: Design and development of a concept-based multi-document summarization system for research abstracts. J. Inf. Sci. **34**(3), 308–326 (2008). https://doi.org/10.1177/0165551507084630

Poulakos, S., Kapadia, M., Schüpfer, A., Zünd, F., Sumner, R., Gross, M.: Towards an accessible interface for story world building. In: AAAI Conference on Artificial Intelligence and Interactive Digital Entertainment, pp. 42–48 (2015). http://www.aaai.org/ocs/index.php/AIIDE/AIIDE15/paper/view/11583

Radev, D.R., Jing, H., Budzikowska, M.: Centroid-based summarization of multiple documents: sentence extraction, utility-based evaluation, and user studies. In: Proceedings of the 2000 NAACL-ANLP-Workshop on Automatic Summarization, NAACL-ANLP-AutoSum 2000, Stroudsburg, PA, USA, pp. 21–30. ACL (2000). https://doi.org/10.3115/1117575.1117578

Raghunathan, K., Lee, H., Rangarajan, S., Chambers, N., Surdeanu, M., Jurafsky, D., Manning, C.: A multi-pass sieve for coreference resolution. In: Proceedings of the 2010 Conference on Empirical Methods in NLP, EMNLP 2010, Stroudsburg, PA, USA, pp. 492–501. ACL (2010). http://dl.acm.org/citation.cfm?id=1870658.1870706

Rehm, G., He, J., Moreno-Schneider, J., Nehring, J., Quantz, J.: Designing user interfaces for curation technologies. In: Yamamoto, S. (ed.) HIMI 2017, Part I. LNCS, vol. 10273, pp. 388–406. Springer, Cham (2017a). https://doi.org/10.1007/978-3-319-58521-5_31

Rehm, G., Moreno-Schneider, J., Bourgonje, P., Srivastava, A., Nehring, J., Berger, A., König, L., Räuchle, S., Gerth, J.: Event detection and semantic storytelling: generating a travelogue from a large collection of personal letters. In: Caselli, T., Miller, B., van Erp, M., Vossen, P., Palmer, M., Hovy, E., Mitamura, T. (eds) Proceedings of the Events and Stories in the News Workshop, Vancouver, Canada, August 2017b, pp. 42–51. Association for Computational Linguistics. Co-located with ACL 2017 (2017b)

Řehůřek, R., Sojka, P.: Software framework for topic modelling with large corpora. In: Proceedings of the LREC 2010 Workshop on New Challenges for NLP Frameworks, Valletta, Malta, May 2010, pp. 45–50. ELRA (2010). http://is.muni.cz/publication/884893/en

Riedl, M.O., Young, R.M.: Narrative planning: balancing plot and character. J. Artif. Int. Res. **39**(1), 217–268 (2010). http://dl.acm.org/citation.cfm?id=1946417.1946422. ISSN 1076–9757

Rishes, E., Lukin, S.M., Elson, D.K., Walker, M.A.: Generating different story tellings from semantic representations of narrative. In: Koenitz, H., Sezen, T.I., Ferri, G., Haahr, M., Sezen, D., Ç atak, G. (eds.) ICIDS 2013. LNCS, vol. 8230, pp. 192–204. Springer, Cham (2013). https://doi.org/10.1007/978-3-319-02756-2_24

Samuel, B., Mateas, M., Wardrip-Fruin, N.: The design of *Writing Buddy*: a mixed-initiative approach towards computational story collaboration. In: Nack, F., Gordon, A.S. (eds.) ICIDS 2016. LNCS, vol. 10045, pp. 388–396. Springer, Cham (2016). https://doi.org/10.1007/978-3-319-48279-8_34

Sasaki, F., Gornostay, T., Dojchinovski, M., Osella, M., Mannens, E., Stoitsis, G., Richie, P., Declerck, T., Koidl, K.: Introducing FREME: deploying linguistic linked data. In: Proceedings of the 4th Workshop of the Multilingual Semantic Web, MSW 2015 (2015)

DBPedia Spotlight. DBPedia Spotlight Website (2016). https://github.com/dbpedia-spotlight/

Srivastava, A., Sasaki, F., Bourgonje, P., Moreno-Schneider, J., Nehring, J., Rehm, G.: How to configure statistical machine translation with linked open data resources. In: Proceedings of Translating and the Computer 38 (TC38), pp. 138–148, London, UK, November 2016

Srivastava, A., Weber, S., Bourgonje, P., Rehm, G.: Different German and English coreference resolution models for multi-domain content curation scenarios. In: Rehm, G., Declerck, T. (eds.) GSCL 2017. LNAI, vol. 10713, pp. 48–61. Springer, Heidelberg (2017)

Strötgen, J., Gertz, M.: Multilingual and cross-domain temporal tagging. Lang. Resour. Eval. **47**(2), 269–298 (2013). https://doi.org/10.1007/s10579-012-9179-y

Turner, S.R.: The Creative Process: A Computer Model of Storytelling and Creativity. Taylor & Francis, Abingdon (2014). https://books.google.gr/books?id=1AjsAgAAQBAJ. ISBN 9781317780625

Wan, X., Yang, J.: Multi-document summarization using cluster-based link analysis. In: Proceedings of the 31st Annual International ACM SIGIR Conference on Research and Development in IR, SIGIR 2008, New York, NY, USA, pp. 299–306. ACM (2008). https://doi.org/10.1145/1390334.1390386. ISBN 978-1-60558-164-4

Wang, C., Paisley, J., Blei, D.M.: Online variational inference for the Hierarchical Dirichlet Process. In: Proceedings of the 14th International Conference on AI and Statistics (AISTATS), vol. 15, pp. 752–760 (2011). http://jmlr.csail.mit.edu/proceedings/papers/v15/wang11a/wang11a.pdf

Wick, M.: Geonames ontology (2015). http://www.geonames.org/about.html

Wood, M.D.: Exploiting semantics for personalized story creation. In: Proceedings of the International Conference on Semantic Computing, ICSC 2008, Washington, DC, USA, pp. 402–409. IEEE Computer Society (2008). https://doi.org/10.1109/ICSC.2008.10. ISBN 978-0-7695-3279-0

Yang, B., Mitchell, T.: Joint extraction of events and entities within a document context. In: Proceedings of the 2016 Conference of the North American Chapter of the ACL: Human Language Technologies, pp. 289–299. ACL (2016)

Twitter Geolocation Prediction Using Neural Networks

Philippe Thomas[(⊠)] and Leonhard Hennig

Language Technology Lab, DFKI GmbH, Alt-Moabit 91c, 10559 Berlin, Germany
{philippe.thomas,leonhard.hennig}@dfki.de

Abstract. Knowing the location of a user is important for several use cases, such as location specific recommendations, demographic analysis, or monitoring of disaster outbreaks. We present a bottom up study on the impact of text- and metadata-derived contextual features for Twitter geolocation prediction. The final model incorporates individual types of tweet information and achieves state-of-the-art performance on a publicly available test set. The source code of our implementation, together with pretrained models, is freely available at https://github.com/Erechtheus/geolocation.

1 Introduction

Data from social media platforms is an attractive real-time resource for data analysts. It can be used for a wide range of use cases, such as monitoring of fire- (Paul et al. 2014) and flue-outbreaks (Power et al. 2013), provide location-based recommendations (Ye et al. 2010), or is utilized in demographic analyses (Sloan et al. 2013). Although some platforms, such as Twitter, allow users to geolocate posts, Jurgens et al. (2015) reported that less than 3% of all Twitter posts are geotagged. This severely impacts the use of social media data for such location-specific applications.

The location prediction task can be either tackled as a classification problem, or alternatively as a multi-target regression problem. In the former case the goal is to predict city labels for a specific tweet, whereas the latter case predicts latitude and longitude coordinates for a given tweet. Previous studies showed that text in combination with metadata can be used to predict user locations (Han et al. 2014). Liu and Inkpen (2015) presented a system based on stacked denoising auto-encoders (Vincent et al. 2008) for location prediction. State-of-the-art approaches, however, often make use of very specific, non-generalizing features based on web site scraping, IP resolutions, or external resources such as GeoNames. In contrast, we present an approach for geographical location prediction that achieves state-of-the-art results using neural networks trained solely on Twitter text and metadata. It does not require external knowledge sources, and hence generalizes more easily to new domains and languages.

The remainder of this paper is organized as follows: First, we provide an overview of related work for Twitter location prediction. In Sect. 3 we describe

G. Rehm and T. Declerck (Eds.): GSCL 2017, LNAI 10713, pp. 248–255, 2018.
https://doi.org/10.1007/978-3-319-73706-5_21

the details of our neural network architecture. Results on the test set are shown in Sect. 4. Finally, we conclude the paper with some future directions in Sect. 5.

2   Related Work

For a better comparability of our approach, we focus on the shared task presented at the 2nd Workshop on Noisy User-generated Text (WNUT'16) (Han et al. 2016). The organizers introduced a dataset to evaluate individual approaches for tweet- and user-level location prediction. For tweet-level prediction the goal is to predict the location of one specific message, while for user-level prediction the goal is to predict the user location based on a variable number of user messages. The organizers evaluate team submissions based on accuracy and distance in kilometers. The latter metric allows to account for wrong, but geographically close predictions, for example, when the model predicts Vienna instead of Budapest.

We focus on the five teams who participated in the WNUT shared task. Official team results for tweet- and user-level predictions are shown in Table 1. Unfortunately, only three participants provided systems descriptions, which we will briefly summarize:

Table 1. Official WNUT'16 tweet- and user-level results ranked by tweet median error distance (in kilometers). Individual best results for all three criteria are highlighted in bold face.

Submission	Tweet			User		
	Acc	Median	Mean	Acc	Median	Mean
FujiXerox.2	0.409	**69.5**	**1,792.5**	0.476	**16.1**	1,122.3
csiro.1	**0.436**	74.7	2,538.2	**0.526**	21.7	1,928.8
FujiXerox.1	0.381	92.0	1,895.4	0.464	21.0	**963.8**
csiro.2	0.422	183.7	2,976.7	0.520	23.1	2,071.5
csiro.3	0.420	226.3	3,051.3	0.501	30.6	2,242.4
Drexel.3	0.298	445.8	3,428.2	0.352	262.7	3,124.4
aist.1	0.078	3,092.7	4,702.4	0.098	1,711.1	4,002.4
cogeo.1	0.146	3,424.6	5,338.9	0.225	630.2	2,860.2
Drexel.2	0.082	4,911.2	6,144.3	0.079	4,000.2	6,161.4
Drexel.1	0.085	5,848.3	6,175.3	0.080	5,714.9	6,053.3

Team *FujiXerox* (Miura et al. 2016) built a neural network using text, user declared locations, timezone values, and user self-descriptions. For feature preprocessing the authors build several mapping services using external resources, such as GeoNames and time zone boundaries. Finally, they train a neural network using the fastText n-gram model (Joulin et al. 2016) on post text, user location, user description, and user timezone.

Team *csiro* (Jayasinghe et al. 2016) used an ensemble learning method built on several information resources. First, the authors use post texts, user location text, user time zone information, messenger source (e.g., Android or iPhone) and reverse country lookups for URL mentions to build a list of candidate cities contained in GeoNames. Furthermore, they scraped specific URL mentions and screened the website metadata for geographic coordinates. Second, a relationship network is built from tweets mentioning another user. Third, posts are used to find similar texts in the training data to calculate a class-label probability for the most similar tweets. Fourth, text is classified using the geotagging tool pigeo (Rahimi et al. 2016). The output of individual stages is then used in an ensemble learner.

Team *cogeo* (Chi et al. 2016) employ multinomial naïve Bayes and focus on the use of textual features (i.e., location indicative words, GeoNames gazetteers, user mentions, and hashtags).

3   Methods

We used the WNUT'16 shared task data consisting of 12,827,165 tweet IDs, which have been assigned to a metropolitan city center from the GeoNames database[1], using the strategy described in Han et al. (2012). As Twitter does not allow to share individual tweets, posts need to be retrieved using the Twitter API, of which we were able to retrieve 9,127,900 (71.2%). The remaining tweets are no longer available, usually because users deleted these messages. In comparison, the winner of the WNUT'16 task (Miura et al. 2016) reported that they were able to successfully retrieve 9,472,450 (73.8%) tweets. The overall training data consists of 3,362 individual class labels (i.e., city names). In our dataset we only observed 3,315 different classes.

For text preprocessing, we use a simple whitespace tokenizer with lower casing, without any domain specific processing, such as unicode normalization (Davis et al. 2001) or any lexical text normalization (see for instance Han and Baldwin (2011)). The text of tweets, and metadata fields containing texts (user description, user location, user name, timezone) are converted to word embeddings (Mikolov et al. 2013), which are then forwarded to a Long Short-Term Memory (LSTM) unit (Hochreiter and Schmidhuber 1997). In our experiments we randomly initialized embedding vectors. We use batch normalization (Ioffe and Szegedy 2015) for normalizing inputs in order to reduce internal covariate shift. The risk of overfitting by co-adapting units is reduced by implementing dropout (Srivastava et al. 2014) between individual neural network layers. An example architecture for textual data is shown in Fig. 1a. Metadata fields with a finite set of elements (UTC offset, URL–domains, user language, tweet publication time, and application source) are converted to one-hot encodings, which are forwarded to an internal embedding layer, as proposed by Guo and Berkhahn (2016). Again batch normalization and dropout is applied to avoid overfitting. The architecture is shown in Fig. 1b.

---

[1] http://www.geonames.org.

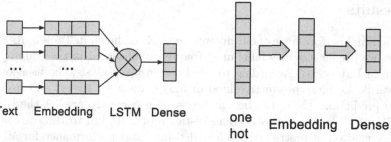

a Example architecture used for textual data. Tokenized text is represented as word embeddings, which are then forwarded to a LSTM. Dropout and batch normalization is applied between individual layers.

b Example architecture used for categorical data. Categorical data is represented as one-hot encodings and internally converted to entity embeddings.

Fig. 1. Architectures for city prediction.

Table 2. Selected parameter settings

Parameter	Property	Parameter	Property
Description embedding dim	100	Text embedding dim	100
Location embedding dim	50	Timezone embedding dim	50
Name embedding dim	100		

Individual models are completed with a dense layer for classification, using a softmax activation function. We use stochastic gradient descent over shuffled mini-batches with Adam (Kingma and Ba 2014) and cross-entropy loss as objective function for classification. The parameters of our model are shown in Table 2.

The WNUT'16 task requires the model to predict class labels and longitude/latitude pairs. To account for this, we predict the mean city longitude/latitude location given the class label. For user-level prediction, we classify all messages individually and predict the city label with the highest probability over all messages.

3.1 Model Combination

The internal representations for all different resources (i.e., *text, user-description, user-location, user-name, user-timezone, links, UTC offset, user lang, tweet-time* and *source*) are concatenated to build a final tweet representation. We then evaluate two training strategies: In the first training regime, we train the combined model from scratch. The parameters for all word embeddings, as well as all network layers, are initialized randomly. The parameters of the full model including the softmax layer combining the output of the individual LSTM– and metadata– models are learned jointly. For the second strategy, we first train each model separately, and then keep their parameters fixed while training only the final softmax layer.

4   Results

The individual performance of our different models is shown in Table 3. As simple baseline, we predict the city label most frequently observed in the training data (Jakarta in Indonesia). According to our bottom-up analysis, the user-location metadata is the most productive kind of information for tweet- and user-level location prediction. Using the text alone, we can correctly predict the location for 19.5% of all tweets with a median distance of 2,190 km to the correct location. Aggregation of pretrained models also increases performance for all three evaluation metrics in comparison to training a model from scratch.

For tweet-level prediction, our best merged model outperforms the best submission (*FujiXerox.2*) in terms of accuracy, median and mean distance by 2.1% points, 21.9 km, and 613.1 km respectively. The ensemble learning method (*csiro*) outperforms our best models in terms of accuracy by 0.6% points, but our model performs considerably better on median and mean distance by 27.1 and 1358.8 km respectively. Additionally, the approach of *csiro* requires several dedicated services, such as GeoNames gazetteers, time zone to GeoName mappings, IP country resolver and customized scrapers for social media websites. The authors describe custom link handling for FourSquare, Swarm, Path, Facebook, and Instagram. On our training data we observed that these websites account for 1,941,079 (87.5%) of all 2,217,267 shared links. It is therefore tempting to

Table 3. Tweet level results ranked by median error distance (in kilometers). Individual best results for all three criteria are highlighted in bold face. Full-scratch refers to a merged model trained from scratch, whereas the weights of the full-fixed model are only retrained where applicable. The baseline predicts the location most frequently observed in the training data (Jakarta).

Model	Tweet			User		
	Acc	Median	Mean	Acc	Median	Mean
Location	0.361	205.6	4,538.0	0.445	43.9	3,831.7
Text	0.195	2,190.6	4,472.9	0.321	263.8	2,570.9
Description	0.087	3,817.2	6,060.2	0.098	3,296.9	5,880.0
User-name	0.057	3,849.0	5,930.1	0.059	4,140.4	6,107.6
Timezone	0.058	5,268.0	5,530.1	0.061	5,470.5	5,465.5
User-lang	0.061	6,465.1	7,310.2	0.047	8,903.7	8,525.1
Links	0.032	7,601.7	6,980.5	0.045	6,687.4	6,546.8
UTC	0.046	7,698.1	6,849.0	0.051	3,883.4	6,422.6
Source	0.045	8,005.0	7,516.8	0.045	6,926.3	6,923.5
Tweet-time	0.028	8,867.6	8,464.9	0.024	11,720.6	10,363.2
Full-scratch	0.417	59.0	1,616.4	0.513	17.8	1,023.9
Full-fixed	**0.430**	**47.6**	**1,179.4**	**0.530**	**14.9**	**838.5**
Baseline	0.028	11,723.0	10,264.3	0.024	11,771.5	10,584.4

speculate that a customized scraper for these websites could further boost our results for location prediction.

As team *cogeo* uses only the text of a tweet, the results of *cogeo.1* are comparable with our text-model. The results show that our text-model outperforms this approach in terms of accuracy, median and mean distance to the gold standard by 4.9% points, 1,234 km, and 866 km respectively.

For user-level prediction, our method performs on a par with the individual best results collected from the three top team submissions (*FujiXerox.2*, *csiro.1*, and *FujiXerox.1*). A notable difference is the mean predicted error distance, where our model outperforms the best model by 125.3 km.

5   Conclusion

We presented our neural network architecture for the prediction of city labels and geo-coordinates for tweets. We focus on the classification task and derive longitude/latitude information from the city label. We evaluated models for individual Twitter (meta)-data in a bottom up fashion and identified highly location indicative fields. The proposed combination of individual models requires no customized text-preprocessing, specific website crawlers, database lookups or IP to country resolution while achieving state-of-the-art performance on a publicly available data set. For better comparability, source code and pretrained models are freely available to the community.

As future work, we plan to incorporate images as another type of metadata for location prediction using the approach presented by Simonyan and Zisserman (2014).

Acknowledgments. This research was partially supported by the German Federal Ministry of Economics and Energy (BMWi) through the projects SD4M (01MD15007B) and SDW (01MD15010A) and by the German Federal Ministry of Education and Research (BMBF) through the project BBDC (01IS14013E).

References

Chi, L., Lim, K.H., Alam, N., Butler, C.J.: Geolocation prediction in Twitter using location indicative words and textual features. In: Proceedings of the 2nd Workshop on Noisy User-generated Text (WNUT), Osaka, Japan, pp. 227–234, December 2016. http://aclweb.org/anthology/W16-3930

Davis, M., Whistler, K., Dürst, M.: Unicode Normalization Forms. Technical report, Unicode Consortium (2001)

Guo, C., Berkhahn, F.: Entity embeddings of categorical variables. CoRR, abs/1604. 06737 (2016)

Han, B., Baldwin, T.: Lexical normalisation of short text messages: makn sens a #Twitter. In: Proceedings of the 49th Annual Meeting of the Association for Computational Linguistics: Human Language Technologies, HLT 2011, Stroudsburg, PA, USA, vol. 1, pp. 368–378 (2011). http://dl.acm.org/citation.cfm?id=2002472.2002520. ISBN 978-1-932432-87-9

Han, B., Cook, P., Baldwin, T.: Geolocation prediction in social media data by finding location indicative words. In: COLING 2012, 24th International Conference on Computational Linguistics, Proceedings of the Conference: Technical Papers, Mumbai, India, pp. 1045–1062, 8–15 December 2012. http://aclweb.org/anthology/C/C12/C12-1064.pdf

Han, B., Cook, P., Baldwin, T.: Text-based Twitter user geolocation prediction. J. Artif. Int. Res. **49**(1), 451–500 (2014). http://dl.acm.org/citation.cfm?id=2655713.2655726. ISSN 1076-9757

Han, B., Rahimi, A., Derczynski, L., Baldwin, T.: Twitter geolocation prediction shared task of the 2016 workshop on noisy user-generated text. In: Proceedings of the 2nd Workshop on Noisy User-generated Text (WNUT), Osaka, Japan, pp. 213–217, December 2016. http://aclweb.org/anthology/W16-3928

Hochreiter, S., Schmidhuber, J.: Long short-term memory. Neural Comput. **9**(8), 1735–1780 (1997). https://doi.org/10.1162/neco.1997.9.8.1735. ISSN 0899-7667

Ioffe, S., Szegedy, C.: Batch Normalization: Accelerating Deep Network Training by Reducing Internal Covariate Shift. CoRR, abs/1502.03167 (2015). http://arxiv.org/abs/1502.03167

Jayasinghe, G., Jin, B., Mchugh, J., Robinson, B., Wan, S.: CSIRO Data61 at the WNUT Geo Shared Task. In: Proceedings of the 2nd Workshop on Noisy User-generated Text (WNUT), Osaka, Japan, pp. 218–226, December 2016. http://aclweb.org/anthology/W16-3929

Joulin, A., Grave, E., Bojanowski, P., Mikolov, T.: Bag of Tricks for Efficient Text Classification. CoRR, abs/1607.01759 (2016). http://arxiv.org/abs/1607.01759

Jurgens, D., Finethy, T., McCorriston, J., Xu, Y.T., Ruths, D.: Geolocation prediction in Twitter using social networks: a critical analysis and review of current practice. In: ICWSM, pp. 188–197 (2015)

Kingma, D.P., Ba, J.: Adam: A Method for Stochastic Optimization. CoRR, abs/1412.6980 (2014). http://arxiv.org/abs/1412.6980

Liu, J., Inkpen, D.: Estimating user location in social media with stacked denoising auto-encoders. In: Proceedings of the 1st Workshop on Vector Space Modeling for Natural Language Processing, Denver, Colorado, pp. 201–210, June 2015. http://www.aclweb.org/anthology/W15-1527

Mikolov, T., Sutskever, I., Chen, K., Corrado, G., Dean, J.: Distributed Representations of Words and Phrases and their Compositionality. CoRR, abs/1310.4546 (2013). http://arxiv.org/abs/1310.4546

Miura, Y., Taniguchi, M., Taniguchi, T., Ohkuma, T.: A simple scalable neural networks based model for geolocation prediction in Twitter. In: Proceedings of the 2nd Workshop on Noisy User-generated Text (WNUT), Osaka, Japan, pp. 235–239, December 2016. http://aclweb.org/anthology/W16-3931

Paul, M.J., Dredze, M., Broniatowski, D.: Twitter improves influenza forecasting. PLOS Currents Outbreaks **6** (2014)

Power, R., Robinson, B., Ratcliffe, D.: Finding fires with Twitter. In: Australasian Language Technology Association Workshop, vol. 80 (2013)

Rahimi, A., Cohn, T., Baldwin, T.: Pigeo: a python geotagging tool. In: Proceedings of ACL-2016 System Demonstrations, Berlin, Germany, pp. 127–132, August 2016. http://anthology.aclweb.org/P16-4022

Simonyan, K., Zisserman, A.: Very Deep Convolutional Networks for Large-Scale Image Recognition. CoRR, abs/1409.1556 (2014). http://arxiv.org/abs/1409.1556

Sloan, L., Morgan, J., Housley, W., Williams, M., Edwards, A., Burnap, P., Rana, O.: Knowing the tweeters: deriving sociologically relevant demographics from Twitter. Sociol. Res. Online, 18 (3) (2013). https://doi.org/10.5153/sro.3001. ISSN 1360-7804

Srivastava, N., Hinton, G., Krizhevsky, A., Sutskever, I., Salakhutdinov, R.: Dropout: a simple way to prevent neural networks from overfitting. J. Mach. Learn. Res. **15**(1), 1929–1958 (2014). http://dl.acm.org/citation.cfm?id=2627435.2670313. ISSN 1532-4435

Vincent, P., Larochelle, H., Bengio, Y., Manzagol, P.-A.: Extracting and composing robust features with denoising autoencoders. In: Proceedings of the 25th International Conference on Machine Learning, ICML 2008, New York, NY, USA, pp. 1096–1103. ACM (2008). https://doi.org/10.1145/1390156.1390294. ISBN 978-1-60558-205-4

Ye, M., Yin, P., Lee, W.-C.: Location recommendation for location-based social networks. In: Proceedings of the 18th SIGSPATIAL International Conference on Advances in Geographic Information Systems, GIS 2010, pp. 458–461, New York, NY, USA. ACM (2010). https://doi.org/10.1145/1869790.1869861. ISBN 978-1-4503-0428-3

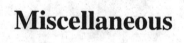

Miscellaneous

Diachronic Variation of Temporal Expressions in Scientific Writing Through the Lens of Relative Entropy

Stefania Degaetano-Ortlieb[1]([✉]) and Jannik Strötgen[2]

[1] Saarland University, Saarbrücken, Germany
s.degaetano@mx.uni-saarland.de

[2] Max Planck Institute for Informatics, Saarland Informatics Campus,
Saarbrücken, Germany
jannik.stroetgen@mpi-inf.mpg.de

Abstract. The abundance of temporal information in documents has lead to an increased interest in processing such information in the NLP community by considering temporal expressions. Besides domain-adaptation, acquiring knowledge on variation of temporal expressions according to time is relevant for improvement in automatic processing. So far, frequency-based accounts dominate in the investigation of specific temporal expressions. We present an approach to investigate diachronic changes of temporal expressions based on relative entropy – with the advantage of using conditioned probabilities rather than mere frequency. While we focus on scientific writing, our approach is generalizable to other domains and interesting not only in the field of NLP, but also in humanities.

1 Introduction

Many types of textual documents are rich in temporal information. A specific type of such information are temporal expressions, which again happen to occur in a wide variety of documents. Thus, during the last years, there has been a growing interest in temporal tagging within the NLP community. While variation of temporal expressions according to different domains has become a well established research area (Mazur and Dale 2010; Strötgen and Gertz 2012; Lee et al. 2014; Strötgen and Gertz 2016; Tabassum et al. 2016), variation of temporal expressions according to time within a domain has received less attention so far.[1] Knowing how temporal expressions might have changed over time within a domain is interesting not only in the field of NLP, e.g., for adaptation of temporal taggers to different time periods, but also in humanities studies in the fields of historical linguistics, sociolinguistics, and the like.

In this paper, we focus on temporal expressions in the scientific domain and study their diachronic development over a time frame of approx. 350 years

---

[1] Note that here *domain* is defined as a group of documents sharing the same characteristics for the task of temporal tagging, cf. (Strötgen and Gertz 2016).

G. Rehm and T. Declerck (Eds.): GSCL 2017, LNAI 10713, pp. 259–275, 2018.
https://doi.org/10.1007/978-3-319-73706-5_22

(from the 1650s to the 2000s). While here we take an exploratory historical perspective, our findings have implications for improving temporal tagging, especially for recall.

Temporal expressions are related to situation-dependent reference (see notably, Biber et al. (1999)'s work), i.e., linguistic reference to a particular aspect of the text-external temporal context of an event (cf. Atkinson (1999, p. 120); Biber and Finegan (1989, p. 492)). While according to Biber et al.'s work, scientific writing has moved towards expressing less situation-dependent reference, to the best of our knowledge, there is no evidence of how this change has been manifested linguistically and whether the types of temporal expressions used in scientific writing have changed over time. To investigate this in more detail, we pose the following questions:

- Do the types of temporal expressions vary diachronically in scientific writing, and if so how is this manifested linguistically?
- What are typical temporal expressions of specific time periods and do these change over time?
- Are different types of temporal expressions, e.g., duration expressions and date expressions referring to points in time, equally affected by a potential change over time?

To process temporal information in scientific research articles, we use HeidelTime (Strötgen and Gertz 2010), a domain-sensitive tagger to extract and normalize temporal expressions according to the TimeML standard (Pustejovsky 2005) for temporal annotation (see Sect. 4). To detect typical temporal expressions of specific time periods, we use relative entropy, more precisely Kullback-Leibler Divergence (KLD) (Dagan et al. 1999; Lafferty and Zhai 2001). By KLD we measure how typical a temporal expression is for a time period vs. another time period (see Sect. 5). The methodology has been adopted from Fankhauser et al. (2016) and successfully used in Degaetano-Ortlieb and Teich (2016) to detect typical linguistic features in scientific writing, Degaetano-Ortlieb and Teich (2017) to detect typical features of research article sections, and Degaetano-Ortlieb (2017) to observe typical features of social variables.

In the analysis, we inspect general diachronic tendencies based on relative frequency and use relative entropy to investigate more fine-grained changes in the use of temporal expressions over time in scientific writing (see Sect. 6). On a more abstract level, we observe that the use of temporal information in scientific writing reflects the paradigm change from observational to experimental science (cf. Fankhauser et al. (2016); Gleick (2010)) and moves further to descriptions of previous work (e.g., *in the last decades*) in contemporary scientific writing.

2   Related Work

Temporal information has been often employed to improve information retrieval (IR) approaches (see Campos et al. (2014) and Kanhabua et al. (2015) for an overview). A prerequisite to exploit temporal information is temporal tagging, i.e., the identification, extraction, normalization, and annotation of temporal expressions based on an annotation standard such as the temporal markup language TimeML (Pustejovsky 2005). While for quite a long time, temporal tagging was tailored towards processing news texts, in the last years, domain-sensitive approaches are being developed, as it has been shown that temporal information varies significantly across domains (Mazur and Dale 2009; Strötgen and Gertz 2016). Domain-sensitive temporal taggers are UWTime (Lee et al. 2014) and HeidelTime (Strötgen and Gertz 2012). We choose HeidelTime as it is being reported to be much faster than UWTime (Agarwal and Strötgen 2017).

Recently, there is also an increasing interest in temporal information in the field of digital humanities. An early approach to operationalize time in narratology has been applied by Meister (2005). Strötgen et al. (2014) show how temporal taggers can be extended for temporal expressions referring to historical dates in the AncientTimes corpus. Fischer and Strötgen (2015) apply temporal taggers to analyze date accumulations in large literary corpora. An analysis of temporal expressions and whether they refer to the future or the past has also been performed on English and Japanese twitter data (Jatowt et al. 2015).

Considering the diachronic aspect of temporal information in scientific writing, it has been mainly investigated by considering temporal adverbs in the context of register studies. Biber and Finegan (1989) and Atkinson (1999), for example, have shown a decrease of temporal adverbs in scientific writing in terms of relative frequencies. Fischer and Strötgen (2015) also studied temporal expressions in a diachronic corpus, but only temporal expressions with explicit day and month information have been considered.

We use temporal tagging tailored at identifying temporal information in scientific writing to obtain a more comprehensive picture of possible diachronic changes. Moreover, besides considering changes in terms of relative frequency, we look at typical temporal expressions and patterns of temporal expressions of specific time periods.

3   Data

As a dataset, we use texts of scientific writing ranging from 1665 to 2007. The first time periods (1665 up to 1869) are covered by the Royal Society Corpus (Kermes et al. 2016a) build from the Proceedings and Transactions of the Royal Society of London – the first periodical of scientific writing – covering several topics within biological sciences, general science, and mathematics. For the later time periods (1966 to 2007), we also use scientific research articles from various disciplines (e.g., biology, linguistics, computer science) taken from the SciTex corpus (Degaetano-Ortlieb et al. 2013; Teich et al. 2013). For comparative purposes, we divide the corpus into fifty year time periods. Table 1 shows the time periods, their coverage and the sub-corpus sizes in number of tokens and documents.

The corpus has been pre-processed in terms of OCR correction, normalization, tokenization, lemmatization, sentence segmentation, and part of speech tagging (cf. Kermes et al. (2016b)).

Table 1. Corpus details.

Period	Coverage	Tokens	Documents
1650	1665–1699	2,589,536	1,326
1700	1700–1749	3,433,838	1,702
1750	1750–1799	6,759,764	1,831
1800	1800–1849	10,699,270	2,778
1850	1850–1869	11,676,281	2,176
1950	1966–1989	18,998,645	3,028
2000	2000–2007	20,201,053	2,111

4   Processing Temporal Information

4.1   Temporal Expressions

Key characteristics. Temporal expressions have three important key character-istics (cf. Alonso et al. (2011); Strötgen and Gertz (2016)). First, they can be *normalized*, i.e., expressions referring to the same semantics can be normalized to the same value. For example, *March 11, 2017* and *the 2nd Saturday in March of this year* point to the same point in time, even though both expressions are realized in different ways. Second, temporal expressions are *well-defined*, i.e., given two points in time X and Y, the relationship between these two points can always be determined, e.g., as X is before Y (cf. Allen (1983)). Third, they can be organized hierarchically on a granularity scale (from coarser to finer gran-ularities and vice versa such as *day, month* or *year*). Relevant in our analysis are normalization and granularity. Normalized values are used to compare tempo-ral expressions across time periods instead of considering only the single lexical realizations. In terms of granularity, we consider granularity scales to determine diachronic changes of temporal expressions.

Types. According to the temporal markup language TimeML (cf. Pustejovsky (2005)), there are four types of temporal expressions (cf. also Strötgen and Gertz (2016)):

- DATE expressions refer to a point in time of the granularity equal or coarser than 'day' (e.g., *March 11, 2017, March 2017* or *2017*).
- TIME expressions refer to a point in time of any granularity smaller than 'day' (e.g., *Saturday morning* or *10:30 am*).
- DURATION expressions refer to the length of a time interval and can be of different granularity (e.g., *two hours, three weeks, four years*).
- SET expressions refer to the periodical aspect of an event, describing set of times/dates (e.g., *every Saturday*) or a frequency within a time interval (e.g., *twice a day*).

In the analysis, we consider all these four types showing how their use has changed diachronically in scientific writing.

4.2   Temporal Tagging

For temporal tagging we use HeidelTime (Strötgen and Gertz 2010), a domain-sensitive temporal tagger. HeidelTime supports normalization strategies for four domains: news, narrative, colloquial, and autonomous. Although HeidelTime has been applied to process scientific documents using the autonomous domain, these scientific documents have been very specific, relatively short (biomedical abstracts) with many so-called autonomous expressions (i.e., expressions not referring to real points in time, but to references in a local time frame).

In contrast, our corpus is quite heterogeneous, containing letters and reports in the earlier time periods and full articles in the later time periods. Thus, we expect that most of the documents are written in such a way that the correct normalization of relative temporal expressions can be reached by using the document creation time as reference time. This makes the documents similar to news-style documents according to HeidelTime's domain definitions. Thus, we apply HeidelTime with its news domain setting. Note, however, that in our analysis we use only normalized values of DURATION and SET expressions, which are normalized to the length and granularity of an expression but not to an exact point in time. Thus, our findings still hold if some of the occurring temporal expressions are not normalized correctly to a point in time.

HeidelTime uses TIMEX3 tags, which are based on TimeML (Pustejovsky et al. 2010), the most widely used annotation standard for temporal expressions. In the following, we briefly explain the *value* attribute of TIMEX3 annotations of DURATION and SET expressions, as we do consider their normalized values for a deeper analysis of the occurring temporal expressions. The value attribute of DURATION and SET expressions contains information about the length of the duration that is mentioned, starting with P (or PT in case of time level durations) followed by a number and an abbreviation of the granularity (e.g., years: Y, month: M, week: W, days: D; hours: H, minutes: M). In addition, fuzzy expressions are referred to by X instead of precise numbers, e.g., *several weeks* is normalized to PXW, *monthly* is normalized to XXXX-XX and *annually* to XXXX.

4.3   Extraction Quality

For meaningful analysis and substantiated conclusions of temporal expressions in our diachronic corpus, the extraction (and normalization) quality of the temporal tagger should be reliable. Although HeidelTime has been extensively evaluated before on a variety of corpora[2], our corpus is quite different from standard temporal tagging corpora as it contains scientific documents from multiple scientific fields published across several centuries. Creation of a proper gold standard with manual annotations covering all scientific fields across all time periods would not be feasible in an appropriate time frame. Instead, for a valuable statement of

---

[2] E.g., on news articles as in the TempEval competitions (Verhagen et al. 2010; UzZaman et al. 2013) and on Wikipedia articles contained in the WikiWars corpus (Mazur and Dale 2010).

temporal tagging quality on our corpus, determining the correctness of expressions tagged by the temporal tagger would be meaningful.

For this, we use precision, i.e., we randomly sample 250 instances for each time period, and manually validate whether the automatically annotated temporal expressions are correctly extracted.[3] Here, we consider correctly extracted instances (RIGHT) and wrongly extracted instances (WRONG). The latter are either cases of ambiguity (e.g., *spring* as 'season' or 'water spring' or *current* meaning 'now' or 'electric current') or wrongly assigned temporal expressions to numbers occurring in the text. On top we differentiate correctly assigned but not relevant instances (OTHER) due to noise in the data itself. These are, e.g., temporal expressions assigned to reference sections (especially in the 1950–2000 periods) or used within tables (mostly in the earlier time periods).

Table 2 presents precision information and the number of instances per assigned category of RIGHT, OTHER, and WRONG. We consider the OTHER instances to be correct in terms of precision of extraction. Across periods, precision achieves 0.89 to 0.96.

Table 2. Precision across time periods.

Period	RIGHT	OTHER	WRONG	Precision
1650	219	13	18	0.928
1700	210	20	20	0.920
1750	218	21	11	0.956
1800	186	37	27	0.892
1850	181	48	22	0.912
1950	116	114	20	0.920
2000	145	96	9	0.964

5 Typicality of Temporal Expressions

To obtain temporal expressions typical of a time period, we use relative entropy, also known as Kullback-Leibler Divergence (KLD) (Kullback and Leibler 1951) – a well-known measure of (dis)similarity between probability distributions used in NLP, speech processing, and information retrieval. In comparison to relative frequency, i.e., the unconditioned probability of, e.g., a word over all words in a corpus, relative entropy is based on conditioned probability.

In information-theoretic parlance, relative entropy measures the average number of additional bits per feature (here: temporal expressions) needed to

---

[3] We chose to use an amount of instances per period rather than an amount of documents, due to possible sparsity of temporal expressions within documents. This also allows us to validate the same amount of instances across time periods, rather than varying amounts of instances across time periods.

encode a feature of a distribution A (e.g., the 1650 time period) by using an encoding optimized for a distribution B (e.g., the 1700 time period). The more additional bits needed, the more distant A and B are. This is formalized as:

$$D(A||B) = \sum_i p(feature_i|A)log_2\frac{p(feature_i|A)}{p(feature_i|B)} \qquad (1)$$

where $p(feature_i|A)$ is the probability of a feature (i.e., a temporal expression) in a time period A, and $p(feature_i|B)$ the probability of that feature in a time period B. The $log_2\frac{p(feature_i|A)}{p(feature_i|B)}$ relates to the difference between both probability distributions ($log_2p(feature_i|A) - log_2p(feature_i|B)$), giving the number of additional bits. These are then weighted with the probability of $p(feature_i|A)$ so that the sum over all $feature_i$ gives the *average* number of additional bits per feature, i.e., the *relative* entropy.

In terms of typicality, the more bits are used to encode a feature, the more typical that feature is for a given time period vs. another time period. Thus, in a comparison of two time periods (e.g., 1650 vs. 1700), the higher the KLD value of a feature for one time period (e.g., 1650), the more typical that feature is for that given time period. In addition, we test for significance of a feature by an unpaired Welch's t-test. Thus, features considered typical are distinctive according to KLD and show a p-value below a given threshold (e.g., 0.05).

To compare typical features across several time periods, the most high ranking features of each comparison are considered. For example, for 1650 we obtain six feature sets typical of 1650 as we have six comparisons of 1650 with each of the other six time periods (i.e., a feature set for features typical of 1650 vs. 1700, of 1650 vs. 1750, etc.). If features are shared across feature sets and are high ranking (e.g., in the top 5), these features are considered to be typical of 1650. In other words, these are features ranking high in terms of KLD, significant in terms of p-value, and typical of a time period across all/most comparisons with other time periods. As in our case we consider seven time periods, features are considered typical which rank high for one time period in 6 to 4 feature sets (i.e., typical in more than half of the comparisons).

6   Analysis

In the following, we analyze diachronic tendencies of temporal expressions from the period of 1650 to 2000 in terms of (1) relative frequency (i.e., unconditioned probabilities), and (2) typicality (i.e., conditioned probabilities of expressions in one vs. the other time periods as described in Sect. 5).

We show how the notion of typicality based on relative entropy leads to valuable insights on the change of temporal expressions in scientific writing w.r.t. more and less frequent expressions.

6.1   Frequency-Based Diachronic Tendencies

Comparing temporal types across fifty years time periods in terms of frequency (see Fig. 1 showing log of frequency per million (pM)), DATE is the most

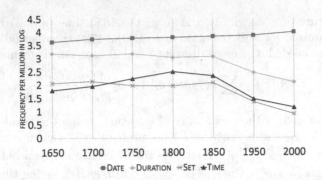

Fig. 1. Diachronic tendencies of temporal expression types in scientific writing.

frequent type, followed by DURATION. SET and TIME expressions are less frequent. In addition, while DATE remains relatively stable over time, expressions of DURATION, SET and TIME drop quite a bit from 1850 onwards, getting relatively rare.

6.2 Diachronic Tendencies of 'Typical' Temporal Expressions

Inspecting diachronic change through the lens of relative entropy (as described in Sect. 5) allows us to consider temporal expressions typical of one time period when compared to the other time periods. We study each type of temporal expression and carefully select the base of comparison.

DATE Considering DATE expressions, instead of comparing single dates (which mostly occur only once in the corpus, such as *June, 3, 1769*), we take a level of abstraction and consider part-of-speech (POS) sequences of annotated DATE expressions to better inspect the types of changes that might have occurred over time. For each DATE expression, we extract POS sequences and use relative entropy to detect typical POS sequences of temporal expressions for each time period.

Table 3 shows POS sequences typical of one time period vs. 6-4 other time periods (see column comp.)[4]. For example, for 1650 the POS sequences DETERMINER-NOUN (DT-NN) and PROPER NOUN (NP) are quite typical, which are all temporal expression referring to seasons in terms of lexical realizations (see Example 1). Both POS sequences are typical of 1650 vs. 1750 to 2000 (i.e., 5 comparisons). If we consider the POS sequences that are typical across time periods and their lexical realizations, there seems to be a development in terms of specificity and interval (see Fig. 2).

---

[4] Note that the examples in Table 3 show most frequent realizations for relatively generic expressions such as seasons (e.g., *in the Spring*) or examples taken randomly from the corpus for specific dates (e.g., *June 3, 1769*) as they occur only once.

Table 3. Typical POS sequences of DATE.

Period	POS sequence	Example	Comp.
1650	DT NN	*in the Spring*	5
	NP	*in Winter*	5
	RB	*now*	4
1700	NP	*in Summer*	5
	NP CD	*March 8*	5
	DT JJ IN NP	*the 6th of March*	4
1750	NP CD, CD	*June 3, 1769*	6
	NP CD	*April 19*	5
	CD NP	*2 June*	4
	DT NN	*the Spring*	4
1800	NP CD, CD	*June 18, 1784*	5
1850	CD	*in 1858*	4
1950	JJ	*current work*	5
	JJ JJ NN	*mid seventeenth century*	4
2000	DT JJ NNS	*the last decades*	5
	DT NNS	*the 1990s*	5
	JJ JJ NN	*late seventeenth century*	5

CD: cardinal number, DT: determiner, IN: preposition, JJ:
adjective, NN: sing. common noun, NNS: pl. common noun,
NP: sing. proper noun, RB: adverb

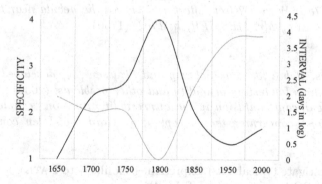

Fig. 2. Specificity (black) and interval (gray) of typical DATE expressions.

To capture the notion of specificity, we consider how many pieces of temporal information are given by a POS sequence to make a temporal expression most specific, with a scale from 1 to 4, where 1 is least specific (e.g., NP denoting seasons such as *Winter* as we do not know of which year etc.) and 4 is most specific (e.g., NP CD, CD such as *June 3, 1769* which gives us an exact date)[5]. For comparison across time periods, Fig. 2 shows the average of the specificity count over all typical POS sequences of a time period (black line). For the interval of typical DATE expressions, the amount of days[6] the expressions refer to is used (shown in log in Fig. 2, gray line). The more specific an expression becomes, the smaller the interval it refers to and vice versa.

Figure 2 also shows how temporal expressions move from relatively unspecific (e.g., *in the Spring* in 1650) to very specific (*June 18, 1784* in 1800) and back to unspecific expressions (e.g., *the last decades* in 2000). The interval moves instead from a wider to a smaller span and back to a wider span in 2000. Investigating the contexts, in which these expressions arise, gives further insights. While in the early time periods, season mentioning is typical, from 1800 to 1850, temporal expressions are typical with exact date, year or month expressions. These expressions are used to present exact dates of observations made by a researcher at several points in time, especially in the field of astronomy (see Example 1). From the 1950 onwards, typical DATE expressions become less explicit, relating to broader (e.g., *the 1970s* in Example 3) and less specific (e.g., *current literature* in Example 3) temporal reference. These expressions are used, e.g., in the context of previous work descriptions in introduction sections of research papers.

Example 1

In **Winter** *it will need longer infusion, than in* **the Spring** *or* **Autumn**. (1650)
The difference between these two plants is this; the papaver corniculatum dies to the root in **the winter**, *and sprouts again from its root in* **the spring**; (1750)

Example 2

March 4, 1783. *With a 7-feet reflector, I viewed the nebula near the 5th Serpentis, discovered by Mr. MESSIER, in 1764.* (1750)

Example 3

In **the 1970s**, *Rabin [38] and Solovay and Strassen [44] developed fast probabilistic algorithms for testing primality and other problems.* (2000)
There is a significant confusion in the **current** *literature on "cellular" or "tessellation arrays" concerning the concept of a "Garden-of-Eden configuration".* (1950)

TIME To investigate typical TIME expressions, similarly to DATE expressions, we consider their POS sequences (see Table 4).

---

[5] On this scale, 1 denotes the mere occurrence of a temporal expression (e.g., NP such as *Winter*), 4 denotes the mere expression plus the inclusion of day, month and year (e.g., NP CD, CD such as *June 3, 1769*, which is a temporal expression + day + month + year resulting in 4 points in total), and 2 and 3 denote an occurrence of a temporal expression plus either two or three combinations of day, month and year (e.g., CD such as *1858* for 2, i.e., temporal expression + year, and NP CD such as *March 8* for 3, i.e., temporal expression + month + day).

[6] We use 'day' because DATE expressions refer to a point in time of the granularity 'day' or coarser, i.e., 'day' is the smallest unit.

Table 4. Typical POS sequences of TIME.

Period	POS sequence	Example	Comp.
1750	NP NN	*Sunday morning*	5
	JJ NN	*next morning*	5
1800	CD NN	*10 A.M.*	5
1850	CD NN	*7 A.M.*	5
	DT NN IN DT JJ IN NP	*the evening of the 28th of August*	4
	IN CD NN	*about 8 A.M.*	4

CD: cardinal number, DT: determiner, IN: preposition, JJ: adjective, NN: sing. common noun, NP: sing. proper noun

It can be seen that only the intermediate time periods show typical expressions (1750 to 1850). In terms of granularity, in the period of 1750, expressions are less granular pointing to broader sections of a day (e.g., *morning*, *evening*) mostly used to describe observations made (see Example 4). In the 1850 period, expressions point to specific hours of a day (e.g., *9 A.M.*) mostly in descriptions of experiments.

Example 4

Monday morning *she appeared well, her pulse was calm, and she had no particular pain.* (1750)
There being usually but one assistant, it was impossible to observe during the whole twenty-four hours; the hours of observation selected were therefore from **3 A.M.** *to* **9 P.M.** *inclusive.* (1850)

DURATION FOR DURATION we consider their TIMEX3 value, as it directly encodes normalized information on the duration length and granularity of temporal expressions. Figure 3 shows typical TIMEX3 values (e.g., P1D for expressions such as *one day*) of specific time periods[7]. The y-axis shows the duration length in seconds on a log scale. In general, duration length gets lower from 1750 to 1850 (with expressions of seconds and hours, which are more granular) and higher in 1950 and 2000 (with expressions of decades, which are less granular).

We then again consider the contextual environments of these typical expressions. In the earlier time periods (1650 and 1700), day and year expressions are typical, mostly relating to observations or experiment descriptions (see Example 5).

---

[7] Note also that typical expressions can either be relatively explicit (e.g., P1D for *24 h*) or fuzzy (indicated by an X in the TIMEX3 value, e.g., *few hours* for the value PTXH).

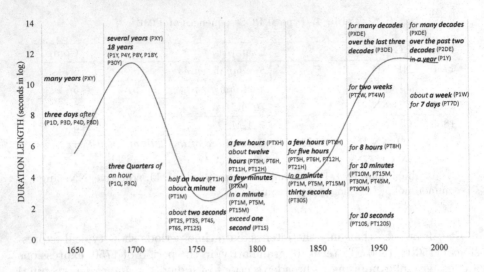

Fig. 3. Diachronic tendencies of typical DURATION expressions.

Example 5

*After the eleven Months, the Owner having a mind to try, how the Animal would do upon Italian Earth, it died **three days** after it had changed the Earth.* (1650)
*[...] the Opium, being cut into very thin slices, [...] is to be put into, and well mixed with, the liquor, (first made luke-warm) and fermented with a moderate Heat for **eight or ten Days**, [...].* (1650)

From the period of 1750 to 1950, duration length is relatively low with expressions of seconds, minutes and hours being typical of these time periods. These expressions are mainly related to observations in the 1750 period and experiment descriptions from 1800 to 1950 (see Example 6).

Example 6

*June 4, the weather continued much the same, and about 9h 30 in the evening, we had a shock of an earthquake, which lasted **about four seconds**, and alarmed all the inhabitants of the island.* (1750)
*[...] the glass produced by this fusion was in **about twelve hours** dissolved, by boiling it in a proper quantity of muriatic acid.* (1800)
*In **a few hours** a mass of fawn-coloured crystals was deposited;* (1850)
*The patient is then switched to the re-breathing system containing 133 Xenon at 5 mCi/1 for a period of **one minute**, and then returned to room air for a period of **ten minutes**.* (1950)

In the 1950, besides weeks and minutes, related to experiment descriptions (see Example 7), expressions of decades are typical. The latter is also true for the 2000. In both periods, expressions relating to decades refer to previous work (see Example 7).

Fig. 4. Diachronic tendencies of typical SET expressions in scientific writing.

Thus, DURATION shifts from being used for purposes of observational to experimental science and finally to previous work references in the latest time periods.

Example 7

*For each speaker, performance was observed across numerous repetitions of the vocabulary set within a single session, as well as across a **2-week** time period.* (1950)

*It constitutes the usual drift-diffusion transport equation that has been successfully used in device modeling for **the last two decades**.* (1950)

*Provably correct and efficient algorithms for learning DNF from random examples would be a powerful tool for the design of learning systems, and **over the past two decades** many researchers have sought such algorithms.* (2000)

SET For SET expressions again their TIMEX3 value is considered. Figure 4 shows typical expressions with the times per year of a SET expression on the y-axis in log[8], mirroring also less granular (*annually*) and more granular (*every day*) expressions. As we have seen from Fig. 1, SET expressions are relatively rare in scientific writing and strongly decrease over time (see Sect. 6.1). This is also reflected in the few temporal expressions typical of each time period in Fig. 4. In terms of granularity, there is a shift from day to month expressions (see Example 8). Interestingly, for the latter, there has been a move from a noun phrase expression (*every/each month*) to an adverb expression (*monthly*). While, in the intermediate periods (1800 and 1850) both expressions are typical, in 1950 only *monthly* is typical. In 1750 to 1850, *every/each month* expressions relate to observations done on a monthly basis of which the mean or average is drawn and the same applies for *monthly* used with *mean* as a term (see Example 8). In 1950, instead, *monthly* is solely used as an adverb. Thus, there is a replacement of longer noun phrase expressions (*every/each month*) by the shorter adverb expression *monthly*.

Example 8

*Besides this, you may there see, that **every day** the Sun sensibly passes one degree from West to East, [...].* (1650)

---

[8] For example, *every day* corresponds to 365 times a year, while *annually* to once a year.

272 S. Degaetano-Ortlieb and J. Strötgen

*In order to determine the annual variations of the barometer, I have taken the mean of the observations in **each month**, [...]. (1800)*

Example 9

*The mean was then taken in **every month** of every lunar hour (attending to the signs), and the **monthly** means were collected into yearly means. (1850)*
*A disk resident file of all current recipient numbers is created **monthly** from the eligibility tape file supplied by Medical Services Administration. (1950)*

7 Discussion and Conclusion

We have presented an approach to investigate diachronic change in the usage of temporal expressions. First, we use temporal tagging to obtain a more comprehensive coverage of possible temporal expressions, rather than investigating specific expressions only, as was the case in previous work. Evaluation of the tagging results showed high precision (approx. 90%) across time periods.

Second, we use relative entropy to detect typical temporal expressions of specific time periods. A clear advantage to frequency-based accounts is that with relative entropy frequent as well as rare phenomena can be investigated in terms of their 'typicality' according to a variable (here: temporal expressions typical of specific time periods). Apart from gaining knowledge on diachronic changes specific to different types of temporal expressions, we also capture more abstract and more fine-grained shifts. On a more abstract level, while our findings confirm the paradigm shift from the more observational to the more experimental character of scientific writing (cf. Fankhauser et al. (2016); Gleick (2010)) for DATE and DURATION expressions, we also show the tendency towards previous work descriptions for these two temporal types in contemporary scientific writing. On a more fine-grained level, for SET (a rarely used temporal type especially towards more contemporary time periods), there is a linguistic shift from longer noun-phrase to shorter adverb expressions.

These findings are not only interesting in historical linguistic terms, but are also relevant to improve adaptation of temporal taggers to different time periods. Especially for recall, gold-standard annotations are needed. Since this is a quite resource and time consuming task, our approach can help in gaining insights on the use of typical temporal expressions in specific contexts across periods. These contexts can then be further exploited in terms of possible temporal expression occurrences to achieve better recall. In addition, temporal expressions might change in terms of linguistic realization as with the SET type in our case. Accounting for shifts in linguistic realization will also improve recall. While this is true for diachronic variation, the approach also generalizes to domain-specific variation. In future work, we plan to work in this direction, further elaborating our methodology for diachronic and domain variation.

Acknowledgments. This work is partially funded by *Deutsche Forschungsgemeinschaft* (DFG) under grant SFB 1102: *Information Density and Linguistic Encoding* (www.sfb1102.uni-saarland.de). We are also indebted to Stefan Fischer for his contributions to corpus processing and Elke Teich for valuable improvement suggestions. Also, we thank the anonymous reviewers for their valuable comments.

References

Agarwal, P., Strötgen, J.: Tiwiki: searching Wikipedia with temporal constraints. In: Proceedings of the 26th International Conference on World Wide Web (WWW 2017), Companion Volume, pp. 1595–1600 (2017)

Allen, J.F.: Maintaining knowledge about temporal intervals. Commun. ACM **26**(11), 832–843 (1983)

Alonso, O., Strötgen, J., Baeza-Yates, R., Gertz, M.: Temporal information retrieval: challenges and opportunities. In: Proceedings of the 1st International Temporal Web Analytics Workshop (TWAW 2011), pp. 1–8 (2011)

Atkinson, D.: Scientific Discourse in Sociohistorical Context: The Philosophical Transactions of the Royal Society of London, 1675–1975. Routledge, New York (1999)

Biber, D., Finegan, E.: Drift and the evolution of English style: a history of three genres. Language **65**(3), 487–517 (1989)

Biber, D., Johansson, S., Leech, G., Conrad, S., Finegan, E.: Longman Grammar of Spoken and Written English. Longman, Harlow (1999)

Campos, R., Dias, G., Jorge, A.M., Jatowt, A.: Survey of temporal information retrieval and related applications. ACM Comput. Surv. **47**(2), 15:1–15:41 (2014)

Dagan, I., Lee, L., Pereira, F.: Similarity-based models of word cooccurrence probabilities. Mach. Learn. **34**(1–3), 43–69 (1999)

Degaetano-Ortlieb, S.: Variation in language use across social variables: a data-driven approach. In: Proceedings of the Corpus and Language Variation in English Research Conference (CLAVIER 2017) (2017)

Degaetano-Ortlieb, S., Teich, E.: Information-based modeling of diachronic linguistic change: from typicality to productivity. In: Reiter, N., Alex, B., Zervanou, K.A. (eds) Proceedings of the 10th SIGHUM Workshop on Language Technology for Cultural Heritage, Social Sciences, and Humanities (LaTeCH 2016), pp. 165–173. ACL (2016)

Degaetano-Ortlieb, S., Teich, E.: Modeling intra-textual variation with entropy and surprisal: topical vs. stylistic patterns. In: Alex, B., Degaetano-Ortlieb, S., Feldman, A., Kazantseva, A., Reiter, N., Szpakowicz, S. (eds.) Proceedings of the Joint SIGHUM Workshop on Computational Linguistics for Cultural Heritage, Social Sciences, Humanities and Literature (LaTeCH and CLfL 2017), pp. 68–77. ACL (2017)

Degaetano-Ortlieb, S., Kermes, H., Lapshinova-Koltunski, E., Teich, E.: SciTex - a diachronic corpus for analyzing the development of scientific registers. In: Bennett, P., Durrell, M., Scheible, S., Whitt, R.J. (eds.) New Methods in Historical Corpus Linguistics. Corpus Linguistics and Interdisciplinary Perspectives on Language - CLIP, vol. 3, pp. 93–104. Narr, Tübingen (2013)

Fankhauser, P., Knappen, J., Teich, E.: Topical diversification over time in the royal society corpus. In: Proceedings of Digital Humanities (DH 2016) (2016)

Fischer, F., Strötgen, J.: When does (German) literature take place? On the analysis of temporal expressions in large corpora. In: Proceedings of Digital Humanities (DH 2015) (2015)

Gleick, J.: At the beginning: more things in heaven and earth. In: Bryson, B. (ed.) Seeing Further. The Story of Science and The Royal Society, pp. 17–36. Harper Press, New York (2010)

Jatowt, A., Antoine, É., Kawai, Y., Akiyama, T.: Mapping temporal horizons: analysis of collective future and past related attention in Twitter. In: Proceedings of the 24th International Conference on World Wide Web (WWW 2015), pp. 484–494 (2015)

Kanhabua, N., Blanco, R., Nørvåg, K., et al.: Temporal information retrieval. Found. Trends Inf. Retr. **9**(2), 91–208 (2015)

Kermes, H., Degaetano-Ortlieb, S., Khamis, A., Knappen, J., Teich, E.: The royal society corpus: from uncharted data to corpus. In: Proceedings of the 10th International Conference on Language Resources and Evaluation (LREC 2016), pp. 1928–1931. ELRA (2016a). ISBN 978-2-9517408-9-1

Kermes, H., Knappen, J., Khamis, A., Degaetano-Ortlieb, S., Teich, E.: The Royal Society Corpus: towards a high-quality corpus for studying diachronic variation in scientific writing. In: Proceedings of Digital Humanities (DH 2016) (2016b)

Kullback, S., Leibler, R.A.: On information and sufficiency. Ann. Math. Stat. **22**(1), 79–86 (1951)

Lafferty, J., Zhai, C.: Document language models, query models, and risk minimization for information retrieval. In: Proceedings of the 24th Annual International ACM SIGIR Conference on Research and Development in Information Retrieval (SIGIR 2001), pp. 111–119. ACM (2001). https://doi.org/10.1145/383952.383970. ISBN 1-58113-331-6

Lee, K., Artzi, Y., Dodge, J., Zettlemoyer, L.: Context-dependent semantic parsing for time expressions. In: Proceedings of the 52nd Annual Meeting of the Association for Computational Linguistics (ACL 2014), pp. 1437–1447. ACL (2014)

Mazur, P., Dale, R.: The DANTE temporal expression tagger. In: Vetulani, Z., Uszkoreit, H. (eds.) LTC 2007. LNCS (LNAI), vol. 5603, pp. 245–257. Springer, Heidelberg (2009). https://doi.org/10.1007/978-3-642-04235-5_21

Mazur, P., Dale, R.: WikiWars: a new corpus for research on temporal expressions. In: Proceedings of the 2010 Conference on Empirical Methods in Natural Language Processing (EMNLP 2010), pp. 913–922. ACL (2010)

Meister, J.C.: Tagging time in prolog: the temporality effect project. Lit. Linguist. Comput. **20**, 107–124 (2005)

Pustejovsky, J.: Temporal and event information in natural language text. Lang. Res. Eval. **39**(2–3), 123–164 (2005)

Pustejovsky, J., Lee, K., Bunt, H., Romary, L.: ISO-TimeML: an international standard for semantic annotation. In: Proceedings of the 7th International Conference on Language Resources and Evaluation (LREC 2010), pp. 394–397. ELRA (2010)

Strötgen, J., Gertz, M.: HeidelTime: high quality rule-based extraction and normalization of temporal expressions. In: Proceedings of the 5th International Workshop on Semantic Evaluation (SemEval 2010), pp. 321–324. ACL (2010)

Strötgen, J., Gertz, M.: Temporal tagging on different domains: challenges, strategies, and gold standards. In: Proceedings of the 8th International Conference on Language Resources and Evaluation (LREC 2012), pp. 3746–3753. ELRA (2012)

Strötgen, J., Gertz, M.: Domain-sensitive Temporal Tagging. Synthesis Lectures on Human Language Technologies. Morgan & Claypool Publishers, San Rafael (2016)

Strötgen, J., Bögel, T., Zell, J., Armiti, A., Van Canh, T., Gertz, M.: Extending HeidelTime for temporal expressions referring to historic dates. In: Proceedings of the 9th International Conference on Language Resources and Evaluation (LREC 2014), pp. 2390–2397. ELRA (2014)

Tabassum, J., Ritter, A., Xu, W.: TweeTime: a minimally supervised method for recognizing and normalizing time expressions in Twitter. In: Proceedings of the 2016 Conference on Empirical Methods in Natural Language Processing (EMNLP 2016), pp. 307–318. ACL (2016). https://aclweb.org/anthology/D16-1030

Teich, E., Degaetano-Ortlieb, S., Kermes, H., Lapshinova-Koltunski, E.: Scientific registers and disciplinary diversification: a comparable corpus approach. In: Proceedings of 6th Workshop on Building and Using Comparable Corpora (BUCC 2013), pp. 59–68. ACL (2013)

UzZaman, N., Llorens, H., Derczynski, L., Allen, J.F., Verhagen, M., Pustejovsky, J.: SemEval-2013 task 1: TempEval-3: evaluating time expressions, events, and temporal relations. In: Proceedings of the 7th International Workshop on Semantic Evaluation (SemEval 2013), pp. 1–9. ACL (2013)

Verhagen, M., Saurí, R., Caselli, T., Pustejovsky, J.: SemEval-2010 task 13: TempEval-2. In: Proceedings of the 5th International Workshop on Semantic Evaluation (SemEval 2010), pp. 57–62. ACL (2010)

A Case Study on the Relevance
of the Competence Assumption for Implicature
Calculation in Dialogue Systems

Judith Victoria Fischer[(✉)]

Sprachwissenschaftliches Institut, Ruhr-Universität Bochum, Bochum, Germany
`fischer@linguistics.rub.de`

Abstract. The *competence assumption* (CA) concerns the estimation of a user that an implicature, derived from an utterance generated in a dialogue or recommender system, reflects the epistemic state of the system about the validity of alternative expressions. The CA can be assigned globally or locally. In this paper, we present an experimental study on the effects of locally and globally assigned competence in a sales scenario. The results of this study suggest that dialogue systems should include means for modelling global competence and that assigning local competence does not improve the pragmatic competence of a dialogue system.

1 Introduction

In order to enhance user acceptance, systems that generate utterances in dialogical scenarios as, e. g., question-answering systems or recommender systems with natural language interfaces, should show some degree of pragmatic behavior. In particular, dialogue systems face the challenge of generating utterances with sufficient scope for the user's possible calculation of implicatures. In the following example, a virtual sales assistant S uses the scalar expression *good*, instead of the semantically stronger *excellent*.

S: *The HP Laptop has a good AMD Radeon graphics coprocessor.*

According to the standard view of quantity implicature calculation, a user, reading this statement, would reason as follows:
The system generated some ψ (*good*) instead of a semantically stronger alternative ϕ (*excellent*). There must be reasons for not generating ϕ: Either the system does not believe ϕ to be true or it believes ϕ not to be true:

$$\neg Bel_S(\phi) \lor Bel_S(\neg\phi)$$

Which interpretation holds, relies on the *competence* or *experthood assumption*: Only if it is shared knowledge of system and user that the system is competent in a way that allows them to predict the truth value of the semantically stronger ϕ, the user will eventually infer the stronger reading (Potts 2015). The inference can be outlined as follows:

G. Rehm and T. Declerck (Eds.): GSCL 2017, LNAI 10713, pp. 276–283, 2018.
https://doi.org/10.1007/978-3-319-73706-5_23

1. ϕ is stronger than ψ, but S didn't claim ϕ.
2. Reason: $\neg Bel_S(\phi)$
3. CA: $Bel_S(\phi) \vee Bel_S(\neg\phi)$
4. 1 – 3 entail $Bel_S(\neg\phi)$

Two different inferences are involved in this overall picture. The first inference concerns the first epistemic assignment (i. e., 2) that just states that the speaker does not know whether or not ϕ holds. This so-called *weak implicature* was derived from the utterance ψ by following the Gricean quantity maxim. The *strong* or *secondary implicature* $Bel_S(\neg\phi)$ strengthens the weak implicature. It can be derived by considering the speaker's competence: Either the speaker believes that ϕ holds or that it doesn't hold. For further details on the relation between implicatures and the competence assumption, see Geurts (2010) and Sauerland (2004).

The competence assumption thus is a crucial component of implicature calculation, because it prevents scalar expressions to be inferred with the stronger implicature reading by default. For the competence assumption to work, certain requirements must be met. Consider the following example:

S: *I've heard a lecture at ComputerCon about that generation of graphics coprocessors. The HP Laptop has a good AMD Radeon graphics coprocessor.*

In this example, the statement that the agent had heard a lecture on graphics coprocessors suggests that S is at least competent with respect to these processors. Hence, the user will infer without doubt that the graphics coprocessor in question is not excellent, which might affect his purchase decision more strongly than the weak implicature that the system just doesn't know whether the coprocessor is excellent or not.

Thus, assuming competence for implicature calculation might have severe effects on the course of the conversation and its outcome. But does the additional information really change the user's assumptions about the system's reliability? Does the sentence *trigger* a competence assumption for the user and thus a stronger implicature reading?

The aim of this paper is to explore this issue in the context of sales and recommender systems: We are interested in positive, neutral, and negative competence triggers and their influence on the competence assumption. Furthermore, we want to know whether the CA should be triggered locally (i. e., for single assertions or themes) or globally (i. e., for the overall dialogue).

2 Related Work

Although there are a number of dialogue systems that deal with various aspects of implicature calculation, the role of the CA in legitimating implicature calculation has not been accounted for. Artificial agents have been utilized before, for either examining various pragmatic reasoning phenomena or maximizing dialogue efficiency, or both. For example, Vogel et al. (2013) use artificial agents to show that they behave in a Gricean manner to maximize their joint utility when faced with reference games or other interactional scenarios. The agent's reasoning about the opponent's belief states, modeled as a variant of the partially observable Markov decision process (POMDPs), to maximize joint utility, results in implicature-rich readings, but the weak–strong distinction has not been accounted for.

Stevens et al. (2016) show that sales dialogue efficiency can be enhanced with pragmatic question answering with indirect answers and consideration of user's requirements by using a game-theoretic model of query answering for their agent: However, the implicature triggered by the indirect answer is, due to the probabilistic model of user types, a weak one only.

Schlöder and Fernández (2015) develop a model for pragmatic rejection by means of implicature(s). Efstathiou and Lemon (2014) consider an account on non-cooperative dialogue in automated conversational systems and teach their agents to behave in this manner. It was shown that in a trading game, non-cooperative behaviour such as deception could increase the agents performance in comparison to a cooperative agent. The CA does not play a role in these models as well.

Insights on the CA originate from linguistic and philosophical analyses (e. g., Geurts (2010)), but these works do not consider requirements for developing computational systems with a generation component in a dialogue setting.

3   Testing the Competence Assumption Locally and Globally

Analogous to the situation in dialogue system research, the CA has not yet been a topic of empirical studies. The pragmatic approach in linguistics assumes that pragmatic reasoning is necessarily global (Sauerland 2004, p. 40), with which it refers to entire speech acts, not embedded sentences. In the context of this paper, *global* refers to an entire conversation, whereas *local* refers to a single speech act, primarily an assertion. It is not yet known on which level the CA is determined or whether it takes into account both levels of interpretation.

In our study, we confine ourselves to the question how the CA may be triggered and whether it will be established globally or locally. We consider the surface forms of the following aspects:

- politeness forms
- personally given indication of competence through additional information
- professionally induced indication of competence through additional information.

3.1   Participants

51 participants were consulted via http://clickworker.com/ and the experiment was distributed by https://www.testable.org/, an open platform for web

experiments. Participation was limited to users from the US and UK. 15 participants were excluded due to failed attention checks, additional 9 participants were excluded, because they had response times below 2500 ms for more than one item. The threshold was set to 2500 ms as a result of a small pretest with three students, where response time minimum was 2900 ms. We assumed here that *clickworkers* are more familiar with the task at hand and that this would justify a lower response time minimum for them.

Thus included in the results are 24 participants.

3.2 Materials

The used items were priorly assigned to one of the categories "positive", "neutral" or "negative".

The examples were considered to be successful competence triggers if they achieved significant ratings within the spectrum of positive (100–70%), neutral (60–40%) or negative (30–0%) competence and in accordance to the priorly assigned categories. The examples were obtained through introspection, prior sales dialogue experiences – online and offline – and media research.

27 statements equally distributed within categories were tested. Of 27 statements, 9 were globally constructed competence triggers like short personal introductions and 18 statements were locally constructed competence triggers like mid-conversational sequences. For all statements, see the appendix; specific statements will hereafter be referred to with their item number (Table 1).

Table 1. Examples for global and local competence triggers

Global competence trigger:	
S:	*My name is Mr. Miller. My shift ends in 5 min, but I'm sure we'll find something for you.*
Local competence trigger:	
S:	*I have to admit I don't know a lot about that.*

The statements and two attention checks were randomized for each participant, which were then asked to "Rate the competence of the sales person on a scale from 0 (not competent) to 100 (very competent)." Answer was given for each statement with a slider from 0 to 100. Other specifics of the sliders grid were hidden from the participants.

3.3 Discussion and Results

First of all, we compared the mean ratings of all items from both groups (local and global) with their priorly assigned categories. As shown in Fig. 1, mean values of positive and neutral items met their categories prerequisites, whereas negative items did not.

We then proceeded to compare the participants ratings for global versus local items with their assigned categories:

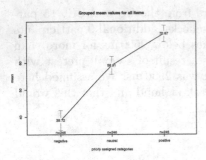

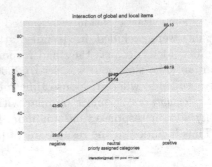

Fig. 1. Items by categories

Fig. 2. Comparison of local and global triggers

As shown, global competence triggers are within their priorly assigned categories, whereas local competence triggers of the categories *positive* and *negative* are not. The results suggest that there might be a significant difference between positive-local and positive-global, negative-local and negative-global items.

The factorial ANOVA allows us to apprehend the effect of groups local and global and categories positive, neutral and negative on the competence ratings simultaneously.

The results of the ANOVA (Table 2) suggest that there is a significant difference within items and in the interaction between items and groups. No significant difference occurs within the groups local and global.

Table 2. Factorial ANOVA results

Analysis of variance table	
Response: competence	
	$Pr(>F)$
Item	$<2.2e{-}16$ ***
Group	0.4973
Item:group	$3.856e{-}13$ ***
Signif. codes: 0 *** 0.001 ***	

This is also supported by the post-hoc Tukey analysis, which gives the analysis of significant differences of different groups and categories compared to each other (e. g., positive-local versus positive-global).

The analysis shows that the following categories are significantly different ($p < .05$):

- neutral-negative
- positive-negative
- positive-neutral.

Furthermore, the following groups and categories are not significantly different:

- neutral:local-neutral:global ($p = 0.972$)
- positive:local-neutral:global ($p = 0.50$)
- positive:local-neutral:local ($p = 0.848$).

This confirms what we have seen before in Fig. 2. Ignoring the groups local and global the categories show a significant difference towards each other. When comparing the interaction between groups and categories, it seems that above all positive-local items and neutral items struggled to differentiate themselves. Also, the findings confirm that negative-local and negative-global items as well as postive-local and positive-global items are significantly different.

4   Implications for Future Work

The results from this study on the role of the CA in dialogue systems suggest that the global competence assumption can be established fairly well. This holds especially true for the competence triggers of the category *positive* and *negative*, which had the most extreme competence ratings. In these categories, polite behaviour and indication of competence with mention of a certain position received the best mean ratings. The negative competence trigger with the lowest mean item number (8) worked with both attributes as well, whereas items number (7) and (9) – direct admission of incompetence and lack of time – didn't compete. Future work thus should pay attention to the factor of polite behaviour and indication of competence. In terms of negative competence triggers, competence should be indicated in an indirect manner, as direct admission of incompetence did not score as well in the mean values.

Local competence triggers of postive-local and neutral-local, as well as neutral-global items failed to distinguish themselves. It should be considered whether neutral competence triggers are beneficial in the first place, or if they can more easily be computed by a lack of competence triggers. Furthermore, local competence triggers did not score as well as their global counterparts. Hence, for computing local competence a more blunt use of polite behaviours or competence indicators should be reviewed.

The global items of the categories *positive* and *negative* were the most promising and therefore suggest that it would be beneficial to integrate a model of CA into the information states of a dialogue system. These first results of our ongoing study on competence triggers will be used for subsequent studies on the role of the CA for deriving weak and strong implicatures. On the whole, they will provide the empirical grounding for a probabilistic model of content determination in a question-answering system in a sales scenario.

Appendix

global-positive: (1) My name is Mr. Miller, I am the assistant Head of the Computer Department at this store. (2) My name is Mr. Miller, I am the designated laptop expert of this store. (3) My name is Mr. Miller and I am here to help you with whatever question you might have.

global-neutral: (4) My name is Mr. Miller. (5) My name is Mrs. Miller. (6) My name is Mrs. Kowalsky.

global-negative: (7) My name is Mr. Miller. Unfortunately I have to inform you, that this is not my department, I am covering this shift for a colleague. (8) My name is Mr. Miller. This is my first day at this store. So please; cut me some slack. (9) My name is Mr. Miller. My shift ends in 5 min, but I'm sure we'll find something for you.

local-positive: (10) Oh, that is easy: (11) I heard a lecture about this at ComputerCon. (12) I'm glad you ask. (13) When I was upgrading my home PC, I came across some information about this: (14) I recently read an article about that. (15) Ah, that is a fascinating piece of hardware.

local-negative: (16) Let me look that up for a second. (17) I have to admit I don't know a lot about that. (18) Sadly, this is not my field of expertise. (19) Well, I don't know much about those to be honest, but what I can tell you is that: (20) I don't think this makes much of a difference, but: (21) I'll have to check in with my colleague about that.

local-neutral: (22) Interesting question. (23) Of course. (24) Well, in regards to this question, I'd have to say: (25) The specifics to that are: (26) As far as I am concerned: (27) On behalf of that, it turns out that.

References

Efstathiou, I., Lemon, O.: Learning non-cooperative dialogue behaviours. In: Proceedings of the SIGDIAL 2014 Conference, pp. 60–68. Association for Computational Linguistics (2014)

Geurts, B.: Quantity Implicatures. Cambridge University Press, New York, NY (2010)

Potts, C.: Presupposition and implicature. In: Lappin, S., Fox, C. (eds.) The Handbook of Contemporary Semantic Theory, 2nd edn, pp. 168–202. Wiley-Blackwell, Oxford (2015)

Sauerland, U.: Scalar implicatures in complex sentences. Linguist. Philos. **27**(3), 367–391 (2004)

Schlöder, J.J., Fernández, R.: Pragmatic rejection. In: Proceedings of the 11th International Conference on Computational Semantics, pp. 250–260. Association for Computational Linguistics (2015)

Stevens, J.S., Benz, A., Reuße, S., Klabunde, R.: Pragmatic question answering: a game-theoretic approach. Data Knowl. Eng. **106**, 52–69 (2016)

Vogel, A., Potts, C., Jurafsky, D.: Implicatures and nested beliefs in approximate decentralized-pompds. In: Proceedings of the 51st Annual Meeting of the Association for Computational Linguistics, pp. 74–80. Association for Computational Linguistics (2013)

Supporting Sustainable Process Documentation

Markus Gärtner[1]([✉]), Uli Hahn[2], and Sibylle Hermann[3]

[1] Institute for Natural Language Processing,
University of Stuttgart, Stuttgart, Germany
markus.gaertner@ims.uni-stuttgart.de
[2] Communication and Information Centre, Ulm University, Ulm, Germany
uli.hahn@uni-ulm.de
[3] University Library, University of Stuttgart, Stuttgart, Germany
sibylle.hermann@ub.uni-stuttgart.de

Abstract. In this paper we introduce a software design to greatly simplify the elicitation and management of process metadata for researchers. Detailed documentation of a research process not only aids in achieving reproducibility, but also increases usefulness of the documented work for others as a cornerstone of good scientific practice. However, in reality, time pressure together with the lack of simple documentation methods makes documenting workflows an arduous and often neglected task. Our method for a clean process documentation combines benefits of version control with integration into existing institutional infrastructure and a novel schema for describing process metadata.

1 Introduction

Lately the term "reproducibility crisis" has frequently been brought up in the context of criticism on research quality in general. However, the German Research Foundation (DFG) correctly emphasizes[1] that reproducibility[2] is only one of many quality measures for good research. Furthermore, it asks for increased attention to questions raised in the close-by area of research data management (RDM). The topic is also in line with recent survey results[3] which stress the insecurity of many researchers when it comes to RDM and the related issues of research data sustainability.

Following the idea of Claerbout and Karrenbach (1992) that articles are only advertisement for the actual scholarship, due diligence should also be exercised in compiling and publishing documentations of research processes alongside the results. This helps especially for tasks that naturally can be reproducible if documented thoroughly, such as computations with strictly deterministic outcomes.

---

[1] Press release by the German Research Foundation (DFG) on "Replicability of Research Results", April 25th 2017.

[2] We do not intend to delve into the deeper semantic discussion of "reproducibility" vs. "replicability" and treat both terms as synonyms here.

[3] By project bwFDCommunities http://bwfdm.scc.kit.edu or Humboldt-University Berlin http://nbn-resolving.org/urn:nbn:de:kobv:11-100213001.

© The Author(s) 2018
G. Rehm and T. Declerck (Eds.): GSCL 2017, LNAI 10713, pp. 284–291, 2018.
https://doi.org/10.1007/978-3-319-73706-5_24

But even if a process conceptually defies reproduction (e.g., results of strictly qualitative analysis) a good documentation can increase usability.

Today's research practice however is dominated by competitive and time pressures. Together with the focus on producing publishable end results combined with lack of gratification for detailed process documentation this leads to a serious neglect of documentation efforts. Our project aims at filling that gap with the software tool described in this paper. The goal is to support scientists in creating process documentation already during the workflow with minimal effort. This concept is being exemplified with use cases from researchers in computational linguistics and digital humanities where a plethora of multifaceted workflows exist.

We contextualize our work in Sect. 2 and introduce our schema for process metadata in Sect. 3. Section 4 provides an overview of our software design and shows possibilities of interconnections to related infrastructure. In Sect. 5 we summarize the findings and hint at possibilities of how this concept can be integrated within a broader context.

2   Related Work

In the context of documenting a research process, systems for two slightly distinct fields are relevant: Workflow management and workflow tracking.

A *Workflow management system* (WMS) is software that helps in setting up workflows as collections of interdependent (executable) steps. The list of (commercial) WMSs for general-purpose or enterprise use is extensive, but their usability for specialized research workflows is limited. For certain research fields dedicated WMS instances have emerged. *GenePattern*[4] (Reich et al. 2006) for instance allows the creation, management and execution of analysis pipelines for genome research.

The task of *workflow tracking* is more reactive and involves documenting steps during or after their execution. Tools similar to *YesWorkflow* (McPhillips et al. 2015) offer the ability to annotate data flows and operations on the code level and to receive a graph visualization of the implicit workflow. They are however primarily aimed at script-heavy workflows. More related to software development, version control systems like Git[5], Apache Subversion (SVN)[6] or others provide sufficient functionality for tracking and documenting complex collaborative development workflows.

In stark contrast to the elaborate solutions listed so far, in practice it is not uncommon for researchers to document their workflow simply by means of a local Word or Excel file. Reasons for this given in surveys include the complexity of many workflow management or tracking systems, the barrier of entry for technically unskilled users and available systems essentially being not specialized enough for a given workflow. In addition many solutions focus on

---

[4]   http://www.genepattern.org.
[5]   https://git-scm.com
[6]   https://subversion.apache.org.

creation and maintenance of executable analysis or processing pipelines, making them unsuited for workflows that involve manual steps, such as annotation or curation.

By splitting process documentation into metadata for *resources* (objects) used in it and the *actions* performed, the need to interface with existing metadata infrastructure becomes evident. Initiatives like CLARIN (Hinrichs and Krauwer 2014) already provide wide coverage of metadata repositories for communities in computational linguistics and digital humanities. While having rich object metadata available is by no means a sufficient replacement of proper process documentation, it does provide a valuable foundation to build on.

3   Process Metadata

We follow the more common convention of modeling workflows as directed acyclic graphs (DAG) (Deelman et al. 2009) where each node represents a single step. As pointed out in Sect. 2, there are already established initiatives and systems for the provision or archiving of object metadata. That is metadata associated with an individual resource, usually created and archived after reaching a mature state in the resource's development or creation lifecycle. Rich in detail, such metadata records are also typically following schemas specific to the field of research they originated from. For example in the context of CLARIN the Component MetaData Infrastructure (CMDI)[7] is used for entries in the CLARIN Virtual Language Observatory (VLO)[8]. Due to such infrastructures also providing means for persistent identification of individual records[9], the metadata schema in this section is focused solely on the process itself. To this end our schema defines the following fields for a single workflow step (with the field's multiplicity in round brackets if it is optional):

Title. User-defined short label for the workflow step. This is meant to be very compact and can also be reduced to simply state the type of the task performed, such as "Annotation".

Description. This more detailed description of the workflow step is meant to be a human readable explanation that should contain enough information for another researcher to understand what was done. It is also the place where basically arbitrary additional notes can be placed. It will help to find and keep track of decisions or expectations and to raise the reusability for others.

Input (0..n). Resources used to perform the action. This includes an extremely diverse spectrum of resources which are in turn highly task-specific. They can range from local resources of arbitrary type (corpora, model files, configuration files, pictures, audio, etc.) to online resources (e.g., public annotation guidelines) to "pointers" at real objects such as books that don't exist in digitized form.

---

[7] http://www.clarin.eu/cmdi.

[8] https://vlo.clarin.eu.

[9] E.g., metadata for the TIGER Corpus: http://hdl.handle.net/11022/1007-0000-0000-8E2D-F.

Output (0..n). Resources generated by the action. Unlike input resources, these usually represent exclusively local files, since the workflow is assumed to take place on a local system.

Tool (0..1). The executable resource used for processing. This is either a local program or a web-service and in both cases command line parameters or settings can and should be recorded.

Person (0..n). Human subjects involved in the workflow step. Similar to `Input`, the content of this field is quite diverse, including, but not limited to, annotators, curators or experiment participants.

Custom properties (0..n). Arbitrary classic textual key-value metadata entries that can be added, primarily to provide machine readable storage options for metadata created by future plugins or if the user wishes to include structured information beyond the free-text in the `Description`.

Complex fields (`Input`, `Output`, `Person`, `Tool`) get assigned one or more typed identifiers. These identifiers take the form of `<type, id>` where `id` is the actual identification string such as a name or web address and `type` defines how that `id` is to be interpreted. These identifiers can also be used to link resources or persons to public repositories or databases (e.g., VLO-Handles for CLARIN resources or an ORCID[10] for registered researchers).

The standard serialization format for the process metadata is JSON[11]. Our tool uses it to store process metadata locally as described in Sect. 4.1 and for exporting (parts of a) workflow documentation. Since JSON is a very simple and also widely used format, it is easily possible to convert the output, or process it with other systems.

4   Architecture

In this section we provide an overview of our architecture, especially the design of the client software which is being developed in our project. The core component is a Java application which bundles the user interactions required for tracking and documenting the workflow. Behind the scenes this client uses a local Git repository for tracking changes to file resources in *workspace* folders designated by the user. The client is also meant to provide a wide range of additional functionality, among others the ability to interact with existing data and/or metadata repositories to store or retrieve resources or metadata. We describe some of these features in the following Sects. 4.1 through 4.3.

4.1   Git and Process Documentation

The Git system allows version control of a local folder's content, that is, it tracks changes to files inside this folder so that individual states or "versions" can be recorded and referenced.

---

[10]   https://orcid.org.
[11]   http://www.json.org.

Storage of Process Metadata. The recording of a workspace's state is triggered by so called Git "commit" operations. In our workflow model every workflow step corresponds to a single Git commit. Each commit is also accompanied by its respective "commit message". Those messages commonly are human-readable descriptions made by the user to explain the nature or reason of performed modifications. However, our application design completely hides the direct Git interface from the end users as to not overwhelm them with unneeded complexity or functions. This way we can use commit messages to store process metadata (cf. Sect. 3) and thereby directly associate physical changes to the data with matching (formal) documentation.

Unfortunately Git can only automatically detect those local files that represent the "output" of a workflow step (since those have been modified or are new). This means that the completeness of a resulting workflow documentation ultimately still relies on the user. We plan assistive functions in the client that try to suggest involved input resources when recording a workflow step to reduce the effort required by the user. Their implementation is at this point however still an open issue and the correctness and usability will have to be evaluated together with the user community at a later stage.

Increasing Documentation Consistency. Manually documenting a workflow is prone to common mistakes, such as forgetting to include changes to a resource in the documentation or introducing inconsistencies in the descriptions. As stated above, Git cannot guarantee completeness when recording all the input resources used in a workflow step. It does on the other hand track reliably all the changes made to files that are under version control. As a result it makes it impossible to miss modification on tracked files and reminds the user to document all of them. Having the entire workflow history available when recording a new workflow step, also enables the client to detect inconsistent documentation, for instance when the user tries to assign a different description to a resource which has been previously used in another step.

Trying Alternatives. Only very few research workflows ever result in a strictly linear concatenation of performed steps. Usually there are situations during a workflow where an assumption was found to be untrue or an evaluation yielded unsatisfactory results. In those cases the researcher typically "goes back" and pursues an alternative approach by performing a different workflow step on the same previous workspace state (e.g., testing new settings for an analysis tool or training systems on different data, etc.). A workflow graph displays this behavior as branches originating from the same node to concurrent new child nodes. Git offers a similarly named functionality where a workflow can split into independent branches and the folder under version control can be changed to reflect any previously recorded state at will.

Backup and Cooperation. Git represents a decentralized or distributed version control system. While every local repository itself allows full version control, one can also import or export changes from and to a remote repository. In the context of our client this allows the "local workflow" to be connected to a remote Git repository (e.g., an institute or university wide GitLab[12] installation). Benefits of this include for example an additional layer of backup for valuable resources used in the workflow or the ability to cooperatively work on the same "shared" workspace from different locations.

While building upon the Git system offers many advantages, there are also limitations and issues to address when using it for workflow documentation, especially in very resource-heavy computational projects. Since Git basically has to duplicate every file that is kept under version control, this leads to a very high storage consumption when used for already big resources, such as web corpora. As a solution the user can exclude files from being tracked, so that they won't affect performance.

4.2 Client Customizability

With the process metadata outlined in Sect. 3 we can model a very broad spectrum of diverse workflows and therefore make the client usable for researchers in many fields. Different research fields and also universities or institutes often already have individual infrastructures for archiving or management of resources and metadata in place. To not create redundancies the following principles are taken into account for the client design:

Independence. In the most basic version our client requires absolutely no external infrastructure or third-party software to work with, besides the libraries it is shipped with. It will provide the full functionality of workflow documentation and also enable the user to create and store object metadata locally in a simple (customizable) schema following Dublin Core (Powell et al. 2005). In this configuration the user can work completely network-independent and also is not reliant on other infrastructure, making the client very light-weight.

Extensibility. To be able to incorporate the client into existing institutional infrastructure we use a plugin-architecture. This allows for example customized implementations for interfacing with additional repositories to be added.

4.3 External Repositories

In addition to workflow documentation in private domains (Treloar et al. 2007), the client also gives the possibility to collaborate in the shared (but not publicly open) domain, and to publish partial or final results in the public domain. To meet both requirements there are two systems with their respective interfaces that will be supported: For publishing within the shared domain, ResourceSpace[13] is being used. The repository software DSpace[14] (Smith 2002) is used

---

[12] https://gitlab.com.
[13] https://www.resourcespace.com.
[14] http://www.dspace.org.

for publishing data with a permanent identifier (DOI) in the public domain. DSpace is a popular software for institutional publication repositories. We plan to interface with ResourceSpace for the shared domain, as it offers a better rights management as well as the possibility to share data within defined communities.

5 Outlook

In this paper we introduced our design of a software supporting process documentation. We have shown the essential benefits of using version control software such as Git as a foundation for workflow tracking with a main focus on computational linguistics and digital humanities. In addition, we propose a simple yet very expressive metadata schema to describe individual steps in a research workflow. Keeping those principles separated – namely the distinction between metadata describing *objects* used in a workflow and the *actions* performed – enables the software to be very flexible. As a result it will be fairly easy to adopt the tool to specific needs (of other disciplines) and also to integrate it into the diverse landscape of existing infrastructures.

Acknowledgments. This work was funded by the Ministry for Science, Research and the Arts in Baden-Württemberg (MWK) via the E-Science project "RePlay-DH".

References

Claerbout, J., Karrenbach, M.: Electronic documents give reproducible research a new meaning. In: Proceedings of 62nd Annual International Meeting of the Society of Exploration Geophysics, pp. 601–604 (1992)

Deelman, E., Gannon, D., Shields, M., Taylor, I.: Workflows and e-science: an overview of workflow system features and capabilities. Future Gener. Comput. Syst. **25**(5), 528–540 (2009). https://doi.org/10.1016/j.future.2008.06.012. ISSN 0167–739X

Hinrichs, E., Krauwer, S.: The CLARIN research infrastructure: resources and tools for e-humanities scholars. In: Proceedings of the 9th International Conference on Language Resources and Evaluation, LREC 2014, pp. 1525–1531, May 2014. URL http://dspace.library.uu.nl/handle/1874/307981

McPhillips, T.M., Song, T., Kolisnik, T., Aulenbach, S., Belhajjame, K., Bocinsky, K., Cao, Y., Chirigati, F., Dey, S.C., Freire, J., Huntzinger, D.N., Jones, C., Koop, D., Missier, P., Schildhauer, M., Schwalm, C.R., Wei, Y., Cheney, J., Bieda, M., Ludäscher, B.: YesWorkflow: a user-oriented, language-independent tool for recovering workflow information from scripts. CoRR, abs/1502.02403, 2015. URL http://arxiv.org/abs/1502.02403

Powell, A., Nilsson, M., Naeve, A., Johnston, P.: Dublin core metadata initiative - abstract model (2005). White Paper, http://dublincore.org/documents/abstract-model

Reich, M., Liefeld, T., Gould, J., Lerner, J., Tamayo, P., Mesirov, J.P.: GenePattern 2.0. Nat. Genet. **38**(5), 500–501 (2006). https://doi.org/10.1038/ng0506-500. ISSN 1061–4036

Smith, M.: Dspace: an institutional repository from the mit libraries and hewlett packard laboratories. In: Agosti, M., Thanos, C. (eds.) ECDL 2002. LNCS, vol. 2458, pp. 543–549. Springer, Heidelberg (2002). https://doi.org/10.1007/3-540-45747-X_40

Treloar, A., Groenewegen, D., Harboe-Ree, C.: The data curation continuum: managing data objects in institutional repositories, vol. 13, no. 9/10. D-Lib Magazine, September/October 2007. https://doi.org/10.1045/september2007-treloar. ISSN 1082–9873

Optimizing Visual Representations in Semantic Multi-modal Models with Dimensionality Reduction, Denoising and Contextual Information

Maximilian Köper[✉], Kim-Anh Nguyen, and Sabine Schulte im Walde

Institut für Maschinelle Sprachverarbeitung, Universität Stuttgart,
Pfaffenwaldring 5B, 70563 Stuttgart, Germany
{maximilian.koeper,kim-anh.nguyen,schulte}@ims.uni-stuttgart.de

Abstract. This paper improves visual representations for multi-modal semantic models, by (i) applying standard dimensionality reduction and denoising techniques, and by (ii) proposing a novel technique *ContextVision* that takes corpus-based textual information into account when enhancing visual embeddings. We explore our contribution in a visual and a multi-modal setup and evaluate on benchmark word similarity and relatedness tasks. Our findings show that NMF, denoising as well as *ContextVision* perform significantly better than the original vectors or SVD-modified vectors.

1 Introduction

Computational models across tasks potentially profit from combining corpus-based, textual information with perceptional information, because word meanings are grounded in the external environment and sensorimotor experience, so they cannot be learned only based on linguistic symbols, cf. the grounding problem (Harnad 1990). Accordingly, various approaches on determining semantic relatedness have been shown to improve by using multi-modal models that enrich textual linguistic representations with information from visual, auditory, or cognitive modalities (Feng and Lapata 2010, Silberer and Lapata 2012, Roller and im Walde 2013, Bruni et al. 2014, Kiela et al. 2014, Kiela and Clark 2015, Lazaridou et al. 2015).

While multi-modal models may be realized as either count or predict approaches, increasing attention is being devoted to the development, improvement and properties of low-dimensional continuous word representations (so-called *embeddings*), following the success of *word2vec* (Mikolov et al. 2013). Similarly, recent advances in computer vision and particularly in the field of deep learning have led to the development of better visual representations. Here, features are extracted from convolutional neural networks (CNNs) (LeCun et al. 1998), that were previously trained on object recognition tasks. For example,

G. Rehm and T. Declerck (Eds.): GSCL 2017, LNAI 10713, pp. 292–300, 2018.
https://doi.org/10.1007/978-3-319-73706-5_25

Kiela and Bottou (2014) showed that CNN-based image representations perform superior in semantic relatedness prediction than other visual representations, such as an aggregation of SIFT features (Lowe 1999) into a bag of visual words (Sivic and Zisserman 2003).

Insight into the typically high-dimensional CNN-based representations is sparse, however. It is known that dimension reduction techniques, such as Singular Value Decomposition (SVD), improve performance on word similarity tasks when applied to word representations (Deerwester et al. 1990). In particular, Bullinaria and Levy (2012) observed highly significant improvements after applying SVD to standard corpus vectors. In addition, Nguyen et al. (2016) proposed a method to remove noisy information from word embeddings, resulting in superior performance on a variety of word similarity and relatedness benchmark tests.

In this paper, we provide an in-depth exploration of improving visual representations within a semantic model that predicts semantic similarity and relatedness, by applying dimensionality reduction and denoising. Furthermore, we introduce a novel approach that modifies visual representations in relation to corpus-based textual information. Following the methodology from Kiela et al. (2016), evaluations are carried out across three different CNN architectures, three different image sources and two different evaluation datasets. We assess the performance of the visual modality by itself, and we zoom into a multi-modal setup where the visual representations are combined with textual representations. Our findings show that all methods but SVD improve the visual representations. This improvement is especially large on the word relatedness task.

2 Methods

In this section we introduce two dimensionality reduction techniques (Sect. 2.1), a denoising approach (Sect. 2.2) and our new approach *ContextVision* (Sect. 2.3).

2.1 Dimensionality Reduction

Singular Value Decomposition (SVD) (Golub and Van Loan 1996.) is a matrix algebra operation that can be used to reduce matrix dimensionality yielding a new high-dimensional space. SVD is a commonly used technique, also refered to as Latent Semantic Analysis (LSA) when applied to word similarity. *Non-negative matrix factorization* (NMF) (Lee and Seung 1999) is a matrix factorisation approach where the reduced matrix contains only non-negative real numbers (Lin 2007). NMF has a wide range of applications, including topic modeling, (soft) clustering and image feature representation (Lee and Seung 1999).

2.2 Denoising

Nguyen et al. (2016) proposed a denoising method (**DEN**) that uses a non-linear, parameterized, feed-forward neural network as a filter on word embeddings to reduce noise. The method aims to strengthen salient context dimensions and to weaken unnecessary contexts. While Nguyen et al. (2016) increase the dimensionality, we apply the same technique to reduce dimensionality.

2.3   Context-Based Visual Representations

Our novel model *ContextVision* (CV) strengthens visual vector representations by taking into account corpus-based contextual information. Inspired by Lazaridou et al. (2015), our model jointly learns the *linguistic* and *visual* vector representations by combining two modalities (i.e., the linguistic modality and the visual modality). Differently to the multi-modal Skip-gram model by Lazaridou et al. (2015), we focus on improving the visual representation, while Lazaridou et al. aim to improve the linguistic representation, without performing updates on the visual representation, which are fixed in advance.

The linguistic modality uses contextual information and word negative contexts, and in the visual modality the visual vector representations are strengthened by taking the corresponding word vector representations, the contextual information, and the visual negative contexts into account.

We start out with describing the Skip-gram with negative sampling (SGNS) (Levy and Goldberg 2014) which is a variant of the Skip-gram model (Mikolov et al. 2013). Given a plain text corpus, SGNS aims to learn word vector representations in which words that appear in similar contexts are encoded by similar vector representations. Mathematically, SGNS model optimizes the following objective function:

$$J_{SGNS} = \sum_{w \in V_W} \sum_{c \in V_C} J_{ling}(w, c) \tag{1}$$

$$\begin{aligned} J_{ling}(w, c) &= \#(w, c) \log \sigma(w, c) \\ &\quad + k_l \cdot \mathbb{E}_{c_N \sim P_D}[\log \sigma(-w, c_N)] \end{aligned} \tag{2}$$

where $J_{ling}(w, c)$ is trained on a plain-text corpus of words $w \in V_W$ and their contexts $c \in V_C$, with V_W and V_C the word and context vocabularies, respectively. The collection of observed words and context pairs is denoted as D; the term $\#(w, c)$ refers to the number of times the pair (w, c) appeared in D; the term $\sigma(x)$ is the sigmoid function; the term k_l is the number of linguistic negative samples and the term c_N is the linguistic sampled context, drawn according to the empirical unigram distribution P. In our model, SGNS is applied to learn the linguistic modality.

In the visual modality, we improve the visual representations through contextual information; therefore the dimensionality of visual representations and linguistic representations needs to be equal in size. We rely on the denoising approach (Nguyen et al. 2016) to reduce the dimensionality of visual representations. The visual vector representations are then enforced by (i) directly increasing the similarity between the visual and the corresponding linguistic vector representations, and by (ii) encouraging the contextual information which co-occurs with the linguistic information. More specifically, we formulate the objective function of the visual modality, $J_{vision}(v_w, c)$, as follows:

$$\begin{aligned}
J_{vision}(v_w, c) = \ & \#(v_w, c)(cos(w, v_w) \\
& + \min\{0, \theta - cos(v_w, c) + cos(w, c)\}) \\
& + k_v \cdot \mathbb{E}_{c_V \sim P_V} [\log \sigma(-v_w, c_V)]
\end{aligned} \qquad (3)$$

where $J_{vision}(v_w, c)$ is trained simultaneously with $J_{ling}(w, c)$ on the plain-text corpus of words w and their contexts c. v_w represents the visual information corresponding to the word w; and term θ is the margin; $cos(x, y)$ refers to the cosine similarity between x and y. The terms k_v, $\mathbb{E}_{c_V}$, and P_V are similarly defined as the linguistic modality. Note that if a word w is not associated with the corresponding visual information v_w, then $J_{vision}(v_w, c)$ is set to 0.

In the final step, the objective function which is used to improve the visual vector representations combines Eqs. 1, 2, and 3 by the objective function in Eq. 4:

$$J = \sum_{w \in V_W} \sum_{c \in V_C} (J_{ling}(w, c) + J_{vision}(v_w, c)) \qquad (4)$$

3 Experiments

3.1 Experimental Settings

We use an English Wikipedia dump[1] from June 2016 as the corpus resource for training the *Context Vision*, containing approximately 1.9B tokens. We train our model with 300 dimensions, a window size of 5, 15 linguistic negative samples, 1 visual negative sample, and 0.025 as the learning rate. The threshold θ is set to 0.3. For the other methods dimensionality reduction is set to 300^2 dimensions. For the resources of image data, we rely on the publically available visual embeddings taken from Kiela et al. (2016)[3]. The data was obtained from three different image sources, namely Google, Bing, and Flickr. For each image source three state-of-the-art convolutional network architectures for image recognition were applied: ALEXNET (Krizhevsky et al. 2012), GOOGLENET (Szegedy et al. 2015) and VGGNET (Simonyan and Zisserman 2014). In each source–CNN combination, the visual representation of a word is simply the centroid of the vectors of all images labeled with the word (mean aggregation). This centroid has 1024 dimensions for GOOGLENET and 4096 dimensions for the remaining two architectures. The size of the visual vocabulary for Google, Bing, and Flickr after computing the centroids is 1578, 1578, and 1582 respectively. For evaluation we relied on two human-annotated datasets, namely the 3000 pairs from MEN (Bruni et al. 2014) and the 999 pairs from SIMLEX (Hill et al. 2015). MEN focuses on relatedness, and SIMLEX focuses on similarity.

---

[1] https://dumps.wikimedia.org/enwiki/latest/enwiki-latest-pages-articles.xml.bz2.

[2] We conducted also experiments with 100 and 200 dimensions and obtained similar findings.

[3] http://www.cl.cam.ac.uk/~dk427/cnnexpts.html.

3.2 Visual Representation Setup

Table 1 shows the results for each of the previously introduced methods, as well as the unmodified image representation (DEFAULT). It can be seen that NMF, DEN and CV increase performance on all settings except for the combination Google & ALEXNET. The performance of SVD is always remarkably similar to its original representations.

Furthermore we computed the average difference for each method across all settings, as shown in Table 2. The performance increased especially on the MEN relatedness task. Here NMF obtains on average a rho correlation of ≈.10 higher than its original representations. Also DEN and CV show a clear improvement, with the latter being most useful for the SIMLEX task.

To ensure significance we conducted *Steiger*'s test (Steiger 1980) of the difference between two correlations. We compared each ouf the methods against its DEFAULT performance.

Table 1. Comparing dimensionality reduction techniques, showing Spearman's ρ on SimLex-999 and MEN. * marks significance over the DEFAULT.

		ALEXNET		GOOGLENET		VGGNET	
		SimLex	MEN	SimLex	MEN	SimLex	MEN
BING	DEFAULT	.324	.560	.314	.513	.312	.545
	SVD	.324	.557	.316	.513	.314	.544
	NMF	.329	**.610***	.341*	**.612***	.330	**.631***
	DEN	.356*	.582*	.342*	.564*	.343*	.599*
	CV	**.364***	.583*	**.358***	.582*	**.357***	.603*
FLICKR	DEFAULT	.271	.434	.244	.366	.262	.422
	SVD	.270	.424	.245	.364	.264	.418
	NMF	.284	.560*	.280*	.556*	.288	**.581***
	DEN	.276	.566*	.273*	.526*	.280	.570*
	CV	**.310***	**.573***	**.287***	**.589***	**.312***	.540*
GOOGLE	DEFAULT	.354	.526	.358	.517	.346	.535
	SVD	**.355**	.527	.359	.518	.348	.536
	NMF	.353	**.596***	**.367**	**.608***	.366	**.609***
	DEN	.343	.559*	.361	.555*	.356	.560*
	CV	.352	.561*	.362	.573*	**.374**	.556*

Table 2. Average gain/loss in ρ across sources and architectures, in comparison to DEFAULT.

	SIMLEX	MEN	BOTH
SVD	0.11	−0.20	−0.05
NMF	1.71	10.49	6.10
DEN	1.63	7.34	4.48
CV	3.23	8.29	5.76

Out of the 19 settings, NMF obtained significant improvements with *=$p <$ 0.001 in 11 cases. Despite having a lower average gain (Table 2), DEN and CV obtained more significant improvements.

In total we observed most significant improvements on images taken from BING and with the CNN GOOGLENET.

3.3 Multi-modal Setup

In the previous section we explored the performance of the visual representations alone.

We now investigate their performance in a multi-modal setup, combining them with a textual representation. Using the same parameters as in Sect. 3.1 we created word representations relying on an SGNS model (Mikolov et al. 2013). We combined the representations by scoring level fusion (or late fusion). Following Bruni et al. (2014) and Kiela and Clark (2015) we investigate the impact of both modalities by varying a weight threshold (α). Similarity is computed as follows:

$$sim(x, y) = \alpha \cdot ling(x, y) + (1 - \alpha) \cdot vis(x, y) \qquad (5)$$

Here $ling(x, y)$ is cosine similarity based on the textual representation only and $vis(x, y)$ for using the visual space.

For the following experiment we focus on ALEXNET, varying the image resource between BING for the SIMLEX task and FLICKR for the MEN task. The results are shown in Fig. 1a for SIMLEX, and in Fig. 1b for MEN.

It can be seen that all representations obtain superior performance on the text-only representation (black dashed line, SIMLEX $\rho = .384$, MEN $\rho = .741$). The highest correlation can be obtained using the DEN or VC representations for SIMLEX. Interestingly these two methods obtain best performance when given equal weight to both modalities ($\alpha = 0.5$) while the remaining methods as well as

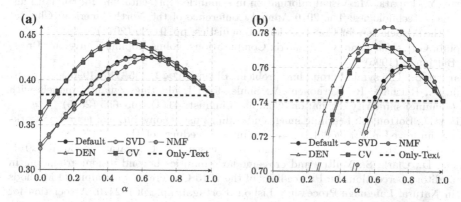

Fig. 1. (a) Comparing multi-modal results on SimLex-999. Image representation from BING using ALEXNET. Y-Axis shows Spearman's ρ. X-axis changes impact of each modality, from only image to the far left to only textual representation (b) Multi-modal results on MEN. Image representation from FLICKR using ALEXNET.

the unmodified default representations obtain a peak in performance when given more weight to the textual representation. A similar picture emerges regarding the results on MEN, where also NMF obtains superior results (.748).

4 Conclusion

We successfully applied dimensionality reduction as well as denoising techniques, plus a newly proposed method *ContextVision* to enhance visual representations within semantic vector space models. Except for SVD, all investigated methods showed significant improvements in single - and multi-modal setups on the task of predicting similarity and relatedness.

Acknowledgments. The research was supported by the DFG Collaborative Research Centre SFB 732 (Maximilian Köper, Kim-Anh Nguyen), the DFG Heisenberg Fellowship SCHU-2580/1 (Sabine Schulte im Walde) and the Ministry of Education and Training of the Socialist Republic of Vietnam (Scholarship 977/QD-BGDDT; Kim-Anh Nguyen). We would like to thank the four anonymous reviewers for their comments and suggestions.

References

Bruni, E., Tran, N.K., Baroni, M.: Multimodal distributional semantics. J. Artif. Intell. Res. **49**, 1–47 (2014)

Bullinaria, J.A., Levy, J.P.: Extracting semantic representations from word co-occurrence statistics: stop-lists, stemming, and SVD. Behav. Res. Methods **44**, 890–907 (2012)

Deerwester, S., Dumais, S.T., Furnas, G.W., Landauer, T.K., Harshman, R.: Indexing by latent semantic analysis. J. Am. Soc. Inf. Sci. **41**, 391–407 (1990)

Feng, Y., Lapata, M.: Visual information in semantic representation. In: Human Language Technologies: The 2010 Annual Conference of the North American Chapter of the Association for Computational Linguistics, pp. 91–99 (2010)

Golub, G.H., Van Loan, C.F.: Matrix Computations. Johns Hopkins University Press, Baltimore (1996)

Harnad, S.: The symbol grounding problem. Physica D **42**, 335–346 (1990)

Hill, F., Reichart, R., Korhonen, A.: Simlex-999: Evaluating semantic models with (genuine) similarity estimation. Comput. Linguist. **41**(4), 665–695 (2015)

Kiela, D., Bottou, L.: Learning image embeddings using convolutional neural networks for improved multi-modal semantics. In: Proceedings of the 2014 Conference on Empirical Methods in Natural Language Processing, Doha, Qatar, pp. 36–45 (2014)

Kiela, D., Clark, S.: Multi - and cross-modal semantics beyond vision: grounding in auditory perception. In: Proceedings of the 2015 Conference on Empirical Methods in Natural Language Processing, Lisbon, Portugal, pp. 2461–2470. Association for Computational Linguistics (2015)

Kiela, D., Hill, F., Korhonen, A., Clark, S.: Improving multi-modal representations using image dispersion: why less is sometimes more. In: Proceedings of the 52nd Annual Meeting of the Association for Computational, Baltimore, USA, pp. 835–841 (2014)

Kiela, D., Verő, A.L., Clark, S.: Comparing data sources and architectures for deep visual representation learning in semantics. In: Proceedings of the Conference on Empirical Methods in Natural Language Processing, pp. 447–456 (2016)

Krizhevsky, A., Sutskever, I., Hinton, G.E.: ImageNet classification with deep convolutional neural networks. In: Pereira, F., Burges, C.J.C., Bottou, L., Weinberger, K.Q. (eds.) Advances in Neural Information Processing Systems 25, pp. 1097–1105. Curran Associates Inc., (2012)

Lazaridou, A., Pham, N.T., Baroni, M.: Combining language and vision with a multimodal skip-gram model. In: The 2015 Conference of the North American Chapter of the Association for Computational Linguistics: Human Language Technologies, Denver, USA, pp. 153–163 (2015)

LeCun, Y., Bottou, L., Bengio, Y., Haffner, P.: Gradient-based learning applied to document recognition. In: Proceedings of the IEEE, pp. 2278–2324 (1998)

Lee, D.D., Seung, H.S.: Learning the parts of objects by nonnegative matrix factorization. Nature **401**, 788–791 (1999)

Levy, O., Goldberg, Y.: Neural word embedding as implicit matrix factorization. In: Proceedings of the 27th International Conference on Advances in Neural Information Processing Systems, Montréal, Canada, pp. 2177–2185 (2014)

Lin, C.-J.: Projected gradient methods for nonnegative matrix factorization. Neural Comput. **19**, 2756–2779 (2007)

Lowe, D.G.: Object recognition from local scale-invariant features. In: Proceedings of the International Conference on Computer Vision, vol. 2, pp. 1150–1157, Washington, DC, USA (1999)

Mikolov, T., Sutskever, I., Chen, K., Corrado, G.S., Dean, J.: Distributed representations of words and phrases and their compositionality. Adv. Neural Inf. Process. Syst. **26**, 3111–3119 (2013)

Nguyen, K.A., im Walde, S.S., Vu, N.T.: Neural-based noise filtering from word embeddings. In: Proceedings of the 26th International Conference on Computational Linguistics, Osaka, Japan, pp. 2699–2707 (2016)

Roller, S., im Walde, S.S.: A multimodal LDA model integrating textual, cognitive and visual modalities. In: Proceedings of the 2013 Conference on Empirical Methods in Natural Language Processing, Seattle, WA, pp. 1146–1157 (2013)

Silberer, C., Lapata, M.: Grounded models of semantic representation. In: Proceedings of the Conference on Empirical Methods in Natural Language Processing, Jeju Island, Korea, pp. 1423–1433 (2012)

Simonyan, K., Zisserman, A.: Very deep convolutional networks for large-scale image recognition. arXiv.org, abs/1409.1556 (2014)

Sivic, J., Zisserman, A.: Video Google: a text retrieval approach to object matching in videos. In: Proceedings of the International Conference on Computer Vision, Nice, France, pp. 1470–1477 (2003)

Steiger, J.H.: Tests for comparing elements of a correlation matrix. Psychol. Bull. **87**, 245–251 (1980)

Szegedy, C., Liu, W., Jia, Y., Sermanet, P., Reed, S., Anguelov, D., Erhan, D., Vanhoucke, V., Rabinovich, A.: Going deeper with convolutions. In: Computer Vision and Pattern Recognition (2015)

Using Argumentative Structure to Grade Persuasive Essays

Andreas Stiegelmayr and Margot Mieskes[✉]

University of Applied Sciences, Darmstadt, Germany
{andreas.stiegelmayr,margot.mieskes}@h-da.de

Abstract. In this work we analyse a set of persuasive essays, which were marked and graded with respect to their overall quality. Additionally, we performed a small-scale machine learning experiment incorporating features from the argumentative analysis in order to automatically classify good and bad essays on a four-point scale. Our results indicate that bad essays suffer from more than just incomplete argument structures, which is already visible using simple surface features. We show that good essays distinguish themselves in terms of the amount of argumentative elements (such as major claims, premises, etc.) they use. The results, which have been obtained using a small corpus of essays in German, indicate that information about the argumentative structure of a text is helpful in distinguishing good and bad essays.

1 Introduction

Writing essays is an essential part of every-day-life of pupils and students. In persuasive essays there is an additional challenge in getting argumentative structures right. Research in automated essay scoring has been looking at a wide variety of features such as text structure, vocabulary, spelling, etc. All of which are important, but considering current research in argument mining, there is a lack of research into the relationship between argument structure and essay quality. In this work, we address how various aspects of arguments (i.e., major claims, premises, etc.) relate to the quality of an essay. Additionally, we use features based on arguments in a classification task using machine learning methods. Our results indicate that persuasive essays can be reliably classified using argument-based features. This work contributes in two ways to research in the area of argument mining and essay scoring: First, we show that the argumentative structure can be used to distinguish good and bad essays in an essay scoring task. Second, to our knowledge, this is the first work to bring these two topics together based on German data.

2 Related Work

As this work is at the intersection of *argument mining* and *essay scoring* we look at relevant previous work in both areas. Reviewing the available literature in detail is beyond the scope of this paper.

G. Rehm and T. Declerck (Eds.): GSCL 2017, LNAI 10713, pp. 301–308, 2018.
https://doi.org/10.1007/978-3-319-73706-5_26

2.1 Argument Mining

Although the topic of argument mining is fairly new, it goes back to acient greece. Habernal and Gurevych (2017) provide a current, extensive overview on the area. We specifically looked at the guidelines presented by Stab and Gurevych (2014): The authors analysed three components for argument structures: Major Claim, Claim and Premise. The basis of an argument is the claim, which relates to one or more premises. This relation has two attributes: support and attack. The Major Claim is the basis for the whole essay and can be found either in the introduction or in the conclusion. In the introduction it serves as a statement, which is related to the topic of the essay. In its conclusion it summarizes the arguments of the author.

Wachsmuth et al. (2016) also based their work on Stab and Gurevych (2014), but they consider Argumentative Discourse Units (ADU). ADUs can be complete sentences or partial sentences, especially in cases where two sentences are connected via "and". The authors defined a set of features, such as n-grams, part-of-speech n-grams, etc., and analysed the flow of ADUs based on graphs.

Work on German data is (compared to English data) rare. One example is by Peldszus and Stede (2013), where artificially constructed short texts were used to determine inter-annotator aggreement on argument annotation. Kluge (2014) used web documents from the educational domain, and Houy et al. (2013) used legal cases. All authors analysed the argumentative structure of their documents.

Work on essays has been carried out for example by Faulkner (2014), but with the aim of identifying the stance of an author towards a specific claim and in the domain of summarization. Stab and Gurevych (2014) also used essays in their study, but focused on the identification of arguments.

2.2 Essay Scoring

Dong and Zhang (2016) present an overview on essay scoring, including commercial tools available. They analysed a range of features for essay scoring and used them in a deep learning approach. The authors used surface features such as the length of characters, words, etc., and linguistic features such as Part-of-Speech (POS) tags and POS-n-grams. They used words and their syonyms based on the prompts for each essay and their appearance in the resulting texts. Additionally, they used uni- and bigrams and corrected for spelling errors. They considered the task as a binary classification task, with good essays defined as "essays with a score greater than or equal to the average score and the remainder are considered as bad scoring essays". The authors report a κ-based evaluation, which achieves results "close to that of human raters".

Using arguments for essay scoring has been done by Ghosh et al. (2016), based on TOEFEL-data. Their results, based on number and length of argument chaines, indicate that essays containing many claims, connected to few or no premises score lower. They also found that length is highly correlated with the scores.

3   Data Set

We collected a corpus containing 38 essays, which are available on the internet[1]. We also tried to get real essays by contacting various schools and teachers. These would also have teachers markings. Unfortunately, this is not a viable path to follow, due to various reasons: Firstly, these essays are subject to a very strict data protection law, which puts a range of obstacles on obtaining such data. Secondly, very few schools use electronic methods and tools for writing essays. So all schools we got in touch with and which would have been willing to grant us access to their essays and markings, provided we agree to the data protection regulations, only had essays which were hand-written on paper. Digitizing them, including proof-reading, would have been beyond the scope of this work. Therefore, we took data that was available on the internet in a machine-readable format. The corpus was manually annotated using the guidelines by Stab and Gurevych (2014) using WebAnno[2]. Figure 1 shows an example structure of the resulting argument tree. The whole data set contains approximately 120,000 words, and slightly over 4,000 sentences. In total, we analysed over 1,000 argument units containing over 1,000 premises and almost 300 claims. 50% of the argument units had more than 15 words. Details can be found in Table 1.

In addition to the argument annotation, we also annotated the quality of the essays, using the German school marking system, which is based on numbers 1 to 4, where 1 represents a very good result and 4 represents a very poor result. We decided to use a reduced version of the German marking systems due to the following reasons: At universities only marks from 1 to 4 are given, with marks >4 being a *fail*[3]. Due to the data set size, using a more fine-grained marking scale would have given us very few data points for each class to train a machine learning system on, especially with respect to the already small data set size.

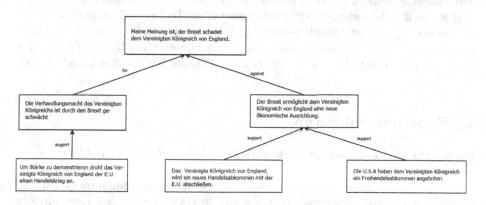

Fig. 1. Example for an argument tree as found in our data.

---

[1] The list of sources can be found at https://b2drop.eudat.eu/s/tR5spZeyRcW20VB.
[2] https://webanno.github.io/webanno/.
[3] One annotator marked one essays with a fail.

Table 1. Statistical information on the corpus.

Element	Count
Words	119,043
Sentences	4,047
Paragraphs	133
Argument units	1,324
Major claims	34
Claims	286
Premises	1,004
Spelling anormaly	130
Median words per argument unit	15.3

We assume, that the quality of the essay corpus is not representative of regular school essays, but rather represent the quality available on the interent. We observe, that the quality of the essays is mediocre, with many authors not explicitly stating their point of view. In some extreme cases the major claim was not detectable. This results in difficulties in deciding whether a sentence contains an argument unit or not. The distribution of the marks is therefore very skewed, with approximately 23.1% of the essays achieving good (mark 2) or very good (mark 1) marks and 77% of the essays achieving poor (mark 3) or very poor (mark 4) results. An additional problem – especially for the later automatic analysis – is the usage of metaphors, which we did not look into in this work.

About one third of the essays (13 out of 38) were graded by two persons. The percentage agreement between the two grades was 0.53. Considering a measure that is specifically designed to evaluate annotations by two coders and correcting for chance agreement (which percentage agreement does not do), we achieve a value of $S = 0.42$, which according to Landis and Koch. (1977) shows a moderate agreement. All values were calculated using DKPro Statistics[4] (Meyer et al. 2014).

4   Experimental Setup

We use DKPro Components[5], such as DKPro Core, DKPro TC and Uby for our experiments.

We defined a range of features, based on the argumentation annotation and previous work. We distinguish between *baseline* features, which have already been used in previous work and *argument* features, which are based on the argumentation annotation. The baseline features contain easy to determine features,

---

[4] https://dkpro.github.io/dkpro-statistics/.
[5] https://github.com/dkpro/.

such as number of tokens, number of sentences, etc. Additionally, we took into account POS-based features, which include nouns, verbs, adjectives, etc.

Based on earlier work, we included information about whether the author used overly long words or short words. We also checked for spelling errors using the LanguageTool[6]. Wachsmuth et al. (2016) observed that questions are not arguments. Therefore, we also extracted the number of questions with and without arguments. According to Stab and Gurevych (2014) one paragraph should only contain one claim. Therefore, we also counted the number of claims and the number of paragraphs in our documents. Additionally, we looked at the number of sentences with and without arguments. Finally, we examined the n-grams found in the annotated arguments. Based on Ghosh et al. (2016) we looked at the graph created by the argument structure over a document. An example can be found in Fig. 1. Tree size and grade show a strong, negative correlation (Pearsons $r = -0.57$[7]), meaning, the larger the tree, the higher the grade. Additionally, we use the argument graph to determine whether it starts with a major claim or not and which arguments are not linked to the major claim. Finally, we determined whether a person consistently uses the correct tense. The full set of features can be found in the respective .arff-files[8].

5 Results and Discussion

We experimented with various machine learning algorithms, using WEKA[9]. As we wanted to gain a qualitative insight into the results obtained through the machine learning methods, we specifically looked into decision trees (J48).

Table 2. Classification result for individual marks using the whole feature set.

Mark	p	r	f1
1	1	0.6	0.75
2	0.833	0.833	0.833
3	0.667	0.6	0.632
4	0.773	0.895	0.829
Avrg.	0.87	0.86	0.858

We observed, that the main features contributing to the results in Table 2 were NrOfMajorClaims, NrOfPremises and RatioSentenceNonQuestion. This supports earlier work, that the number of major claims and premises allows for detecting good essays.

---

[6] https://languagetool.org/de/.
[7] This correlation is significant on $\alpha = 0.01$.
[8] https://b2drop.eudat.eu/s/mTyUTJrCNyO4I3e
[9] http://www.cs.waikato.ac.nz/ml/weka/.

Table 3. Classification result for individual marks using baseline features only.

Mark	p	r	f1
1	0	0	0
2	0	0	0
3	0.545	0.6	0.571
4	0.721	0.838	0.775
Avrg.	0.476	0.545	0.508

Using only the baseline features we observed that the lower marks (3 and 4) were still classified fairly reliably, but the better marks (1 and 2) performed very poorly. Looking into the results in detail revealed that essays marked as "2" were mostly confused with essays marked as "1", which indicates, that not so good essays suffer from more than just a lack of good argumentative structure, which is already visible with the surface features. This becomes very prominent looking at the resulting tree, where the most important features for 3 and 4 were a combination of fewer characters and a high amount of spelling errors. In order to reduce the importance of the spelling errors, we artificially introduced spelling errors to the good essays (marked 1 and 2). We tried to achieve a similar ratio as for the bad essays (marked 3 and 4). Thereby, we managed to reduce the importance of the spelling feature in the feature ranking. But the overall results (including the observations concerning the usage of major claims and premises in connection to the resulting grade) were similar to those presented in Table 2 ($p = 0.86$; $r = 0.85$ and $f1 = 0.85$) and the discussion above (Table 3).

Table 4. Classification result for individual marks using custom features only.

Mark	p	r	f1
1	1	0.6	0.75
2	0.833	0.833	0.833
3	0.700	0.700	0.700
4	0.810	0.895	0.850
Avrg.	0.87	0.86	0.858

The argumentative features allow us to clearly identify and distinguish between the various essays. A closer look at the resulting tree indicates, that good essays use premises cautiously and also keep the major claims low, which is in line with observations from previous work. Bad essays have a higher number of major claims, but also a high number of disconnected arguments (Table 4).

Overall, our results indicate that poor essays suffer from more than just poor argumentation and authors should address issues such as spelling, usage of tense, number of conjunctions and length of words. Once these issues are considerably improved, the argumentative elements of the essays should be considered, such as a high number of major claims. For authors who already achieve good results, the focus can be put on argumentative elements, such as the number of premises, which is higher than in very good essays.

6 Conclusion and Future Work

We presented work on using argumentative structures and elements in identifying the quality of persuasive essays. We found that argumentative elements support the identification of good essays. Bad essays can be classified reliably using traditional features, indicating that these authors need to address issues such as spelling errors before improving on argumentative elements in their writing.

The next step would be to increase the data set size in order to solidify our findings. More data would also allow us to use more sophisticated machine learning methods. Additionally, we would like to incorporate a range of features previously used in the area of essays scoring, such as latent semantic analysis. Finally, we would like to have a closer look at the issue of metaphors in argumentative essays and their contribution to arguments and essay quality.

Acknowledgements. We would like to thank the German Institute for Educational Research and Educational Information (DIPF), where this work was carried out. Additionally, we would like to thank the reviewers for their helpful comments.

References

Dong, F., Zhang, Y.: Automatic features for essay scoring - an empirical study. In: Proceedings of the 2016 Conference on Empirical Methods in Natural Language Processing, pp. 1072–1077 (2016)

Faulkner, A.R.: Automated Classification of Argument Stance in Student Essays: A Linguistically Motivated Approach with an Application for Supporting Argument Summarization. Ph.D. thesis, Graduate Center, City University of New York (2014)

Ghosh, D., Khanam, A., Han, Y., Muresan, S.: Coarse-grained argumentation features for scoring persuasive essays. In: Proceedings of the 54th Annual Meeting of the Association for Computational Linguistics. Short Papers, vol. 2, pp. 549–554. Association for Computational Linguistics (2016). https://doi.org/10.18653/v1/P16-2089. http://aclweb.org/anthology/P16-2089

Habernal, I., Gurevych, I.: Argumentation mining in user-generated web discourse. Comput. Linguist. **43**(1), 125–179 (2017)

Houy, C., Niesen, T., Fettke, P., Loos, P.: Towards automated identification and analysis of argumentation structures in the decision corpus of the German Federal Constitutional Court. In: The 7th IEEE International Conference on Digital Ecosystems and Technologies (IEEE DEST-2013) (2013)

Kluge, R.: Searching for Arguments - Automatic analysis of arguments about controversial educational topics in web documents. AV Akademikerverlaug, Saarbrücken (2014)

Landis, R.J., Koch, G.G.: The measurement of observer agreement for categorical data. Biometrics **33**(1), 159–174 (1977)

Meyer, C.M., Mieskes, M., Stab, C., Gurevych, I.: DKPro Agreement: an open-source Java library for measuring inter-rater agreement. In: Proceedings of the 25th International Conference on Computational Linguistics (COLING), Dublin, Ireland, pp. 105–109 (2014)

Peldszus, A., Stede, M.: Ranking the annotators: an agreement study on argumentation structure. In: Proceedings of the 7th Linguistic Annotation Workshop and Interoperability with Discourse, pp. 196–204 (2013)

Stab, C., Gurevych, I.: Identifying argumentative discourse structures in persuasive essays. In: Proceedings of the 2014 Conference on Empirical Methods in Natural Language Processing (EMNLP), pp. 46–56. Association for Computational Linguistics (2014). https://doi.org/10.3115/v1/D14-1006. http://aclweb.org/anthology/D14-1006

Wachsmuth, H., Al-Khatib, K., Stein, B.: Using argument mining to assess the argumentation quality of essays. In: Proceedings of the 26th International Conference on Computational Linguistics (COLING 2016), pp. 1680–1692, December 2016

Author Index

Printed in the United States
By Bookmasters